Introduction to Computational Engineering with MATLAB®

Introduction to Computational Engineering with MATLAB® aims to teach readers how to use MATLAB® programming to solve numerical engineering problems. The book focuses on computational engineering with the objective of helping engineering students improve their numerical problem-solving skills. The book cuts a middle path between undergraduate texts that simply focus on programming and advanced mathematical texts that skip over foundational concepts, feature cryptic mathematical expressions, and do not provide sufficient support for novices.

Although this book covers some advanced topics, readers do not need prior computer programming experience or an advanced mathematical background. Instead, the focus is on learning how to leverage the computer and software environment to do the hard work. The problem areas discussed are related to data-driven engineering, statistics, linear algebra, and numerical methods. Some example problems discussed touch on robotics, control systems, and machine learning.

Features:

- Demonstrates through algorithms and code segments how numeric problems are solved with only a few lines of MATLAB® code.
- Quickly teaches the basics and gets readers started programming interesting problems as soon as possible.
- No prior computer programming experience or advanced math skills required.
- Suitable for undergraduate students who have prior knowledge of college algebra, trigonometry, and are enrolled in Calculus I.
- MATLAB® script files, functions, and datasets used in examples are available for download from http://www.routledge.com/9781032221410.

Tim Bower is an Associate Professor of Robotics and Automation Engineering Technology and Computer Systems Technology at Kansas State University Salina. He received the B.S. Electrical Engineering degree from Kansas State University (K-State) in 1987 and the M.S. Electrical Engineering degree from the University of Kansas in 1990. He was a Senior Member of the Technical Staff at Sprint's Local Telephone Division from 1989 to 1998. From 1998 to 2003, he was a systems administration manager and instructor at Kansas State University in Manhattan Kansas while taking graduate course work in Computer Science. He joined the faculty of K-State's campus in Salina Kansas in 2004. He teaches undergraduate courses related to programming in C, Python, and MATLAB®, robotics programming, machine vision, numerical computation, operating systems, data structures and algorithms, and systems administration.

Away from teaching, he enjoys spending time with his wife, three grown children, and five grandchildren.

Numerical Analysis and Scientific Computing Series
Series Editors:
Frederic Magoules, Choi-Hong Lai

About the Series
This series, comprising of a diverse collection of textbooks, references, and handbooks, brings together a wide range of topics across numerical analysis and scientific computing. The books contained in this series will appeal to an academic audience, both in mathematics and computer science, and naturally find applications in engineering and the physical sciences.

Computational Methods for Numerical Analysis with R
James P Howard, II

Numerical Techniques for Direct and Large-Eddy Simulations
Xi Jiang, Choi-Hong Lai

Decomposition Methods for Differential Equations
Theory and Applications
Juergen Geiser

Mathematical Objects in C++
Computational Tools in A Unified Object-Oriented Approach
Yair Shapira

Computational Fluid Dynamics
Frederic Magoules

Mathematics at the Meridian
The History of Mathematics at Greenwich
Raymond Gerard Flood, Tony Mann, Mary Croarken

Modelling with Ordinary Differential Equations: A Comprehensive Approach
Alfio Borzì

Numerical Methods for Unsteady Compressible Flow Problems
Philipp Birken

A Gentle Introduction to Scientific Computing
Dan Stanescu, Long Lee

Introduction to Computational Engineering with MATLAB®
Timothy Bower

For more information about this series please visit: https://www.crcpress.com/
Chapman--HallCRC-Numerical-Analysis-and-Scientific-Computing-Series/book-series/
CHNUANSCCOM

Introduction to Computational Engineering with MATLAB®

Timothy Bower

Kansas State University Salina, USA

CRC Press

Taylor & Francis Group

Boca Raton London New York

CRC Press is an imprint of the
Taylor & Francis Group, an **informa** business

A CHAPMAN & HALL BOOK

First edition published 2023
by CRC Press
6000 Broken Sound Parkway NW, Suite 300, Boca Raton, FL 33487-2742

and by CRC Press
4 Park Square, Milton Park, Abingdon, Oxon, OX14 4RN

CRC Press is an imprint of Taylor & Francis Group, LLC

Library of Congress Cataloging-in-Publication Data

A catalog record for this title has been requested.

ISBN: 978-1-032-22178-6 (hbk)
ISBN: 978-1-032-22141-0 (pbk)
ISBN: 978-1-003-27143-7 (ebk)

DOI: 10.1201/ 9781003271437

Typeset in Latin Modern
by KnowledgeWorks Global Ltd.

Publisher's note: This book has been prepared from camera-ready copy provided by the authors.

Access the Support Material: http://www.routledge.com/9781032221410

To Pam,
Joe, Beth, and Naomi

Contents

Contents   xv

Preface

Like a dog on a country road, the mind must poke into as many holes as it can.

Frank W. Boreham
1919

This book is for people that are curious about technology. It is for people that must always ask, "How does that work?", "What is under the hood?", and "Can I come up with a better solution?". Such people are naturally drawn to the fields of mathematics, science, and engineering. When the mind of these people is in hot pursuit of a fresh idea, time is of the essence. Efficient tools, proficiency, and knowledge are needed to prototype, simulate, visualize, and analyze. Perhaps, the most obvious tool that they will deploy is the computer. With the speed and storage capacity of modern computers along with advanced computational software, your personal computer is a powerful engineering tool. My hope in writing this book is that readers will press their computers into the service of problem solving. We call this *computational engineering*.

Computational engineering uses numerical computing, data analysis, visualization, and software tools to model, analyze, and solve a variety of science and engineering problems. Writing computer programs is merely a component of the process toward the goals of computational engineering. Mathematics, scientific knowledge, abstract reasoning, logic, and common sense are also components of the process.

Successful professionals in all branches of science and engineering use computers to their best advantage. But computers have strict rules about how they are programmed. However, that does not mean that effective engineers need to also be computer scientists. A few software development environments have become particularly popular with engineers that allow users with limited programming experience to quickly solve a variety of numerical problems and plot data. Modern software tools for engineers are not yet able to relax the strict rules of programming, but they can accomplish some amazing things with just a few lines of code. The primary software tool that we will use is MATLAB [®1] from The MathWorks, Inc.

Think of the material in this book in the same way that you think about long division. I'm glad that I learned how to do long division in elementary school. But when presented with a division problem, I either do a rough estimate in my head or I reach for a calculator. I seldom use pencil and paper to do long division. In the same manner, engineers look to computers as a tool for solving engineering problems as much or more than pencil, paper, or even hand held calculators. Computers allow us to work quickly, accurately, and produce impressive results. Just as it was important to learn about long division in elementary school, we need to discuss some mathematics concepts that are used in computational engineering. But the mathematics that we cover will directly apply to problems and will be solved by computers.

Some of the algorithms that we will take advantage of are quite advanced, but are pre-developed and available for our use. To affectively apply these algorithms, we need to understand the basic concepts of what they do and the relationship of the input variables to the output. So we strive for a balance. Some basic knowledge about the algorithms is needed. If I can write a simple implementation of an algorithm, then I really understand it and can properly apply it to problems. However, it is not necessary to develop implementations of algorithms at the level of robustness, numerical accuracy, and speed as the functions found in MATLAB.

Origins and Strategy

The material in the book started as my notes for an undergraduate course that I teach to engineering students. The course is intended to teach computer programming concepts; but more importantly, it is intended to help students improve their ability to solve numerical problems.

In my preparation to teach the course, I could not find a textbook with the focus that I was looking for. Several good undergraduate textbooks provide a gentle introduction to MATLAB. However, I wanted the course to focus more on the mathematics and algorithms of engineering problems than writing computer programs or the specific details of MATLAB. Most computer programming textbooks take the approach of providing a survey of programming semantics along with the features and syntax of the language. Then some examples from science, engineering, and business are used to illustrate the programming. I wanted to take the opposite approach. Beyond an introduction to MATLAB, I wanted to focus on technical topics and then bring in the mathematics and programming concepts as is needed to solve the

---

[1]MATLAB is a registered trademark of The MathWorks, Inc.

problems. The books that I found that address the material I wanted to cover are at the graduate school level, which would frustrate undergraduate students. So I decided to write notes for the course that would cut a middle path. Although the course covers some advanced topics, students do not need prior computer programming experience or advanced math skills. Instead, the focus is on learning how to leverage the computer and software environment to do the hard work.

Since the origin of book came from my course notes, the style of writing is more casual than formal. I try to explain what is needed while cutting through advanced details that undergraduate level readers might find confusing. The secret to achieving this objective is the source code. In graduate school, one of my professors often told his students, "If you want to know how things really work, read the source code". That is the model that I have tried to apply. MATLAB code examples illustrate nearly every observation. In some cases, the examples are a few commands typed in the MATLAB Command Window. In many cases, the examples are complete scripts and functions that feature plots showing the data relationships. I find that examples with real numbers always help when trying to understand an algorithm or math equation. I feel that source code examples often teach concepts best because details can not be omitted or assumed.

Content Comments

Determining what content to include is perhaps the question that has required the most thought in my course development. The question that I asked myself was, "What do the students need?". What do they need to learn to get through more advanced courses and before they graduate? An introduction to programming was, of course, considered obligatory. Then visually displaying data with plots and graphs was quickly deemed essential. Beyond the prerequisites, I tried to pick topics where a computer and a bit of computational engineering knowledge might pay the best dividends.

Chapter 1 offers a fast–paced but sufficient introduction to MATLAB. The goal is to quickly get students programming with a some basic knowledge. As a continuation of chapter 1, chapter 2 covers how to make effective two and three dimensional data plots.

Chapter 3 provides an introduction to computational statistics. Although coverage is given to basic statistical metrics, calculations, and common probability distributions, the focus is on computation and graphical display of statistical information. The statistics chapter could have easily grown to be much larger. The intent, however, is to provide a supplement to what students learn in a course on statistics and probability.

Chapter 4 is a quick introduction to using MATLAB's *Symbolic Math Toolbox*, which is the only toolbox covered that requires an additional purchase beyond MATLAB itself. I hope that the chapter shows how the computer can help with complex analytic mathematics problems as well as numeric problems.

Chapter 5 introduces the topic of linear algebra and its application to computational engineering. Finding solutions to systems of equations is central to the discussion in chapter 5. I feel confident in making the following statements about linear algebra.

- Vectors, matrices, and systems of equations have application to every science and engineering discipline.

- The focus of mathematics in most current engineering and science curriculums is on algebra, trigonometry, calculus, and to a lesser extent differential equations. Linear algebra is often not given the attention that it merits.

- Linear algebra is uniquely suited to computational engineering. Applying linear algebra to real problems requires tedious matrix and vector calculations that don't fit well in analytic math courses. The obvious solution is to let computers handle the calculations. Thus, it is best to learn linear algebra in the context of computational engineering where computers and programming are central to the course of study.

Chapter 6 continues the discussion of linear algebra, but shifts the focus from systems of equations to applications that take advantage of the eigenvalues and eigenvectors of a matrix. Since this is a new concept to many undergraduate students, particular attention is given to what eigenvalues and eigenvectors are and the significance of the relationship $\mathbf{A}\boldsymbol{x} = \lambda\boldsymbol{x}$.

Computational numerical methods are discussed in chapter 7. This chapter started as a short chapter focused entirely on understanding and using the efficient functions provided by MATLAB. However, the limited coverage did not give an adequate appreciation of the algorithmic considerations. So the coverage was extended to include more detailed algorithmic descriptions and algorithm alternatives.

The appendices describe supporting material that is more complex and is not strictly required for applying the material in the previous chapters, but will provide curious readers with proofs, more detailed definitions, and descriptions of some algorithms used by the MATLAB functions. Most of the material in the appendices relate to the linear algebra topics.

Acknowledgments

I wish to thank the administration, faculty, staff, and students at Kansas State University Salina for their support and encouragement. Thank you to my colleague Kaleen Knoop and the Writing Center for making the Grammarly program available. It helped me numerous times. I owe a particular debt of gratitude to my colleague Gayan Samarasekara for reviewing and offering corrections to chapter 3 on statistical data analysis.

I wish to thank the Taylor & Francis Group for giving me the opportunity to fulfill a lifelong goal in publishing this work. Special thanks to editors Mansi Kabra and Callum Fraser for their encouragement, patience, and advice.

I am deeply blessed and thankful for the emotional support and encouragement from my wife, Pam. I would not have been able to complete the writing without her. I'm also grateful for the encouragement and joy that comes from our son, daughters, their spouses, and our grandchildren.

Downloadable Support Material

All MATLAB script files, functions, and datasets used in examples are part of the support material available for download from the book's website at http://www.routledge.com/9781032221410.

Tim Bower
Salina, Kansas

List of Program Files

Chapter 1

MATLAB Programming

MATLAB[1] (name taken from MATrix LABoratory) is a high-level programming language and development environment well suited for numerical problem solving, data analysis, prototyping, and visualization. MATLAB has become very popular with engineers, scientists, and researchers in both industry and academia. Three features of MATLAB make it stand out:

1. Matrix and vector support

2. Data visualization tools

3. Huge library of functions covering nearly every area of mathematics, science, and engineering.

MATLAB has extensive built-in documentation. It contains descriptions of the main functions, sample code, relevant demonstrations, and general help pages. MATLAB documentation can be accessed in a number of different ways including command-line help, a graphical documentation browser, and web pages.

MATLAB code that users write is interpreted, rather than compiled, so it is not necessary to declare and allocate variables in memory prior to using them. Although MATLAB programs are interpreted, MATLAB has features that make it still execute fairly fast.

1. When the code is vectorized (explained later), instead of using loops, the performance is better than expected for interpreted programs.

2. MATLAB has an extensive library of functions to do most of the work in your program. When MATLAB functions are run, the computer is executing highly optimized, compiled code.

This chapter covers the basics of MATLAB programming. The intent is to start at the very beginning to let readers start programming simple programs right away. In later chapters, we will apply MATLAB to solve computational engineering problems while also introducing additional MATLAB features and applying them with examples. Special attention is given in this chapter to vectorized code using element-wise arithmetic, logical arrays, and vector aware functions to operate at the vector and matrix level with limited need for loops operating on the individual values contained in arrays.

---

[1]MATLAB is a registered trademark of The MathWorks, Inc.

DOI: 10.1201/9781003271437-1

Additional Resources For Beginners

MathWorks MATLAB product page
    https://www.mathworks.com/products/matlab.html

MATLAB Onramp A free training course from MathWorks.
    https://matlabacademy.mathworks.com/

MathWorks Documentation and Videos MathWorks has produced
    many pages of documentation and videos about the usage of MAT-
    LAB. So make a habit of searching their website. A good introduction
    video showing the capabilities of MATLAB is at [41].
    https://www.mathworks.com/

Other Books If you would benefit from a slower paced introduction to
    programming and MATLAB, then you might consider an undergrad-
    uate level textbook such as [4], [11], or [32]. For a more advanced
    reference book with a strong focus on the available commands and
    functions, the *MATLAB Guide* [34] is recommended.

1.1   The MATLAB Development Environment

The MATLAB integrated development environment (IDE) is conducive for
rapid program development. It is shown in figure 1.1 and includes the following.

MATLAB Desktop: This typically contains at least five subwindows: the
    Command Window, the *Workspace Browser*, the *Current Directory Win-
    dow*, the *Command History Window*, and one or more *Figure Windows*,
    which are visible when displaying plots or images. The *Layout* menu on
    the *HOME* tab lists other subwindows that users may wish to try.

MATLAB Editor: This is used to create and edit M-files. It includes a num-
    ber of useful functions for saving, viewing, and debugging M-files.

Help System: This includes the Help Browser, which displays HTML doc-
    uments and contains a number of search and display options.

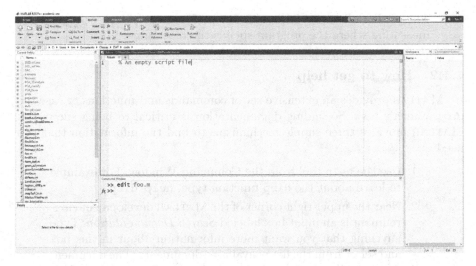

FIGURE 1.1: The MATLAB integrated development environment.

1.1.1 Using the IDE

The interaction between the *Command Window*, *Workspace Browser*, *Command History Window*, and the *Editor* make a convenient environment for software development.

- In the Command Window, users may interactively type commands. Simple calculations may be entered using only the command window. It is also nice for testing commands and syntax before adding those commands to a file in the editor.

- The value and size of variables can be viewed either from the Command Window or the Workspace Browser.

- The Command History Window shows previous commands entered. Click on a previously entered command with the mouse. Hold down the shift key and use the arrow keys to select multiple previous commands. With commands selected, press the F9 keyboard key to run the selected commands. Commands from the Command Window and the Editor may also be highlighted with the mouse and run by pressing F9.

- In the Editor, users type code into scripts and functions so that they will be saved for future use.

  When writing a script file, commands in the script can be run as though they were typed in the command window.

- The debugger is used to set breakpoints that stop the execution at set places. With the execution stopped, a user can use the command window to determine what changes are needed to fix a problem.

- The profiler allows for improving the performance of larger programs by measuring where the program spends the most time.

1.1.2 How to get help

MATLAB provides an extensive set of commands and functions to ease the programmer's task. So finding documentation is critical to being successful. MATLAB provides three simple mechanisms to find the information that you need.

1. Use the `help` tool from the Command Window. For example to learn about the `disp` function type, `help disp`.

2. Near the upper right corner of the MATLAB development environment is an input box labeled *Search Documentation*. Type anything that you want more information about in this box and select from the list of available documents. The documentation is displayed in a separate document browser window. The information found will usually be more detailed than that given by the help tool from the Command Window. It will often include useful examples. The `doc` command in the Command Window also launches the document browser and displays the same information.

3. The same documentation available from the built-in search tool can be found on MathWorks' web pages. Using your web browser and favorite Internet search engine, just add the word 'matlab' to what you are looking for.

1.2 Variables and Values

As with math equations, we use variables in computer programs. They allow us to associate a storage location in the computer's memory with a name. We use the variable name to store and retrieve values from the storage location.

1.2.1 Command Window Calculator

The characters >> in the Command Window is a prompt where commands are given to MATLAB.

```
>> 2 + 1
ans =
    3
```

We can save a result to a variable for future reference:

```
>> x = 2 + 1
x =
    3
```

Tip: The keyboard arrow keys may be used to retrieve previously entered commands.

Variable

When a value is given a name, it becomes a *variable*. Variables are used to store data for display or future computations. The name given to a variable is called an *identifier*, which just means that we choose the name. The rules for what is a valid identifier are covered later.

Assignment Statement

Storing data in a variable with an equal sign (=) is formally called an *assignment statement* in computer science terminology. A data value is assigned to be held in a variable.

ans

MATLAB knows that users might perform a calculation without saving the result to a variable but later want to use the result. So any result not saved to a variable is automatically stored in a variable called **ans**.
NOTE: Only the most recent unsaved result is saved in **ans**.

The mathematics operators used with simple variables and constant numbers are:

- **+** Addition
- **−** Subtraction
- ***** Multiplication
- **/** Division
- **^** Exponent

The order of operator precedence is what you would expect from basic algebra.

1. parentheses, brackets
2. exponents, roots
3. multiplication, division
4. addition, subtraction

When operations are at the same level, then the order execution is left to right.

```
>> 2 * (3-4) / 5
ans =
    -0.4000
```

Caution: A common error for beginning programmers is leaving off the * for multiplication.

1.2.2   Identifier Names

A few simple rules govern the names (called identifiers) for variables and functions. Table 1.1 shows examples of common mistakes in choosing identifiers and how they might be corrected.

- Identifiers can only contain letters, numbers, and underscores.

- Identifiers can only start with letters.

- Identifiers are case-sensitive.

- Identifiers can be between 1 and 63 characters long.

- Identifiers can not be one of MATLAB's keywords.

TABLE 1.1: Incorrect and Corrected Identifiers

Incorrect	Problem	Correct
X-Y	– not allowed	XY
month&year	& not allowed	monthYear
1987value	begins with a number	value1987
CO2 conc	space character not allowed	CO2_conc

1.2.3   Calling Functions

MATLAB has functions for any calculation that you can imagine and a lot more.

```
>> sin(1)
ans =
    0.8415
```

What is a keyword?

Look up the function `iskeyword` in the MATLAB documentation. You will find information about the boolean function and also a list of keywords that may not be used as identifiers.

```
break        else         global       return
case         elseif       if           spmd
catch        end          otherwise    switch
classdef     for          parfor       try
continue     function     persistent   while
```

This command is an example of a function call. The name of the function is `sin`, which is the usual abbreviation for the trigonometric sine. The value in parentheses is called the *argument*. All the trigonometry functions in MATLAB work in radians, but there is a set of trigonometry functions ending with a **d** that use angles in degrees. These include `sind`, `cosd`, `tand`, etc.

Some functions take more than one argument, in which case they are separated by commas. For example, `atan2` computes the inverse tangent, which is the angle in radians between the positive x axis and the point with the given y and x coordinates.

```
>> atan2(1,1)
ans =
    0.7854
```

Caution: A common error is to leave out the commas between the arguments of a function.

1.2.4 Numeric Data Types

The default data type for numeric variables is *double* precision floating point (64 bits). It supports very large positive or negative numbers and numbers that are very close to zero.

```
>> big1 = 9.578942e50
big1 =
    9.5789e+50
>> small = 1.07124e-122
small =
    1.0712e-122
```

Other numeric data types, such as signed integers and unsigned integers are available if needed. Look up the documentation for the `cast` function for a detailed list of data types.

MATLAB treats all numeric data as a matrix (discussed later). Scalar values are thus reported as having a size of 1×1.

```
>> a = 5
a =
     5
>> whos
  Name      Size              Bytes  Class     Attributes

  a         1x1                   8  double
```

`format` Command

Look up the documentation for the `format` command. How can it be applied? Does it impact the number of digits that MATLAB uses to store variables? Note that the command `format compact` removes extra blank lines from results displayed in the command window.

1.2.5   Simple Arrays

We use arrays to hold a list of values. Arrays and matrices are central to using MATLAB. We will discuss arrays in more detail in section 1.7. For now, let us learn how to use a simple array, which MATLAB calls a row vector. Calling it a row vector just means that data is stored in a single row. Consider the following lines of code.

In the following code example, the `zeros(1, 5)` function call creates a row vector consisting of one row and five columns that are filled with zeros. In the next two commands, assignment statements change the values stored in the first and forth positions. Notice two things about the use of an index number that are different than most other programming languages. First, we use a pair of parenthesis around the index number. Secondly, the indices are from 1 to N, where N is the size of the row vector.

```
>> x = zeros(1, 5)
x =
     0     0     0     0     0
>> x(1) = 3
```

```
x =
     3     0     0     0     0
>> x(4) = 6
x =
     3     0     0     6     0
>> x(1)
ans =
     3
```

1.2.6 Clearing Variables

The `clear` command removes all variables from the Workspace. Specific variables may also be removed:

```
>> clear x
```

In some cases, it may be desirable, particularly while developing a program, to clear all variables and close any open plots when the program starts. Commands `clear; close all` achieves this.

1.2.7 Some Pre-defined Constants

Since MATLAB is primarily used for numerical calculations, there are a few math constants that we need to know about. The irrational number π is perhaps the most obvious, but MATLAB also includes constants for working with the imaginary part of complex numbers ($\sqrt{-1}$) and even infinity. Table 1.2 lists some commonly used constants and their meaning.

TABLE 1.2: Math Constants

Name	Meaning
i	Imaginary number
j	Imaginary number
Inf	Infinity
pi	Ratio of circle's circumference to its diameter
NaN	Not-a-Number

You might expect to see the irrational number $e \approx 2.718281828459046$. Instead, there is a function, `exp(x)`, that returns e^x. The usage of the number e almost always involves an exponent. If the exponent were always a real number, then an expression such as `e^x` would suffice, but the exponent may be a complex number. Thus a function gives the flexibility needed.

```
>> exp(1)                    >> exp(i*pi/2)
ans =                        ans =
     2.7183                      0.0000 + 1.0000i
>> exp(i*pi)
ans =
   -1.0000 + 0.0000i
```

The complex exponent demonstrates Euler's formula for complex exponentials: $e^{i\,x} = \cos x + i \sin x$ (appendix B.1.2).

1.3   MATLAB Scripts

Scripts are files with a `.m` file name extension containing the same type of commands that one might enter in the Command Window. Scripts simplify the task of developing a correct program because they allow code to be saved, modified and run repeatedly; whereas, to re-run code typed into the Command Window, the code would need to be retyped with any needed corrections.

You will see in the development environment that MATLAB also supports a new type of script file called a *Live Script*. Live scripts can be more interactive with output optionally displayed inline within the code. The code examples in this book may be used in either traditional scripts or live scripts. MathWorks has considerable documentation and demonstration videos about the use of live scripts.

Here are few important points to know about script files.

- Like the Command Window, scripts access variables from the global Workspace.

- Create a script with the command `edit myscript.m`, or from the Home tab, select either **New Script**, or **New -> Script**.

- It is possible to run just one or two lines of code at a time to verify the expected results.

  - Use the mouse to highlight a few lines of code, then press the F9 key on the keyboard to execute just those lines of code.

  - Another way to execute a few lines is to make those lines a *section*. The percent sign (%) is the beginning of a comment line. Two percent signs and a space character (%%) at the beginning of a line mark a section boundary. Select a section by clicking anywhere in the section and then, in the editor tab ribbon, click on *run section*.

1.3.1 Displaying Results

As you read MATLAB code, you will quickly notice that sometimes a semicolon (;) is placed at end of commands, but not always.

- The semicolon suppresses output.

- If you want to see the result of a calculation, leave off the semicolon. The output will be displayed in the Command Window.

- If the output would be too much to display on the screen; or you just don't want to see it, then add a semicolon after the command.

- Many commands, such as plotting commands, do not produce output to the Command Window, so there is no difference between using a semicolon or not.

In the Command Window, just enter a variable name without a semicolon to see its value.

```
>> b
b =
     5
```

In a script, you may want to use `disp` or `fprintf` to show results. We discuss them in more detail in section 1.4.2. You may also enter a variable name without a semicolon to see its value in a script, but MATLAB will display a warning telling you that you should terminate a statement with a semicolon in a script.

```
>> disp(b)
     5

>> fprintf('The value of b = %d\n', b)
The value of b = 5
```

1.3.2 Adding Sections

The overall structure of the scripts you create can naturally be divided into different sections. This is especially nice for larger files where you may just want to focus on one area, such as adjusting a plot.

Sections allow you to modify and reevaluate the code in just that section to see the impact.

To create a section, add `%% Section Title` at the beginning of a line. Note the space character between the second % and the section title.

```
%% Title of a Section

% The code of the sections goes here.

%% The start of a new section.
```

Tip:　An optional window in the IDE called Panel Titles displays the section titles and allows a quick way to move to and select a section.

1.3.3　Comments

A few words of explanation can make it much easier to reuse a program after you have forgotten the details. Adding text descriptions that are not executed, called comments, allows you to explain the functionality of your code. Text following a percent sign, %, is a comment. Comments can appear on lines by themselves, or they can be appended to the end of a line.

The first contiguous block of comments after a function declaration (section 1.8) is reserved for help information which is displayed when you use the `help` or `doc` commands. Comments after the first blank line are not treated as part of the help.

1.4　Input and Output

1.4.1　Input Function

Scripts share the global Workspace variables with the Command Window, and functions receive input data from passed arguments, so the need to prompt users to input data is not as prevalent in MATLAB as is the case for most other programming languages. But, of course, there is a function for cases when it is needed.

The function `result = input('PROMPT')` displays the prompt on the screen and returns the number entered by the user. The returned data is of type double as expected, but may be cast to another data type.

```
>> num = input('Enter a number: ')
Enter a number: 15
num =
     15
>> intNum = int32(input('Enter an integer number: ' ))
Enter an integer number: 8
intNum =
   int32
     8
```

The input function *evaluates* what the user enters, so more complex data such as a vector and matrix can also be entered.

```
>> vect = input('Enter a vector: ')
Enter a vector: [1 2 3]
vect =
     1     2     3
>> vect = input('Enter a vector: ')
Enter a vector: 1:4
vect =
     1     2     3     4
```

If a string result is desired, pass the 's' option to input.

```
>> name = input('What is your name? ', 's' )
What is your name? Tim
name =
    'Tim'
```

1.4.2   Output Functions

Two MATLAB functions are commonly used to display results in the Command Window. The disp function is generally easier to use because it formats the output in the same way as you see it when you enter a variable name in the Command Window without the semicolon. The fprintf function allows you to customize how the output is formatted.

1.4.2.1   disp

The disp function takes only one argument and generally formats the output in an acceptable manner. If you want to display multiple outputs, such as a text string and a number, there are two options. You can use separate calls to disp, or pass one item to disp that is an array (row vector) where each item in the array has the same data type. In the following example, the num2str function converts a number to a string, thus both items in the array (notice the square brackets) are strings.

```
>> disp(['The number is: ', num2str(num)])
The number is: 15

>> disp('Here is the 2x2 identity matrix')
Here is the 2x2 identity matrix
>> disp(eye(2))
     1     0
     0     1
```

1.4.2.2   fprintf

The fprintf function doesn't provide as much help in formatting the output, but offers the programmer complete flexibility to customize the appearance of the output. When the output is to be displayed in the Command Window, the first argument to fprintf is a string called a format specifier. The string may contain text to display and also information about the data type and location for variables to be displayed in the output. Additional special characters (called escape characters) are often used in the format specifier. A new line is achieved with \n. A tab character is inserted in the output with \t. To display an actual backslash, use two backslashes \\.

Any number of variables may be passed to fprintf, but they should match the variable references in the format specifier. Each variable reference begins with a percent sign (%) and uses a letter code to indicate the data type. In addition, the variable reference can also specify information such as the number of characters to use when displaying the variables (called the field width) and the justification (left or right) within the character field. Programming beginners may need practice using fprintf. The examples below can serve as a guide for how to use fprintf. Be sure to read MATLAB's documentation for fprintf.

The data type codes in the format specifier are listed in table 1.3 followed by some examples. Data type codes 's', 'd', 'g', 'f', and 'e' are the most frequently used codes.

TABLE 1.3: fprintf format specifiers

Letter	Meaning
s	string
c	character
d	integer
u	unsigned integer
f	floating point
g	floating point, but fewer digits
e	scientific notation
o	octal
x	hexadecimal

```
>> fprintf('Number is %g. Integer number is %d.\n', ...
    num, intNum)
Number is 15. Integer number is 8.
>> fprintf('Number is %f. Integer number is %d.\n', ...
    num, intNum)
Number is 15.000000. Integer number is 8.
```

```
>> fprintf('Pi is %f.\n', pi)
Pi is 3.141593.
>> fprintf('Pi is %g.\n', pi)
Pi is 3.14159.
>> fprintf('Pi is %e.\n', pi)
Pi is 3.141593e+00.

>> big = 32.8997898e20;
>> fprintf('A big number is %e.\n', big)
A big number is 3.289979e+21.
>> fprintf('A big number is %g.\n', big)
A big number is 3.28998e+21.
>> fprintf('A big number is %f.\n', big)
A big number is 3289978979999999852544.000000.
```

1.5 For Loops

Here we introduce our first programming *control construct*. Control constructs determine which, if any, code is executed, and how many times the code is executed.

The `for` loop is considered a *counting loop* because the loop's declaration explicitly states the number of times that the loop will execute.

1.5.1 Code Blocks

A sequential collection of commands are often grouped together as a set that together perform some task. This set of commands is called a *code block*. A code block may be a single command or it may be a large section of code. Different programming languages use different strategies to identify a code block. In MATLAB, a code block is all of the code between the initial line of the control construct (in this case `for`) and the keyword `end`.

1.5.2 For Loop Syntax

Here is the syntax of a `for` loop.

```
for idx = sequence
    code block
end
```

Here, *sequence* is a series of values (usually numbers). The sequence could also be called a row vector. The variable `idx` sequentially gets the next value of the sequence each time the code block runs.

Here is an example `for` loop. The variable k is 1 the first time through the loop. On the second iteration, k is 2. During the third a final execution of the loop, k is 3.

```
for k = 1:3
    disp(['Iteration: ',num2str(k)])
    disp(2*k)
end
```

The output from this loop is:

```
Iteration: 1
    2
Iteration: 2
    4
Iteration: 3
    6
```

1.5.3   Colon Sequences

The colon operator, as used in the previous example, is frequently used to create a sequence of numbers. They are used in `for` loops and also for several other purposes in MATLAB.

The simplest colon operator usage takes two arguments and counts with a step size of one between the two arguments.

```
>> 1:5
ans =
    1    2    3    4    5
>> 3:7
ans =
    3    4    5    6    7
>> 1.5:4.5
ans =
    1.500    2.500    3.500    4.500
```

With three arguments, the first and third argument specify the range as before, while the second argument gives the step size between items of the sequence.

```
>> 1:2:5
ans =
    1    3    5
>> -12:4:12
ans =
```

```
     -12    -8    -4     0     4     8    12
>> 0:0.5:3
ans =
     0    0.500   1.000   1.500   2.000   2.500   3.000
>> 3:-1:0
ans =
     3     2     1     0
```

1.5.4 Application of For Loops in MATLAB

A for loop is often used to iterate through a sequence of values or items in an array, which is a cornerstone of numerical computing. Although there are syntax differences, for loops are a major component of every computer programming language. However, as will be explained in sections 1.7 and 1.9, MATLAB does not always need a for loop to iterate through an array. We will see how element-wise arithmetic and functions that operate on all of the values in a vector can be used instead of a for loop.

In MATLAB, for loops are needed to execute algorithms with different parameters or data sets, but not usually to apply a calculation to a single set of data. For example, one might use a for loop to plot a series of data curves in a chart. Whereas, the data for each curve on the chart might be generated using element-wise arithmetic and vector aware functions.

1.5.5 Fibonacci Sequence

The Fibonacci sequence of numbers is an exception to what was said earlier about not usually needing loops to generate data values in MATLAB. This is because each value is derived from previously calculated values.

This is the first example where we will use an array (vector) to save the results. We will discuss arrays more in section 1.7. Notice that we preallocate the array with the zeros function.

The definition of the Fibonacci sequence is $F_1 = 0$, $F_2 = 1$, and $F_i = F_{i-1} + F_{i-2}$ for $i \geq 3$.

```
n = 50;  % number of terms
F = zeros(1,n);
% F(1) = 0 -- already set
F(2) = 1
for i = 3:n
    F(i) = F(i-1) + F(i-2);
end
```

Fibonacci sequence

The Fibonacci sequence is interesting because it is a series of numbers that naturally occurs in nature. It is also a sequence that the computer science education world has latched onto because it can be implemented various ways to teach programming concepts and to illustrate programming strategies. A simple recursive function runs very slow because the program recalculates values many times. Using dynamic programming greatly improves the performance, and we will see in section 6.6.1 that there is also a closed form (not iterative) equation for calculating Fibonacci sequence values.

1.5.6   First Plot

Plotting data will be discussed several times in following chapters. Here we will use a `for` loop to plot a sequence of points.

Start by entering the following in the Command Window:

```
>> plot(1, 2, 'o')
```

You should see a plot with a small circle at point $(x = 1, y = 2)$. If you plot another point, the first plot is replaced by the new one.

```
>> plot(2, 3, 'o')
```

If we want multiple plots on the same figure, we want to use **hold on** to retain the same axis for all of the plots. When finished plotting, we issue the **hold off** command so that future plots start over. It is common, to make the first **plot** to generate the graph and then use the **hold on** command before adding new plots, but we can also use a loop to make all the plots after the **hold on** command.

Copy the code from figure 1.2 into a MATLAB script. The appearance of the data points here are specified by the `'r*'` option, which calls for red asterisks with no connecting line.

A peak ahead

A better way to code figure 1.2 is to pass a sequence of points to one plot command as follows. We will discuss how to find the sequence of y axis data points in section 1.7.3.

```
k = 0:5;
plot(k, k.^2, 'r*');
```

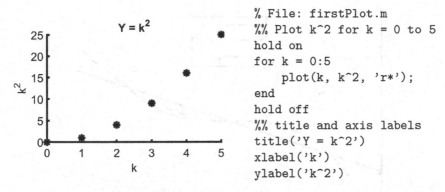

```
% File: firstPlot.m
%% Plot k^2 for k = 0 to 5
hold on
for k = 0:5
    plot(k, k^2, 'r*');
end
hold off
%% title and axis labels
title('Y = k^2')
xlabel('k')
ylabel('k^2')
```

FIGURE 1.2: Simple plot from a for Loop

1.5.7 A Multi-line Plot

Code 1.1 shows an example of using a for loop to plot multiple lines. Note how MATLAB automatically uses a different color for each curve. Also note the legend displayed at the top of the plot. We discuss the details of plotting in the next chapter. Your plot should look like figure 1.3, except you should see colored plot curves.

```
% File: multiline.m
% Multiline plot with a for loop
x = -1.5:0.1:1.5;
style = ["-", "--", "-."];
hold on
for k = 1:3
    plot(x, x.^k, style(k), 'LineWidth', 2)
end
axis tight
legend('y = x', 'y = x^2', 'y = x^3', 'Location', 'North')
hold off
```

CODE 1.1: Using a loop to plot multiple data curves on a plot.

1.6 Control Constructs

Here we continue to present MATLAB control constructs. Whereas, the for loop is considered a *counting* loop because the number of loop executions is predetermined; the constructs here are *conditional*. The evaluation of a logical

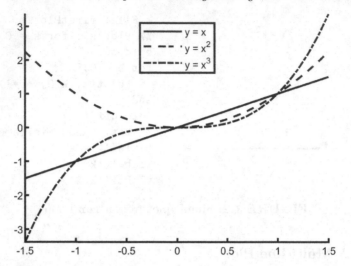

FIGURE 1.3: Plot where each iteration of a `for` loop adds a curve to the plot.

(Boolean) expression determines which code block to execute and, in the case of `while` loops, the number of executions. A logical expression is a statement that evaluates to either *true* (1) or *false* (0). Each logical value is displayed in MATLAB as 1 or 0. To write logical expressions, we need relational and logical operators.

Relational operators compare the relationship between two items and return either a *true* or *false* verdict. For example, using variables x and y we might write a logical expression to see if x is greater than y as x > y. We usually think of relational operators as comparing numeric values, but this is not always the case. For example, the alphabetical order of text strings could be compared as in "Bob" > "Bill". The relational operators are listed below.

- `==` Equal to each other (note: x == y, NOT x = y)
- `~=` Not equal
- `>` Greater than
- `<` Less than
- `<=` Less than or equal
- `>=` Greater than or equal

Logical operators allow us to combine other logical expressions to form a composite expression. Here we are using the symbols a and b to represent individual logical expressions. Symbol a, for example, might represent x > y and b might represent x > z, where x, y, and z are variables. The logical operators are listed in table 1.4.

TABLE 1.4: Logical Operators

Operator	Meaning	Example
&&	AND	a && b
||	OR	a || b
~	NOT	~a

Parenthesis, ( ), may be used to control the order in which logical expressions are evaluated.

Note that as soon as MATLAB determines if a statement is true or false, it stops evaluating. For example, if the logical expression is a && b and a is found to be false, then b is not evaluated because the overall statement will be false. Similarly, if the expression is a \|\| b, b is not evaluated if a is true. This behavior is commonly called *short circuit evaluation*.

Note: In addition to the && and \|\| operators for AND and OR, MATLAB also has (& and \|) operators. These are not used with control constructs, but with logical vectors, which are discussed in section 1.9.3.

1.6.1 Selection Statements

Selection statements determine which, if any, code will run.

1.6.1.1 If Construct

If the logical condition of an if statement is true, the code block runs. If the condition is false, the code block is skipped.

```
if condition
    code block
end
```

1.6.1.2 Else

Add an else statement to the if construct when there is an alternate code block that should run when the condition is false.

```
if condition
    code block 1
else
    code block 2
end
```

1.6.1.3   Elseif

Selection between multiple code blocks is achieve with any number of `elseif` statements.

The final `else` statement is optional. Its code block runs when all of the other logical condition statements are false.

```
if condition1
    code block 1
elseif condition2
    code block 2
elseif condition3
    code block 3
else
    code block 4
end
```

Note: Only the code block for the first true condition will run, even if multiple conditions are true.

What is the output of the following code?

```
a = 1;
if a < 2
    disp(2)
elseif a < 3
    disp(3)
elseif a < 4
    disp(4)
else
    disp(1)
end
```

1.6.1.4   Switch–Case Construct

Switch is a multi-branching selection statement. The switching value can be a variable, expression, or function that evaluates to either a scalar (usually integer values) or a character vector. The code below the first matching case statement is ran. Notice that in the third `case` statement that brackets surround two values. This is a cell array and it is used to match any of multiple values. The code following the `otherwise` statement is run when none of the `case` statements match the switching value.

```
switch x
    case value_1            % x == value_1
        code block 1
    case value_2            % x == value_2
```

```
            code block 2
        case {value_3, value_4} % x == value_3, or value_4
            code block 3
        otherwise               % x is none of the above
            code block 4
    end
```

1.6.1.5   Example Selection Statements

The example in code 1.2 illustrates a nested `if` construct. It also demonstrates the `input` and `error` functions that are used for interacting with the user of the program. It determines the interest rate for a loan based on the loan amount.

Code 1.2 uses the `error` function. The program terminates with a message in the Command Window when the `error` function is called.

```
% File: ifElse.m
loan = input('Enter the loan amount: ');

if loan >= 1e6
    error('Loans must be less than one million dollars')
else
    if loan < 100
        rate = 0.08;
    elseif loan < 1000
        rate = 0.06;
    elseif loan < 10000
        rate = 0.05;
    elseif loan < 100000
        rate = 0.04;
    else
        rate = 0.03;
    end
    disp(['The interest rate is ',num2str(rate*100),'%.'])
end
```

CODE 1.2: Example of multi-branch selection statements.

1.6.2   While Loop

The `while` loop will execute a code block until the logical condition is false, so it may not run, run once, or run many times. Use a `while` loop instead of a `for` loop whenever the number of loop executions can not be pre-determined. This might be because an algorithm needs to run a different number of times depending on the value of a variable, or because of interactions from users.

The loop will evaluate the condition and run the code block if it is true. Each time after running the code block, the condition is re-evaluated for a possible additional run. The loop stops when the condition is false.

```
while condition
    code block
end
```

1.6.3   Example Control Constructs—sinc

MATLAB has an built-in constant called `eps`, which is the smallest, positive, nonzero value that can be noticed when added to the number 1. A fun illustration of a `while` loop is to find this value with a program. When we add a number to 1 and the computer thinks that the sum is 1, then we have a number smaller than `eps`.

We will use our calculated `myeps` variable to prevent a divide by zero error. Our sinc function should plot a smooth curve at $sinc(0) = 1$. The code and plot are found in code 1.3 and figure 1.4.

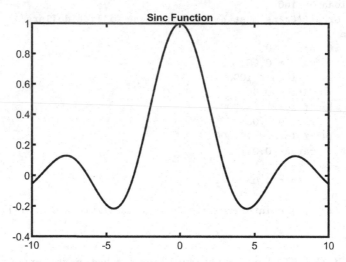

FIGURE 1.4: Sinc function $(\sin(x)/x)$.

1.6.4   Continue and Break

MATLAB has two addition commands related to looping control constructs. These commands provide mechanisms for altering the execution flow in the middle of a loop. Both of these commands are normally placed inside an `if` construct that is there to catch special conditions.

```
%  File: sinc1.m
%% Manually find eps (myeps) using a while loop,
%  then use myeps to prevent divide by zero and plot
%  the sinc function.

epsilon = 1;
while (1 + epsilon) ~= 1
    myeps = epsilon;
    epsilon = epsilon / 2;
end
fprintf('1 + %9.5g is the same as 1\n', epsilon)
fprintf('myeps = %9.5g\n', myeps);


%% Use myeps with a sinc function sinc(x) = sin(x)/x

t = -10:0.1:10;
y = t;              % just create y array for efficiency sake
% This could be vectorized, but this code illustrates a for
% loop and a selection statement.
for k = 1:length(t)
    if t(k) == 0
        x = myeps;  % prevent a divide by zero error
    else
        x = t(k);
    end
    y(k) = sin(x)/x;
end
plot(t, y)
title('Sinc Function')
```

CODE 1.3: Manual find of `eps` and sinc $(\sin(x)/x)$ function plot.

1.6.4.1 Continue

The `continue` keyword causes execution of the current loop iteration to skip past the remainder of the loop's code block. Control returns to the beginning of the loop where the loop condition is evaluated again to either advance to the next loop iteration or exit the loop.

In the following pseudocode example, if the `special_condition` is true, code block 2 is skipped and control moves back to evaluating `loop_condition`.

```
while loop_condition
    code block 1
    if special_condition
        continue
    end
    code block 2
end
```

Round–off Errors

The topic of eps, also called *machine epsilon*, brings up the important topic of round–off errors and numerical stability. We saw here that if a number smaller than eps is added to 1, the result is 1. We can think of the difference between 1 and 1 + eps as the round–off error of the digital numbers between 1 and 2. Because floating point numbers are stored in the computer using a binary scientific notation, the round–off error doubles for each increment of the exponent.

$$(-1)^{sign} \times 1.N \times 2^{exponent}$$

A number smaller than $2 \times$ eps is lost in addition when added to numbers between 2 and 4. Similarly any number less than $4 \times$ eps is lost in addition when added to numbers between 4 and 8. Algorithms that are *numerically stable* avoid addition and subtraction between large and small numbers. We will consider this topic again in sections 5.5.4 and 5.6.

1.6.4.2 Break

The **break** keyword causes execution of the current loop to stop. Control advances to the code after the loop.

In the following pseudocode example, if the **special_condition** is true, code block 2 is skipped and the loop is finished.

```
while loop_condition
    code block 1
    if special_condition
        break
    end
    code block 2
end
```

1.6.4.3 Continue and Break Example

Code 1.4 lists a short script program using **continue** and **break** statements. The program prompts the user to enter numbers one at a time. The user chooses how many numbers to enter. Since we don't know how many numbers will be entered, the loop is programmed with a **while true** statement to run until the user enters a value of zero. When zero is entered, the **break** statement stops the loop. If the user makes a mistake and enters a negative number, we don't want to include that number in the average, so a **continue** statement skips the rest of the loop's code and the user is prompted again.

```
% File: continueBreakDemo.m
% Prompt the user for the length of items and
% calculate the average.

disp('Enter item lengths, enter zero when finished')
sum = 0;
n = 0;
while true
    item_len = input("Enter the next length: ");
    if item_len == 0
        break;
    elseif item_len < 0
        disp('Items must have a length greater than 0.')
        continue;
    end
    n = n + 1;
    sum = sum + item_len;
end
average = sum/n;
disp(['The average length is ', num2str(average)])
```

CODE 1.4: Example use of `continue` and `break` statements.

1.7 Vectors and Matrices in MATLAB

MATLAB (name taken from MATrix LABoratory) is designed to work with variables that actually contain multiple numeric values. This allows us to perform calculations on every value in the variable with just one expression. It also allows MATLAB programs to conveniently solve problems using linear algebra. We will put off the linear algebra aspect for now and focus on *vectorized* equations that can do calculations on many values at once without the need to write a `for` loop.

The numeric variables that we have used so far have only one value. In math terminology, we call such variables *scalars*. MATLAB reports that the size of scalars are 1×1. We will now introduce vectors and matrices, but focus on vectors. Vectors can also be called *arrays*, but note that MATLAB has both row vectors and column vectors.

scalar

A scalar is a variable with size 1×1. That is, it has only one value. The `isscalar` function tests if a variable is a scalar.

row vector

A row vector is a variable of size $1 \times n$, where $n > 1$. It has one row containing n values. Row vectors can be created with the colon operator, manually entered, or created with one of several MATLAB functions. The `isvector` and function tests if a variable to is a vector, while the `isrow` function more specifically tests if a variable is a row vector.

```
>> C = 1:6          % C is a 1x6 row vector
C =
     1     2     3     4     5     6
>> D = [3 5 7 9]    % D is a 1x4 row vector
D =
     3     5     7     9
```

column vector

A column vector is a variable of size $m \times 1$, where $m > 1$. It has one column containing m values. Column vectors can be created with the transpose of a row vector, manually entered, or created with one of several MATLAB functions. The `iscolumn` function tests if a variable is a column vector. We usually use column vectors to hold geometric points and vectors.

```
>> E = (1:6)'   % ' is the transpose operator.
E =             % A transpose operation is used to change
     1          % a row vector into a column vector.
     2
     3
     4
     5
     6
>> F = [3; 5; 7; 9]
F =
     3
     5
     7
     9
```

Vectors in our physical 3-dimensional world are said to be in $\mathbb{R}^3$, that is they consist of three real numbers defining their magnitude in the x, y, and z directions. Vectors on a plane, like a piece of paper, contain two elements and are said to be in $\mathbb{R}^2$. Vectors may also be used for applications not relating to geometry and may have higher dimension than 3. Generally, we call this

$\mathbb{R}^n$. For some applications, the coefficients of vectors may also be complex numbers, which is denoted as $\mathbb{C}^n$.

Row vectors are usually used for sequences of numbers, such as is used for plotting. However, when referring to a geometric vector in either $\mathbb{R}^2$ or $\mathbb{R}^3$ space, then the term *vector* usually refers to a column vector. For convenience of notation, the elements of a column vector may be written as a row of numbers separated by commas and between parenthesis.

These are both column vectors in $\mathbb{R}^3$.

$$\begin{bmatrix} 1 \\ 2 \\ 3 \end{bmatrix} = (1, 2, 3)$$

matrix

A matrix is a variable of size $m \times n$, where $m > 1$ and $n > 1$. It has m rows and n columns of data values. Matrices can be manually entered, or created with one of several MATLAB functions. The `ismatrix` function tests if a variable is a matrix.

Note that the `ismatrix` function considers vectors ($m = 1$ or $n = 1$) to be matrices; although, we will attempt to distinguish between them since vectors have some unique properties.

Matrices may also be formed by concatenating row vectors or column vectors.

```
>> G = [1 2 3; 4 5 6; 7 8 9]    >> D = [1; 1; 2];
G =                             >> E = [0; 1; 3];
                                >> J = [D E]
     1     2     3              J =
     4     5     6
     7     8     9                   1     0
                                     1     1
>> A = [1 2 4];                      2     3
>> B = [2 1 6];
>> H = [A; B]
H =

     1     2     4
     2     1     6
```

Matrices of different sizes may be combined as long as the resulting matrix is rectangular or square. Figure 1.5 and the following code sample show how matrices of size 4×5, 4×1, and 2×6 are concatenated with the command D = [[A B] ; C] to make a 6×6 matrix.

```
>> A = ones(4, 5)
A =
```

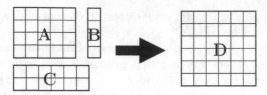

FIGURE 1.5: Matrices concatenated, D = [[A B] ; C]

```
         1       1       1       1       1
         1       1       1       1       1
         1       1       1       1       1
         1       1       1       1       1
>> B = 2*ones(4, 1)
B =
         2
         2
         2
         2
>> C = 3*ones(2, 6)
C =
         3       3       3       3       3       3
         3       3       3       3       3       3
>> D = [[A B];C]
D =
         1       1       1       1       1       2
         1       1       1       1       1       2
         1       1       1       1       1       2
         1       1       1       1       1       2
         3       3       3       3       3       3
         3       3       3       3       3       3
```

Higher dimension data

Data arrays containing more than two dimensions are also possible. Color images have size $m \times n \times 3$, as they contain three monochrome images for red, green, and blue.

1.7.1   Matrix Generating Functions

One example of when it is desired to create a simple matrix or vector of pre-determined size is when a loop is used to set or modify the vector data. For example, consider the following lines of code:

```
for k=1:5
    A(k) = k;
end
```

To this code, MATLAB gives a warning that says:

```
The variable A appears to change size with each loop
iteration (within a script). Consider preallocating for
speed.
```

The warning is fixed by first creating (preallocating) the vector as follows:

```
A = zeros(1,5);
for k=1:5
    A(k) = k;
end
```

MATLAB has several functions that will create matrices and vectors. The zeros and ones functions create vectors or matrices with either all zero or one values.

```
>> A = zeros(1,5)          >> C = ones(3,1)
A =                        C =
   0   0   0   0   0          1
>> B = ones(2)                1
B =                           1
   1   1                   >> D = zeros(3,2)
   1   1                   D =
                              0   0
                              0   0
                              0   0
```

1.7.2 Scalar—Vector Arithmetic

When an arithmetic operation occurs between a scalar and a vector or matrix, the operation is applied to each element of the vector.

```
>> A = zeros(1,5)
A =
   0   0   0   0   0
>> A = A + 2
A =
   2   2   2   2   2
>> A = A*3
A =
   6   6   6   6   6
>> A = A/2
A =
   3   3   3   3   3
>> A = A - 1
A =
   2   2   2   2   2
```

1.7.3   Element-wise Arithmetic

Addition and subtraction arithmetic between vectors and matrices happens as expected as long as the two matrices are the same size.

```
>> A = 2*ones(1,4)          >> A + B
A =                         ans =
    2    2    2    2             3    3    3    3
>> B = ones(1,4)
B =
    1    1    1    1
```

When we want to perform multiplication, division, or use an exponent *element-wise*, that is between each element of the vectors, we use the .*, ./, and .^ element-wise operators. These operators take two vectors or matrices of the same size and perform the operation between pairs of values at the same position in the vectors or matrices. In reference to element-wise multiplication, division, or exponent operations the data might be called an *array* rather than a vector or matrix.

```
>> a = [2 3 4];             >> c ./ a
>> b = [5 6 7];             ans =
>> c = a.*b                      5    6    7
c =
    10    18    28
```

The element-wise exponent can be used between vectors and between a scalar and a vector.

```
>> 2.^a                     >> b.^a
ans =                       ans =
    4    8    16                 25       216      2401
>> a.^2
ans =
    4    9    16
```

Note: Matrix and vector multiplication (A*B) and division like operations (A\B and A/B) are covered in chapter 5.

1.7.4   Vector and Matrix Indices

You can select elements of a vector with parentheses around an index number or range of indices. The range of valid indices for a vector of length N is 1 to N.

Warning: Most other programming languages count indices from 0 to N-1.

MATLAB produces an error when a vector is indexed outside of its range. However, when an assignment statement adds elements beyond the current range then the vector is expanded.

```
>> A = [2 4 6 8]
A =
    2    4    6    8
>> A(1)
ans =
    2
>> A(4)
ans =
    8
>> A(5)
Index exceeds the number of array elements (4).
>> A(5) = 10
A =
    2    4    6    8   10
```

1.7.4.1 Ranges of Indices

The colon operator can be used to select a subset of a vector. MATLAB keeps a special identifier called **end** for the last element of a vector. Ranges of indices can be used when reading from a vector and when assigning values to it.

```
>> A = 1:10;
>> A(1:5)
ans =
    1    2    3    4    5
>> A(6:end) = A(1:5)
A =
    1    2    3    4    5    1    2    3    4    5
>> A(2:2:end-2) = [10 9 8 7]
A =
    1   10    3    9    5    8    2    7    4    5
```

1.7.4.2 Accessing Data in Matrices

You can extract values from a matrix by indexing the row number and column number.

```
>> x = A(rowNum,colNum);
```

You can use the MATLAB keyword **end** as either a row or column index to reference the last element. The following line of code accesses the last element in A.

```
>> x = A(end, end)
x =
    92
```

The following reads the data value that is in the row before the last row and the column that is two columns before the last column.

```
>> x = A(end-1,end-2)
x =
    72
```

To extract an entire row or column from an array, you can use a colon (:). Think of the colon as saying *all rows* or *all columns*. This extracts a column from A:

```
>> x = A(:,colNum);
```

To index a range of rows or columns use a vector containing those index values. For example, both lines of the following code will extract the first three rows of A.

```
>> firstThree = A(1:3,:);
>> firstThree = A([1 2 3],:);
```

What index values will return all four corners of A?

```
>> x = A([1 end],[1 end])
```

1.7.5   Delete Vector or Matrix Data

Assigning vector or matrix elements to take values of empty square brackets ([]) deletes them. The selection of the elements to delete can use either array indexing or logical vectors, which are covered in section 1.9.3. The empty square brackets will also create an empty vector.

Note: The isempty function tests if a vector or matrix is empty, that is of size 0×0.

```
>> w = []
w =
     []
>> isempty(w)
ans =
  logical
   1
>> v = 1:6
v =
     1.00     2.00     3.00     4.00     5.00     6.00
>> v(2:4) = []
v =
     1.00     5.00     6.00

>> B = randi(20, 1, 5)   % random numbers
B =
    19.00    14.00    16.00    15.00     8.00
>> B(B > 15) = []        % delete based on a logical vector
B =
    14.00    15.00     8.00
```

1.7.6   Linear and Logarithmic Spaced Vectors

An alternative to the colon operator for creating vectors that span a range of number is the `linspace` function. The `linspace` function returns a row vector with a fixed number of elements. Use `linspace` when you want a certain number of evenly spaced elements in the vector.

The first and second arguments to `linspace` give the range. The third argument, which is optional, gives the number of elements. The default for the third argument is 100, which is a nice value for creating a smooth line plot.

The `logspace` function also generates a sequence of numbers, but the numbers are evenly spaced on a logarithmic scale. The first two arguments to `logspace` specify the range as an exponent to 10 ($1 \rightsquigarrow 10^1 = 10; 2 \rightsquigarrow 10^2 = 100$). Like `linspace`, the third argument gives the number of elements desired. The default number of elements is 50.

```
>> x = linspace(0,10,5)
x =
        0    2.5000    5.0000    7.5000   10.0000
>> y = linspace(-pi, pi);
>> whos
Name      Size              Bytes  Class     Attributes

x         1x5                  40  double
y         1x100               800  double
```

```
>> logspace(0,2,5)
ans =
    1.0000    3.1623   10.0000   31.6228  100.0000
```

1.8 MATLAB Functions

Scripts as discussed in section 1.3 are nice because they let us edit our code until we get it right and save the code in files for future use. But we often need something more. We have used many functions that are part of MATLAB's library of functions, now we need to write our own functions.

- Functions have a private workspace, so there is less concern of name conflicts.

- A function provides a simple way to invoke code that is called multiple times. Functions allow common code to be called with different parameters or data, which is easier than replicating code.

- Knowing when to divide a program into separate functions versus keeping the code together is somewhat an art form which is best perfected by practice. Two questions will help with the design decisions.

  1. Does the code need to be executed more than once, especially with different values?

  2. Does the code provide a specific task? If so, putting it in a function may help to draw attention to the task and thus improve the clarity of the program.

- Functions may be placed either inside a script file or in their own file. Functions inside a script are good for functions that would only be used in that script. Place a function in its own file to make it generally available for other scripts and functions that you might develop.

1.8.1 Syntax of a Function

In MATLAB, the returned value(s) from a function are specified by their local variable name on the function definition line. The returned values are the values of those variables at the end of the function. If there is more than one return value, they are listed like a vector.

Note: The word `function` is a keyword.

Here is the syntax for defining a function with one return values and two input arguments.

```
function outPut = my1stFunction(input1, input2)
% MY1STFUNCTION - Help documentation
    your code
    outPut = <return value>;
end
```

This function has two return values and two input arguments.

```
function [outPut1, outPut2] = my2ndFunction(input1, input2)
% MY2NDFUNCTION - Help documentation
    your code
    outPut1 = <one return value>;
    outPut2 = <another return value>;
end
```

Note: Create help text by inserting comments at the beginning of your program. If your program includes a function, position the help text immediately below the function definition line (the line with the function keyword). In the help comments, we usually start by the typing the name of the function in capital letters, which causes the help command to display the name of the function with bold letters.

Functions are either private or public. The code for private functions is in the same file as a script, and are only ran from the script or other private functions in the same file. Private functions may not be run directly from the Command Window. In contrast, public functions may be run from the Command Window or from scripts and functions contained in other files. Public functions have two important constraints.

1. A public function must be the only function in its file.

2. The name of the function must match the name of the file.

Note: Private functions are also called *helper* functions.

1.8.2 Calling a Function

```
[out1, out2] = my2ndFunction(in1, in2);
```

The values of the input arguments and outputs are copied between the Command Window, calling script, or function and the called function. The variable names between the two are completely independent. They may be the same name with no name space conflict or they may be completely different names. The only important thing about the values and variables is the order in the list of arguments and return values.

Self Quiz

(Select all that apply) Which of the following are valid ways to call a function called `myFunction` that takes three inputs arguments and returns two output variables?

1. [out1, out2] = myFunction(in1, in2, in3)

2. out1 = myFunction(in1, in2, in3)

3. myFunction(in1, in2, in3)

Answer: All of the syntaxes shown are valid. The function call returns only outputs that are requested. If no outputs are requested, the function still runs and the first output is saved to `ans`.

1.8.3   Example Function

Code 1.5 lists a short example function. Notice how the return value is set as the last value of the return variable at the end of the function. The function calculates the sum of a sequence of numbers according to the equation that Johann Carl Friedrich Gauss reportedly discovered when he was a young boy. His teacher wanted to keep the students busy with a tedious assignment so the students were instructed to find the sum of the digits from 1 to 100. Gauss surprised the teacher by finding the answer very quickly. He recognized that pairs of numbers had the same sum. The sum of 1 and 100 is the same as 2 and 99, as for 3 and 98, etc. In the sequence from 1 to 100, there are 50 such pairs. In this example, we generalize the equation to work with any sequence with consistent spacing between the numbers.

```
function ssum = sequenceSum(sequence)
% SEQUENCESUM - The sum of a sequence of numbers. The
%                spacing between numbers in the sequence
%                must be consistent.
    start = sequence(1);
    stop = sequence(end);
    n = length(sequence);
    if mod(n, 2) == 0    % n is even
        ssum = n*(start+stop)/2;
    else                 % n is odd
        ssum = (n-1)*(start+stop)/2 + sequence(ceil(n/2));
    end
end
```

CODE 1.5: Function to find the sum of a sequence of numbers.

We can test the function including the help information from the Command Window. We can use the `sum` function to verify the results.

```
>> help sequenceSum
sequenceSum - The sum of a sequence of numbers. The
              spacing between numbers in the sequence
              must be consistent.
>> sequenceSum(1:100)
ans =
        5050
>> sum(1:100)
ans =
        5050

>> sequenceSum(50:3:100)
ans =
        1258
>> sum(50:3:100)
ans =
        1258
```

1.8.4 Function Handles

MATLAB has a shortcut way of expressing simple functions that can be expressed with one single statement. In the following example, f is the name of the *function handle*. The argument(s) passed to the function are inside the parenthesis after the @ symbol. The two primary applications of function handles are to create data points according to an equation and to pass a function as an argument to another function.

```
%% An example of a function handle

f = @(x) exp(-x) .* sin(x);

t = linspace(0, 3*pi);
plot(t, f(t));
```

Function Handles and Anonymous Functions

These two terms are often confused and interchanged, which as a practical matter is not really a problem. An anonymous function is the definition of the function that is not given a name. It is usually an argument to a function, such as fplot(@(x) x.^2, [0 5]). The fplot function makes a plot of a function over the specified range of the x axis. An anonymous function becomes a function handle when it is assigned to a variable name, as in f = @(x) x.^2.

1.9   Functions Operating on Vectors

1.9.1   Replacing Loops with Vectorized Code

As mentioned in sections 1.5 and 1.7, MATLAB has the ability to operate on whole vectors and matrices with a single arithmetic expression. Code that takes advantage of this feature is said to be *vectorized*. Such code has two benefits over code that uses loops to iterate over the elements of vectors. First and most importantly, it can be simpler and faster to write vectorized code than to write code using loops. Secondly, vectorized code runs faster because MATLAB's interpreter runs less and optimized, compiled code runs more to perform the calculations.

The functions `tic` and `toc` are used in code 1.6 to illustrate the difference. More complex expressions yield more speed improvements for vectorized code.

FYI, calculating π

The equation in the next example is the Maclaurin series for $\tan^{-1} 1$.

$$\tan^{-1} x = x - \frac{x^3}{3} + \frac{x^5}{5} - \frac{x^7}{7} + \cdots \qquad -1 \le x \le 1$$

When we let $x = 1$, we get $\frac{\pi}{4}$, which is a series that could be used to estimate the value of π.

$$\frac{\pi}{4} = \tan^{-1} 1 = 1 - \frac{1}{3} + \frac{1}{5} - \frac{1}{7} + \cdots$$

The three code sections in code 1.6 find the same estimate for π, but the vectorized code is faster.

1.9.2   Vectors as Input Variables

Functions that you write should assume that they may be called with vectors as input variables, even if your intent is that it will work with scalar inputs. Thus, unless the function is performing matrix operations or working with a scalar constant, use element-wise operators (section 1.7.3).

1.9.3   Logical Vectors

Logical vectors are formed by testing a vector with a logical expression. The resulting vector is 1 (true) for each value in the original vector where the logical expression is true and 0 (false) in the other positions. Notice the single &

```
% File: piEstimate.m
% Series calculation of pi via pi/4 (0.7849)
N = 1003;
%% For Loop Code
tic
mysum = 1;
sign = 1;
for n = 3:2:N
    sign = -sign;
    mysum = mysum + sign/n;
end
toc
disp(['For Loop Code: ', num2str(4*mysum)])

%% Vectorized Code 1
tic
denom1 = 1:4:N;
denom2 = 3:4:N;
mysum = sum(1./denom1) - sum(1./denom2);
toc
disp(['Vectorized Code 1: ', num2str(4*mysum)])

%% Vectorized Code 2
tic
n = 1:4:N;
mysum = sum(1./n - 1./(n+2));
toc
disp(['Vectorized Code 2: ', num2str(4*mysum)])
```

CODE 1.6: Comparison of looped code and vectorized code.

symbol used as an AND operator between logical vector expressions. The |
symbol is the logical vector OR operator.

```
>> A = 1:10
A =
    1   2   3   4   5   6   7   8   9  10
>> A > 3 & A <8
ans =
  1x10 logical array
   0   0   0   1   1   1   1   0   0   0
```

Next we use a logical vector to modify the vector A. The mod function gives
the result of modulus division (the remainder after integer division). The first
expression below shows a 1 where the vector is odd and a 0 where it is even.
The second expression sets A to 0 for each odd value. It does so by first creating

a temporary logical vector. Then, each element in A where the logical vector is true is given the new value of 0.

```
>> mod(A,2)
ans =
     1   0   1   0   1   0   1   0   1   0
>> A(mod(A,2) == 1) = 0
A =
     0   2   0   4   0   6   0   8   0  10
```

Table 1.5 lists a few functions that operate on logical vectors.

TABLE 1.5: Logical Vector Functions

Function	Purpose	Output
any	Are any of the elements true?	true/false
all	Are all the elements true?	true/false
nnz	How many elements are true?	double
find	What are the indices of the elements that are true?	double

find

The find function takes a logical vector as input and returns a vector holding the indices where the logical vector is true.

The following example uses an integer random number generator to make the sequence of numbers.

```
>> A
A =
     8   4   6   2   7   3   7   7   8   5
>> find(A > 6)
ans =
     1   5   7   8   9
```

nnz

Use the function nnz to count the number of nonzero or true values in a logical array.

```
>> nnz(A > 6)
ans =
     5
```

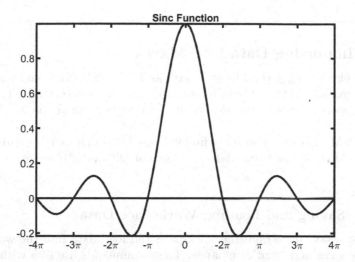

FIGURE 1.6: Sinc function $(\sin(x)/x)$ with logical vector indexing.

1.9.4   Sinc Revisited

When we looked at the sinc function in section 1.6.3, we commented that it would be better to use vectorized code. Notice how a logical vector is used in code 1.7 to index row vector t. The plot is shown in figure 1.6.

```
% File: sinc2.m
% Sinc function using a logical vector to prevent divide by 0.

t = linspace(-4*pi,4*pi,101);
t(t == 0) = eps;
y = sin(t)./t;
figure, plot(t, y)
hold on
plot([-4*pi 4*pi],[0 0], 'r')
xticks(linspace(-4*pi,4*pi, 9))
xticklabels({'-4\pi','-3\pi','-2\pi','-\pi','0',...
        '\pi','2\pi','3\pi','4\pi'});
title('Sinc Function')
axis tight
hold off
```

CODE 1.7: The sinc $(\sin(x)/x)$ function implemented with logical vector indexing.

Note: In code 1.7, we took care to set and label the x axis tick marks appropriately for a trigonometry function. See chapter 2 for more details about controlling the appearance of plots.

1.10   Importing Data Into MATLAB

Data files having several formats such as TXT, CSV, XLS, and XLSX can be imported into MATLAB both interactively and from commands. The latest list of supported import and export file formats is available on MathWorks' website [42].

Search MathWorks' website to find videos related to importing spreadsheet data into MATLAB. In particular, the videos at [22] and [61] are quite helpful.

1.10.1   Saving and Loading Workspace Data

You can save data from the workspace and load data into the workspace using the `save` and `load` commands. These commands use files with a `.mat` file name extension.

```
>> save myData
```
Creates `myData.mat` containing current workspace variables.

Instead of saving the entire MATLAB workspace, you can selectively save individual variables.
```
>> save someFile x y
```
Saves `x` and `y` to `someFile.mat`.

```
>> load myData
```
Loads variables from `myData.mat` into the workspace.

1.10.2   Import Tool

The Import Tool is a convenient tool for importing data from files with an intuitive graphical user interface. If the file has labels in the first row for each column, they will appear at the top of each column. If there are no labels, then you can type them in, see figure 1.7. The labels are used for naming the column vectors or table fields.

The Import Tool is available from the Home tab, labeled as "Import Data". Notice the pull down menu specifying "Output Type". The choice of "Column vectors" yields similar vector data as we have previously used. Each column of data from the file is imported as a column vector.

Notice on the pull down menu on the right side of the Import Tool labeled "Import Selection". The two choices "Generate Script" and "Generate Function" can be useful to automate the task for future use. Importing the data as column vectors is one of the more complex import tasks, so the Import Tool helps out quite a bit.

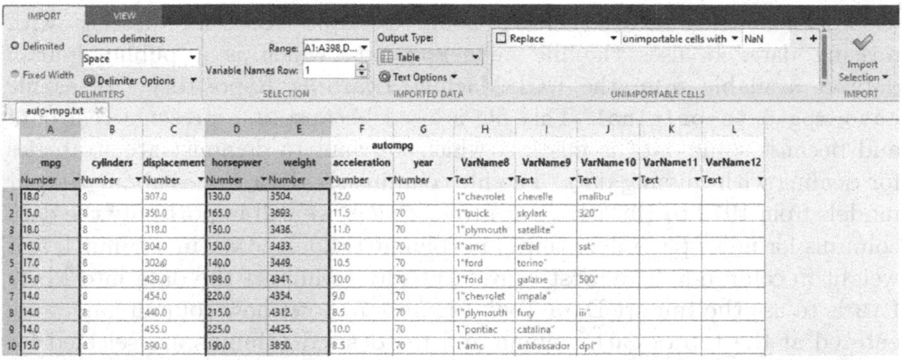

FIGURE 1.7: The Import Data Tool is used to import desired data columns.

Notice also how missing data will be handled. A pull–down menu offers choices. Replacing missing data with NaN (not a number) is often a good choice.

1.10.3 Reading Tables

Column vectors is what we are most familiar with at this point, but it may not always be the best choice. Importing data as a table is simpler to perform from commands. The columns of a table can be converted to column vectors if needed. It is not necessary for each column of a table to be of the same data type. The data can be thought of as a series of observations (rows) over a set of different variables (columns). The `sortrows` function is used to sort the data based on any table column and the data in each row is kept together.

The command for importing a table from a file is `readtable('filename')`. The columns of a table may be accessed as column vectors using the notation of *tableName.columnName*.

1.10.4 Dealing with Missing Data

Data values that are missing show up as NaN (Not a Number). The MAT-LAB function `ismissing` returns a logical vector showing where data is missing. There are three basic ways to deal with missing data.

1. **Ignore it:** Some MATLAB statistics functions have an option `'omitnan'` that will ignore any NaN values.

2. **Delete it:** Setting a table row to [] removes the data.

3. **Interpolate it:** MATLAB has a function called `fillmissing` that will interpolate missing data in a vector between its neighbor data. Section 7.2 has more information about data interpolation.

The following tutorial shows some example code for dealing with missing data. It uses the file `auto-mpg.txt`, which is a public domain dataset available from the UCI Machine Learning Repository [20] as file `auto-mpg.data-original`. This file is used because it represents real data and because some data is missing, which we want to demonstrate strategies for dealing with missing data. The file contains six data values about 398 car models from 1970 to 1982. For our investigation, we will import only the data columns for miles per gallon (mpg) in column 1, horsepower in column 4, and weight in column 5. The most convenient way to import the data into MATLAB is to use the Import Data Tool. Figure 1.7 shows how column names are entered at the top of each column and the desired columns are selected for import. The data will be imported into a table and any missing values will be replaced with NaN (not a number).

After the data is imported into a table, we verify the names of the table fields. We will also save the data so that it is easier to load later. The saved file is `autompg.mat`.

```
>> autompg.Properties.VariableNames
ans =
  1x3 cell array
    'mpg'    'horsepower'    'weight'
>> save autompg
```

We will put our commands into a script, but use the "Run Section" button from the Edit tab so that we can take things one step at a time.

First, we will check the three data fields to see if any data is missing. The `ismissing` function will return a logical array and the `nnz` function will count how may values are missing.

```
% File: carTable.m
%% Dealing with missing data
%   Run this code one section at time.

% take a look at the part of the table
% autompg(1:4,:)

% count missing data
disp('missing MPG')
nnz(ismissing(autompg.mpg))
disp('missing Horsepower')
nnz(ismissing(autompg.horsepower))
disp('missing Weight')
nnz(ismissing(autompg.weight))
```

We see that the `horsepower` field has 6 missing values.

```
missing MPG
ans =
     0
missing Weight
ans =
     0
missing Horsepower
ans =
     6
```

If we just want to calculate statistics of the data, we can ignore the missing data with the `'omitnan'` option. Most of the statistics functions have this option. For more on statistics functions, see section 3.2 and check the documentation of functions that you want to use.

```
%% Ignore missing data for mean

disp('mean Horsepower')
mean(autompg.horsepower, 'omitnan')
```

We may want to delete the rows with missing values.

```
%% Delete rows with missing data

%   idx holds index of any rows with missing (NaN) values
idx = ismissing(autompg.horsepower);
autompg(idx,:) = [];    % remove rows missing values
disp('missing Horsepower')
nnz(idx)
```

The above code also displays a count of how many rows were deleted.

```
missing Horsepower
ans =
     6
```

We would expect a negative correlation between horsepower and mpg values, a quick scatter plot verifies this. We could use MATLAB's `corrcoef` function as shown in section 3.6.3 to calculate the correlation, but a scatter plot verifies what we already expected. We just want to see the plot for our own use, so we didn't bother to label the axes or give the plot a title. Figure 1.8 shows the scatter plot.

```
%% scatter plot to observe correlation between MPG
%   and Horsepower

scatter(autompg.horsepower, autompg.mpg)
```

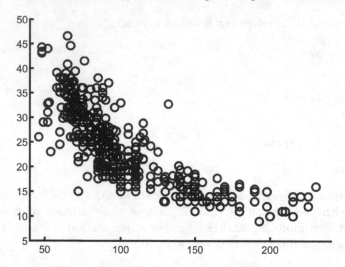

FIGURE 1.8: Scatter plot of car horsepower against mpg.

Next after re-loading the data, we will use linear data interpolation to fill in the missing data. Since we know that the horsepower data is correlated to the mpg data, we can get reasonable results if we sort the data based on the mpg field. The rather long output (not shown) of this step shows the values before and after the filled in values and verifies that the estimates are reasonable.

Whether it is better to delete rows with missing values or use interpolation depends on the application, the data available, and a preferences of individuals.

```
%% Interpolate missing data
% re-load data with missing values
load autompg

% sort rows by MPG to group cars.
autompg = sortrows(autompg, 'mpg');
% find missing data
idx = find(ismissing(autompg.horsepower));
disp('missing data')
for i = idx'
    disp(autompg.horsepower(i-2:i+2))
    disp(' ')
end

% use linear interpolation
autompg.horsepower = ...
    fillmissing(autompg.horsepower, 'linear');
```

```
disp('interpolated data')
for i = idx'
    disp(autompg.horsepower(i-2:i+2))
    disp(' ')
end
```

1.10.5 Exporting Table Data

Use the `writetable` function to export a table to a delimited text file or a spreadsheet.

```
>> writetable('myTable.csv');
```

By default, `writetable` uses the file name extension to determine the appropriate output format.

Note: There is a lot more information about MATLAB tables that are not included here. The tables tutorial in exercise 1.7 shows examples of additional MATLAB commands related to working with tables.

1.11 Text Strings in MATLAB

We have already passed textual data to functions. Actually, what we have mostly used so far are called character arrays where each string is inside a pair of single quote marks, `'Hello world'`. Character arrays have been a part of MATLAB from its beginning, but a new data type of string arrays was introduced with version R2016b and enhanced in version R2017a. String arrays are created with either double quote marks `"My String"` or with the `string()` function. String arrays have many more functions to aid in manipulating textual data.

We might want to store a sequence of individual strings in a single array. Look at the following code to see a problem with doing so with character arrays and how strings and cell arrays provide solutions to the problem. Each element of a cell array holds a reference to data, rather than actual data.

```
>> months1to3 = ['Jan', 'Feb', 'Mar']
months1to3 =
    'JanFebMar'
>> months3 = ["January", "February", "March"];
>> months3
months3 =
  1x3 string array
    "January"    "February"    "March"
```

```
>> CellMonths1to3 = {'January', 'February', 'March'}
CellMonths1to3 =
  1x3 cell array
    'January'    'February'    'March'
```

MATLAB arrays of character arrays concatenate the string elements, which in some cases is fine, but often is a problem. For this reason, we often include spaces as part of character arrays, such as when we use `disp` to combine textual and numeric data displayed to the user:

```
disp(['The answer is: ', num2str(x)])
```

For more information about character arrays and strings, see MathWorks' online documentation [45].

Cell Arrays

One solution for passing textual data to functions is to use a cell array to store the data. Cell arrays are created with curly brackets { } instead of square arrays used for numeric arrays [ ].

A good application for using cell arrays is to set the tick labels on the axis of a plot, which is demonstrated in section 2.2.9.

Cell arrays can store references to many types of data besides strings. Cell arrays hold references to data rather than the actual data. A cell array does not require homogeneous data types. So, for example, a string, character array, data structure, numeric vector, and a MATLAB object may reside together in the same cell array.

1.12   Exercises

Exercise 1.1   Entering Basic Equations

Simple numeric equations may be entered into either the MATLAB Command Window or into a script in the same way. In this exercise you will practice entering simple equations into MATLAB Scripts. If you are not certain about how to enter an expression, you might experiment with entering the equations in the Command Window.

1. **Area and circumference of a circle**
   The size of a circle is determined by the radius (distance from the center to edge of the circle). We will use variable r for the radius. The area of the circle is: $area = \pi r^2$. The circumference of the circle is: $circ = 2\pi r$. In MATLAB, the value of π is found with the built-it

variable called `pi`. Write equations for the area and circumference of a circle. Store the results in variables named `area` and `circ`.

```
r = 4.5;   % radius

% Calculate the area and circumference
area =
circ =
```

2. Geometry of a rectangle

The area of a rectangle is the product of the lengths of the two sides a and b. The perimeter of a rectangle is the sum of all sides. Remember that the lengths of a and b are each on two sides. Store the area and perimeter in variables called `area` and `perim`.

```
% sides of a rectangle
a = 10.7;
b = 8.2;

% Calculate the area and perimeter
area =
perim =
```

3. Area of a trapezoid

Calculate the area of a trapezoid. Store the area in a variable called `area`. A trapezoid is a four sided shape with two parallel sides of different length. The area is the product of the average length of the parallel sides and the height. $area = \left(\frac{a+b}{2}\right) h$.

```
% trapezoid dimensions
a = 7.4;
b = 4.8;
h = 5;

% Calculate the area of a trapezoid
area =
```

4. Right triangle hypotenuse and area

Given a right triangle with perpendicular sides $a = 5$ and $b = 8$, Find the length of the hypotenuse, c, and area of the triangle. Recall the Pythagorean theorem, $c^2 = a^2 + b^2$. The area of a right triangle is $area = \frac{ab}{2}$.

```
% Right triangle sides
a = 5;
b = 8;

% Calculate the hypotenuse and area of a right triangle.
c =
area =
```

Exercise 1.2   Fourier for Loop

In this exercise, we will take a sneak peek at using arrays in MATLAB and we will also practice writing a simple `for` loop.

First, enter the following commands into the MATLAB command window. In these lines of code, you can see that the variables `t` and `y` are arrays (also called vectors). You can also see that MATLAB has the ability to operate at the array level, which eliminates the need to write a `for` loop to assign each value of the `y` array.

```
>> T = 5;
>> t = 0:15/100:15;
>> y = sin(2*pi*t/T);
>> plot(t, y);
```

Did you see three cycles of a sine wave?

Next we create a sum of sine waves, which is called a Fourier series. A square wave with a period of T can be expressed with the following Fourier series.

$$y(t) = \frac{4}{\pi} \sum_{n=1,3,5...}^{\infty} \frac{1}{n} \sin(n\, 2\, \pi\, t/T)$$

Below are instructions to create the Fourier series of a square wave in MATLAB.

1. Create a new script.

2. Copy the code for creating the variables `t` and `T` into the script.

3. Add the line: `y = zeros(1,101);`. This lines gives the `y` array initial values of zero.

4. Write a `for` loop where a variable called `n` has the values of 1, 3, 5, and 7 during the successive iterations of the loop.

5. In the code block of the `for` loop enter the fourier series sum from above. **Hint:** Add a new sine term to the existing `y` array each time through the loop.

6. After the loop code, multiply `y` by $4/\pi$.

7. Plot the `y` array against `t`. What could you change in your code to make the plot look more like a square wave?

Exercise 1.3   Save Your Money

For this exercise, we will practice writing a `while` loop with a selection construct.

Since you will be well paid after graduating from college, let's say that you invest $1,000 at the beginning of each month. At the end of each month, you are paid interest on your investment. The interest rate paid depends on the account balance. The annual rate is

$$\begin{cases} 0 < balance \le 10,000: 1\% \\ 10,000 < balance \le 15,000: 2\% \\ 15,000 < balance \le 30,000: 3\% \\ 30,000 < balance \le 100,000: 5\% \\ 100,000 < balance \qquad\quad : 7\% \end{cases}$$

Note that these are annual interest rates, not monthly rates. Write a program that tracks the balance of the savings until it is at least 1 million dollars. Make a plot of the savings with years on the x axis and dollars on the y axis.

Exercise 1.4   Close Enough

The MATLAB function `isequal` compares two arrays and returns *true* if they are identical and *false* otherwise. In some situations, two numbers which are slightly different could be treated as close enough for the purposes of a particular application. For example, a variable with value 2.1 and another variable with value 2.100000009 differ by a small value on the order of 10^{-9}.

Write a function that takes three scalar inputs, x, y, and a tolerance, `tol`. It should return *true* if x and y differ by a number less than `tol` and *false* otherwise. Make it a private function by adding the function to a script file that tests the function.

Exercise 1.5   Practice with Vectors and Functions

For this exercise will practice writing some simple functions that use vectors.

1. Write a function that has one input argument, a row vector, v and one output argument, a row vector w that is of the same length as v. The vector w contains all the elements of v, but in the exact opposite order. For example, if v is equal to [1 2 3] then w must be equal to [3 2 1]. You are not allowed to use the built-in function `flip`.

2. Write a function that takes as input a row vector of distances in kilometers and returns two row vectors of the same length. Each element of the first output argument is the time in minutes that light would take to travel the distance specified by the corresponding element of the input vector. To check your math, it takes a little more than 8 minutes for sunlight to reach Earth which is 150 million kilometers away. The second output contains the input distances converted to miles. Assume that the speed of light is 300,000 km/s and that one mile equals 1.609 km.

3. Write a function that is called as: `amag = accelerate(F1,F2,m)`. `F1` and `F2` are three-element column vectors that represent two forces applied to a single object. The argument `m` equals the mass of the object in units of kilograms. The three elements of each force equal the x, y, and z components of the force in Newtons. The output variable `amag` is a scalar that is equal to the magnitude of the object's acceleration. The function calculates the object's acceleration vector by using Newton's law: $F = ma$, where F is the sum of `F1` and `F2`. Then it returns the magnitude of `a`.

4. Write a function called `income` that takes two row vectors of the same length as input arguments. The first vector, `rate` contains the number of various products a company manufactures per hour simultaneously. The second vector, `price` includes the corresponding sale price per item. The function must return the overall income the company generates in a week assuming a 5-day work week and two 8-hour long shifts per day.

Exercise 1.6   Element-wise Arithmetic

Write a function called `poly_val` that is called as `p = poly_val(c0,c,x)`, where `c0` and `x` are scalars, and `p` is a scalar. If `c` is an empty vector, then `p = c0`. If `c` is a scalar, then `p = c0 + c*x`. Otherwise, `p` equals the polynomial,

$$p = c_0 + c_1 x + c_2 x^2 + \cdots + c_n x^n$$

where `n` is the length of the vector `c`.

Hints:

1. The functions `isempty`, `isscalar`, `iscolumn`, and `length` will tell you everything you need to know about the vector `c`.

2. When `c` is a vector, use an element-wise exponent to determine the vector $\begin{bmatrix} x & x^2 & x^3 & \cdots & x^n \end{bmatrix}$.

3. When `c` is a vector, use the `sum` function with element-wise multiplication. Matrix multiplication could also be used, but we have not yet covered that.

Here are some example runs:

```
>> p = poly_val(-17,[],5000)
p =
    -17
>> p = poly_val(3.2,[3,-4,10],2.2)
p =
    96.9200
```

```
>> p = poly_val(1,[1;1;1;1],10)
p =
   11111
>> p = poly_val(8,5,4)
p =
     28
```

Exercise 1.7   Working with Tables

This exercise is formatted as a tutorial to demonstrate commands and functions used with tables. Enter the listed commands into the MATLAB Command Window and answer the questions.

The file `tallest_bldgs.txt` contains a public domain dataset with information about the world's 200 tallest buildings as of the year 2010. The columns of data are named and described as follows.

bldg_name building name

city city in which the building is located

country country in which building is located

stories the number of stories

year the year in which the building was structurally completed.

height_m height in meters

1. From the command window, import the data in the file `tallest_bldgs.txt` and save it to a table named `buildings`.

   ```
   >> buildings = readtable('tallest_bldgs.txt');
   ```

2. The dot notation (`tableName.VariableName`) may be used to create a MATLAB column vector from a table variable. Create a numeric vector named `height_feet` which contains the heights of all the buildings converted into feet. (1 meter = 3.28084 feet)

   ```
   >> height_feet = buildings.height_m * 3.28084;
   ```

   Using the `max` and `min` functions with the `height_feet` array, what is the difference in feet between tallest and shortest building listed in the table?

3. Modify the existing buildings table to include an additional variable called `height_feet` containing the height data you just calculated. Notice the use of curly brackets, {} as one way to add a variable to a table. The curly brackets are also a good way to work with a subset of a table.

   ```
   >> buildings{:,'height_feet'} = height_feet;
   ```

4. The dot notation is simplest when working with all of the data from a table variable. Remove the `height_m` table variable.

```
>> buildings.height_m = [];
```

5. The sorting capability is a good reason for using tables to hold data. Sort the values in the buildings table in order of decreasing height.

```
>> buildings = ...
      sortrows(buildings, 'height_feet', 'descend');
```

   (a) In what country is the tallest building?
   (b) Which country has two buildings that are the same height at the fourth and fifth tallest buildings?

6. Indexing a table using parenthesis can create a table from a portion of the original table. If curly brackets were used, a vector or matrix would be created from the table data. Create a table that contains the data of the five tallest buildings.

```
>> fiveTallest = buildings(1:5,:)
```

7. Write the contents of `fiveTallest` to a file named `tallBldgs.txt`.

```
>> writetable(fiveTallest, 'tallBldgs.txt');
```

8. Create a logical vector of the buildings over 1000 feet tall.

```
>> over1k = buildings.height_feet > 1000;
```

9. Find the number of buildings that are over 1000 feet tall. Store the result in `n1k`.

```
>> n1k = nnz(over1k);
```

How many buildings are over 1000 feet tall?

10. Create a table of the buildings over 1000 feet tall.

```
>> tallest = buildings(over1k,:);
>> tallest(1:5,:)
>> tallest(end-4:end,:)
```

11. Sort the tallest buildings by age. The default sorting order is ascending.

```
>> oldtall = sortrows(tallest,'year');
>> oldtall(1:5,:)
```

Which country has the oldest building that is over 1000 feet tall?

12. Using the table dot notation, table variables may be used like column vectors with results saved to a new table variable. Determine which buildings have the most and least head room on each floor (story).

```
>> buildings.feet_per_story = ...
    buildings.height_feet./buildings.stories;
>> buildings = ...
    sortrows(buildings, 'feet_per_story', 'descend');
>> buildings(1:5,:)
>> buildings(end-4:end,:)
```

In which country is the building with the most feet per story?

Chapter 2

Graphical Data Analysis

Visualizing data is critical to understanding it. What does a function, $f(x)$, do when x is near zero or very large? Are there any discontinuities? Where does $f(x)$ cross the x axis? If the data is statistical in nature, are there any outliers? Do we know enough about our data to make data driven decisions? We need to make effective two and three-dimensional data plots so that we and others can understand the data.

We made several simple plots in chapter 1 for the purpose of learning about the features of MATLAB. Here we direct our attention specifically to plotting. Keep in mind that MATLAB has a large variety of plotting options, so we can only cover the essentials of plotting.

2.1   Using the Plot Tool

The Plot Tool is convenient when you don't know what type of plot that you need or how to create it. Click on the plot tab to browse through a list of plot types. As shown in figure 2.1, when you select the variables that you want to plot from the Workspace, the plot types suitable for the data will be available to select. Click on a plot type, and the plot will be created. As shown in figure 2.2, the pull–down menus are used to add annotations such as a title and labels for the x– and y–axes. Note that the annotation tools are also available to plots created from the Command Window, scripts, or functions.

One factor that discourages the use of the Plot Tool is that plots are created manually with the graphical interface; whereas, plotting commands entered in a script or function are easy to fine-tune with editing and can be re-run as desired. However, the Plot Tool can create code for a script or function to preserve the steps used in making a plot with the Plot Tool.

DOI: 10.1201/9781003271437-2

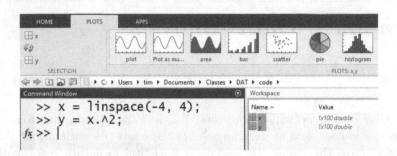

FIGURE 2.1: The Plot Tool lets user select the data and plot type with a click of the mouse.

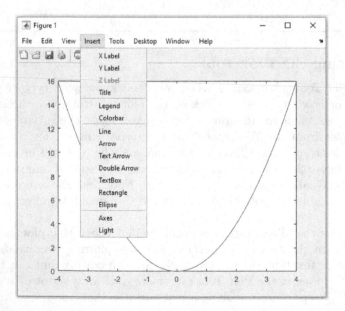

FIGURE 2.2: The Plot Tool pull down menus provide a graphical interface to annotate a plot.

Video Tutorial

MathWorks has made a tutorial video that demonstrates how to use the Plot Tool [31].

2.2 Basic Line Plots

Resources from MathWorks

MathWorks' website has a lot of documentation and videos related to plotting. Some recommendations follow.

- An introductory video on how to use basic plotting functions [59].

- Documentation on how to create 2-D line plots [38].

- Documentation on 2-D and 3-D plots [43].

- Documentation on properties of plot lines [40].

- MathWork's *MATLAB Fundamentals* course has a good introduction to creating plots [44].

2.2.1 Basic Plotting Notes

The Plot Tool is easy to use, however, using commands and functions allows for easier reproduction of plots and more automation. Use the `plot(x, y)` function to plot vector y against vector x. If only a single vector is passed to `plot`, then the index values of the vector are used for the x axis.

Use an additional argument to the `plot` function specifying a color code to change the color of the data curve. The following command will plot y versus x in magenta.

```
>> plot(x, y, 'm')
```

Some other color codes are:

- 'k' black
- 'b' blue
- 'c' cyan
- 'g' green

- 'r' red
- 'w' white
- 'y' yellow

Line style codes are used if you want something other than a solid line. The following command will plot y versus x with a dotted line.

```
>> plot(x, y, ':')
```

Some other line style codes are:

- '-' solid
- '--' dashed
- '-.' dash-dot

You can combine these codes. Generally, the order does not matter, so both of the following commands make a plot using a red dotted line.

```
>> plot(x, y, 'r:')
>> plot(x, y, ':r')
```

In addition to color and line style, you can specify a marker style. The following command will plot y versus x using asterisks.

```
>> plot(x, y, '*')
```

Some other marker style codes are:

- '.' points
- 'o' circles
- 'x' crosses
- '+' plus sign

- 's' square
- 'd' diamond
- '^' upward-pointing triangle
- 'v' downward-point triangle

If the only plot specification is a marker, the default line style is none. Add a line style, such as a dash (-), to the marker style to also plot a line.

The **grid** command adds or removes a grid to your plot:

```
>> grid on
```

```
>> grid off
```

2.2.2 Annotating Plots

Labels are added to plots using plot annotation functions such as `xlabel`, `ylabel`, and `title`. The input to these functions must be text. Traditionally, the textual data was entered as a *character array* made with characters and digits inside a pair of `'apostrophes'` (single quotation marks). MATLAB now allows text labels that are the newer *string arrays* made with characters and digits inside a pair of `"quotation marks"`. The text may be a simple textual constant or be held in a variable. See section 1.11 for more information on textual data.

```
>> xlabel('Text Label for X Axis') % a text constant

>> ylabel(yAxisName)               % a text variable
```

If you need more than one line in an axis label or title, use a cell array created with a pair of curly brackets `{}`.

```
>> title({'first line', 'second line'})
```

LATEX markup formatting may be used in the annotations. Thus, x^2 will become x^2 and x_2 becomes x_2. LATEX markup is a default standard for mathematical equations. You can search the Internet and find good documentation on formatting LATEX math equations. Other symbols, such as Greek letters, can also be added, such as Δ, γ, and π. See MathWorks documentation on *Greek Letters and Special Characters* [39]. Try the following code to see what happens.

```
>> xlabel('\sigma \approx e^{\pi/2}')
```

If you need to use a symbol such as an underscore (_) or a backslash (\), then preface it with the escape character \. Thus, to display \, you must type \\.

The MATLAB `text` function adds a text annotation to a specified position on the plot. See the MATLAB documentation for `text`.

```
>> text(3, 5, 'Best result')
```

2.2.3 Starting a New Plot

You may notice your plot being replaced when you enter a new plot command after making a previous plot. Use the `figure` command to keep the previous plot and start over in a new window.

2.2.4   Multiple Plots in the Same Figure

Several small plots may be put in a figure. The `subplot(m, n, p)` function accomplishes this by making a grid of $m \times n$ plots. The variable `p` is the plot number from 1 to $m \times n$. The plot for the following code is shown in figure 2.3.

```
>> x = linspace(-5, 5);
>> subplot(2, 2, 1), plot(x, x), title('y = x')
>> subplot(2, 2, 2), plot(x, x.^2), title('y = x^2')
>> subplot(2, 2, 3), plot(x, x.^3), title('y = x^3')
>> subplot(2, 2, 4), plot(x, x.^4), title('y = x^4')
```

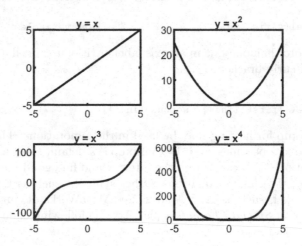

FIGURE 2.3: Plots created using subplots.

Enhanced subplots with the `tiledlayout` function were introduced in MATLAB release R2019b. As with `subplot`, a fixed layout grid of tiled plots is simple to create.

```
>> tiledlayout(2, 2);
```

The `nexttile` command is used to advance to the next plot. Several new features were added beyond the capabilities of subplots.

- Figures may now have a global `title`, `xlabel`, and `ylabel` for the set of plots.

- The size and spacing between tiles is adjustable. The following commands reduce the spacing between tiles and the padding between the edges of the figure and the grid of tiles. The net effect is to increase the size of the individual plots.

```
t = tiledlayout(3, 3);
t.TileSpacing = 'compact';
t.Padding = 'compact';
```

- Specifying the size and location of tiles can create a custom layout. This is done by passing a tile number and an optional range of tiles to be used to `nexttile`. The example in figure 2.4 shows a 3×3 grid of tiles. Three plots are added across the first row. Starting at tile 4, the whole second row is filled with the next plot, which spans a 1×3 set of tiles. The third row has a single tile plot and a 1×2 tile plot.

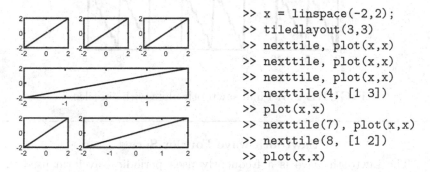

```
>> x = linspace(-2,2);
>> tiledlayout(3,3)
>> nexttile, plot(x,x)
>> nexttile, plot(x,x)
>> nexttile, plot(x,x)
>> nexttile(4, [1 3])
>> plot(x,x)
>> nexttile(7), plot(x,x)
>> nexttile(8, [1 2])
>> plot(x,x)
```

FIGURE 2.4: Tiled plots with a custom layout.

- The layout of the plots may be adjusted with each `nexttile` command. This is done with the `'flow'` option to `tiledlayout`. The first plot fills the whole figure. After the second plot, two stacked plots are shown. As each tile is added, the layout of tiles is adjusted to accommodate the new tile. This can make for a nice demonstration showing successive changes to data. Run the code from the `sawTooth.m` file (code 2.1) to see this. Successive plots show the Fourier series of a sawtooth wave as each new sine wave is added. The user presses the `Enter` key when they are ready to see the next plot. The final tiled figure with six plots is shown in figure 2.5.

2.2.5 Multiple Plots on the Same Axis

Visualizations are often more informative when multiple data sets are plotted on the same axis. To add new plots onto an existing axis without replacing the previous plot, use the `hold on` command. When you are done adding plots, issue the `hold off` command. Then, unless a new figure is started, the next plotting command issued will replace what is currently in the figure.

Another way to plot multiple data sets onto a single axis is to put the y axes values in a matrix. Each column of the matrix will be a different plot.

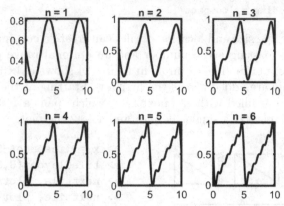

FIGURE 2.5: Final sawtooth plots with six tiles.

Sawtooth wave Fourier Series

The sawtooth wave is a frequently used periodic waveform used in various electrical circuits including music synthesizers. The Fourier series for a sawtooth wave with amplitude of 1 and period of L is given by:

$$f(t) = \frac{1}{2} - \frac{1}{\pi} \sum_{n=1}^{\infty} \frac{1}{n} \sin(2\pi n t/L)$$

The following simple examples illustrate both methods. See figure 2.6 for the plot.

```
>> x = linspace(0,10); Y = zeros(100,3);
>> Y(:,1) = x; Y(:,2) = 2*x; Y(:,3) = 3*x;
>> plot(x, Y)
```

```
>> y1 = x; y2 = 2*x; y3 = 3*x;
>> hold on
>> plot(x, y1, 'k'), plot(x, y2, 'k:'), plot(x, y3, 'k-.')
>> hold off
>> xlabel('x'), ylabel('y')
>> legend('y = x', 'y = 2x', 'y = 3 x', 'Location', 'northwest')
>> title('Multi-line Plot')
```

```
% File: sawTooth.m
% Plots of the Fourier series of a sawtooth function
% displayed using a tiledlayout. Each time through
% the loop, a new plot is added.

clear; close all
L = 5;
N = 200;
t = linspace(0, 2*L, N);
s = zeros(1, N);
plt = tiledlayout('flow');
title(plt, 'Sawtooth Fourier Series')
for n = 1:6
    in = input('Press enter for next plot');
    nexttile
    s = s + sin(2*pi*n*t/L)/n;
    f = 0.5 - s/pi;
    plot(t, f, 'k', 'LineWidth', 2)
    title(['n = ', num2str(n)])
end
```

CODE 2.1: Tiled plots of the Fourier series of a sawtooth function.

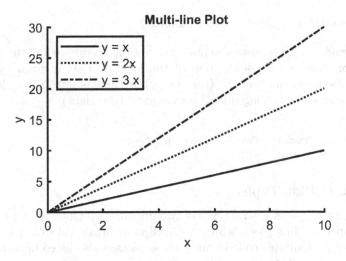

FIGURE 2.6: Plot with multiple data lines.

2.2.6 Adding a Plot Legend

A legend helps identify what data each plot in the graph corresponds to. The order of legend labels matters—the inputs to legend should be in the same order in which the plots were drawn.

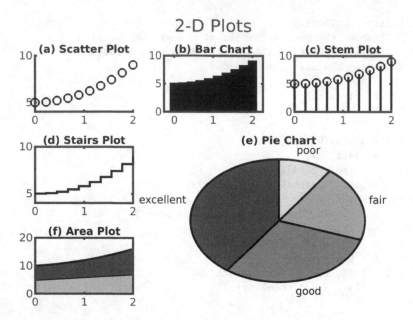

FIGURE 2.7: Common types of 2-D plots other than line plots.

```
legend('first label','second label')
```

The `legend` function also accepts two optional arguments: the keyword `'Location'` and a text description of the location that is given by a compass direction, such as `'north'` (top center) or `'southwest'` (lower left). The default location for the legend is `'northeast'` (top right).

```
legend('Trucks','Cars','Location','west')
```

2.2.7   2-D Plot Types

MATLAB supports several types of two dimensional plots. Six types of plots other than basic line plots, which we have used in several examples, are described below. Example code to make these plots is also listed in code 2.2 and figure 2.7 shows what the plots look like.

Scatter Plot

`scatter(x, y)`: Scatter plot with variable marker size and color. Example shown in figure 2.7 (a).

```
% File: Plots2D.m
% Some 2-D plotting examples

x = linspace(0, 2, 10);
y = 5 + x.^2;
y2 = 5 + x;
plts = tiledlayout(3, 3);
plts.TileSpacing = 'compact';
plts.Padding = 'compact';
title(plts, '2-D Plots');
nexttile, scatter(x, y, 'ko'), title('(a) Scatter Plot');
nexttile, bar(x, y, 'k'), title('(b) Bar Chart');
nexttile, stem(x, y, 'k'), title('(c) Stem Plot');
nexttile, stairs(x, y, 'k'), title('(d) Stairs Plot');
labels = {'excellent', 'good', 'fair', 'poor'};
nexttile(5, [2 2]), pie([0.4 0.3 0.2 0.1], labels)
title('(e) Pie Chart')
newcolors = [0.7 0.7 0.7; 0.2 0.2 0.2]; % make grayscale
nexttile(7), area(x, [y2; y]'), title('(f) Area Plot');
colororder(newcolors)
```

CODE 2.2: Code to make common types of 2-D plots other than line plots.

Bar graph

bar(x, y): Bar graph (vertical and horizontal). Example shown in figure 2.7 (b).

Stem plot

stem(x, y): A discrete sequence plot—each point is plotted with a marker and a vertical line to the x axis. Example shown in figure 2.7 (c).

Stair step plot

stairs(x, y): The lines on the plot are either horizontal or vertical, resembling the run and rise of stairs. Example shown in figure 2.7 (d).

Pie chart

pie([0.4 0.3 0.2 0.1], labels): The values of the data are passed as a vector that often sums to 1; otherwise, MATLAB will calculate the percentages. The labels of the pie slices should be in a cell array. Example shown in figure 2.7 (e).

Filled area plot

area(x, [y1; y2]): Plot multiple data sets with the area between the x axis and each data curve shaded a different color. Each data set is a column of a matrix passed as the y data. Example shown in figure 2.7 (f).

2.2.8   Axis Control

The `axis` command returns a 4 element row vector containing the x and y axis limits.

```
>> v = axis
v =
   10 20 1 5
```

Use the functions `xlim` and `ylim` to set the x and y axis limits. Both functions take a vector of two elements as input. For example, the following command sets the lower y axis limit to 2 and the upper y axis limit to 4.

```
>> ylim([2 4])
```

When the lower or upper range of the data falls between the limits that MAT-LAB picked, the data may not fill the x axis. Use the `axis tight` command to change the limits based on the range of the data.

For plots depicting geometry, we would like the x, y, and z axes to have the same scale. There are two ways to accomplish this. The `axis` command has an `equal` option that sets the scale of each axis to be the same. We can also set the aspect ratio to be the same for each axis with the `daspect` function, which takes a vector with three values, even for 2-D plots.

```
>> axis equal        % Set the axes scales to be the same
>> %                 % or
>> daspect([1 1 1])  % Another way to accomplish the same
```

2.2.9   Tick Marks and Labels

MATLAB generally does a good job of setting the tick marks on a plot. The tick labels are by default the numeric values of the tick marks. We sometimes want to change the labels and locations of the tick marks because certain points on the axis have significance. A good example of this is when plotting equations that use trigonometry functions.

The `xticks` and `yticks` functions take vectors of numeric locations for the tick marks. The `xticklabels` and `yticklabels` functions take cell arrays of textual data. See section 1.11 for more information on cell arrays. The following code illustrates custom tick marks and labels on the x axis. Notice also that the title uses the `texlabel` function to format strings with LATEX to match an equation. The plot is shown in figure 2.8.

```
x = linspace(pi/4, 5*pi/4);
figure, plot(x, 4 - 4*sin(2*x - pi/2))
txt = texlabel('f(x) = 4 - 4*sin(2*x - pi/2)');
```

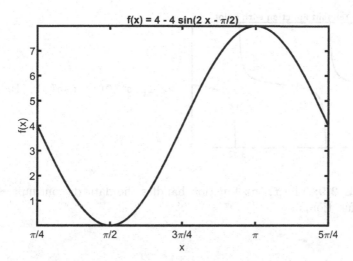

FIGURE 2.8: Sine wave with TeX formatted x axis tick marks and title.

```
title(txt), ylabel('f(x)'), xlabel('x')
xticks(linspace(pi/4, 5*pi/4, 5))
xticklabels({'\pi/4', '\pi/2', '3\pi/4','\pi', '5\pi/4'});
axis tight
```

2.2.10   Fplots

MATLAB includes another 2-D plotting function that comes in handy when working with either function handles, anonymous functions, or symbolic math functions (sections 1.8.4 and 4.4). The plot, annotations, and options are the same as the `plot` function. However, instead of passing vectors for the x and y data, we give `fplot` a function and a range of values for the x axis, or accept the default x axis range of -5 to 5.

In addition to being convenient when working with function handles or symbolic math functions, `fplot` also correctly handles plotting difficulties, such as data discontinuities. A good example of this, as shown in figure 2.9, comes from the trigonometric tangent function.

2.3   3-D Plots

MATLAB has several functions for plotting data in three dimensions. Try the code in figure 2.10. The `membrane` data is built into MATLAB. Do you recognize the shape of the data?

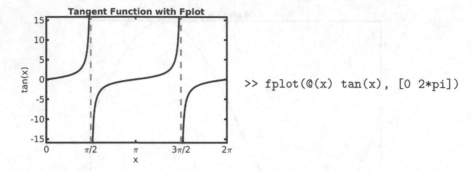

```
>> fplot(@(x) tan(x), [0 2*pi])
```

FIGURE 2.9: The `fplot` function handles the data discontinuities of the tangent function.

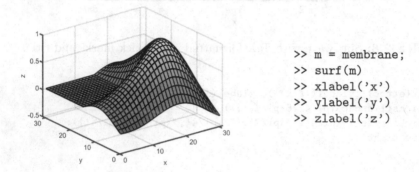

```
>> m = membrane;
>> surf(m)
>> xlabel('x')
>> ylabel('y')
>> zlabel('z')
```

FIGURE 2.10: The `surf` plot is the first plot to try for plotting 3-D surfaces.

Given only a matrix as input, the 3-D surface plotting functions will show the matrix values as the z axis values and use the matrix indices for the x and y axis.

The `meshgrid` function is used to create matrices of the x and y axis values that cover the range of data. Using data created from `meshgrid`, the z axis values are simple to create from an equation.

```
>> x = -3:3;
>> y = -3:3;
>> [X, Y] = meshgrid(x,y);
>> X
X =
        -3      -2      -1      0       1       2       3
        -3      -2      -1      0       1       2       3
        -3      -2      -1      0       1       2       3
        -3      -2      -1      0       1       2       3
        -3      -2      -1      0       1       2       3
        -3      -2      -1      0       1       2       3
```

```
        -3    -2    -1     0     1     2     3
>> Y
Y =
        -3    -3    -3    -3    -3    -3    -3
        -2    -2    -2    -2    -2    -2    -2
        -1    -1    -1    -1    -1    -1    -1
         0     0     0     0     0     0     0
         1     1     1     1     1     1     1
         2     2     2     2     2     2     2
         3     3     3     3     3     3     3
>> Z = X.^2 + Y.^2
Z =
        18    13    10     9    10    13    18
        13     8     5     4     5     8    13
        10     5     2     1     2     5    10
         9     4     1     0     1     4     9
        10     5     2     1     2     5    10
        13     8     5     4     5     8    13
        18    13    10     9    10    13    18
```

2.3.1   3-D Plot Functions

Here are descriptions of some of the 3-D plots available in MATLAB. Examples of the plots are shown in figures 2.11 to 2.15. The data for the plots is from the following code. Except for the plot3 function, the X, Y, and Z data is a two dimension grid of values, such as produced by the meshgrid function.

```
x = -8:0.25:8;
y = -8:0.25:8;
y1 = 8:-0.25:-8;
t = x.^2 + y1.^2;
z = (y1-x).*exp(-0.12*t); % for plot3
[X, Y] = meshgrid(x,y);
T = X.^2 + Y.^2;
Z = (Y-X).*exp(-0.12*T); % for surface plots
```

surf

surf(X, Y, Z): A 3-D surface plot, which is probably the most often used 3-D plot function. The surface is color coded to the z axis height. The surface is covered with grid lines parallel to the x and y axis. Examples are shown in figure 2.10 and figure 2.11.

surfc

surfc(X, Y, Z): A contour plot under a surface plot. Example is shown in figure 2.12.

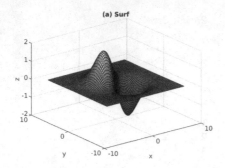

FIGURE 2.11: surf plot

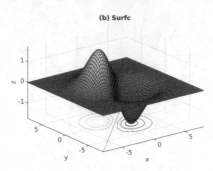

FIGURE 2.12: surfc plot

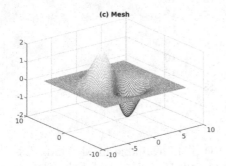

FIGURE 2.13: mesh plot

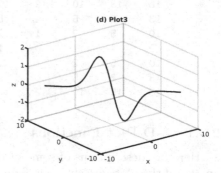

FIGURE 2.14: plot3 plot

mesh

mesh(X, Y, Z): A wireframe mesh with color determined by Z, so color is proportional to surface height. Example is shown in figure 2.13.

plot3

plot3(x, y, z): A line plot, like the plot function, but in 3 dimensions. Example is shown in figure 2.14.

contour

contour(X, Y, Z): A contour plot displays isolines to indicate lines of equal z axis values, like found on a topographic map. Example is shown in figure 2.15 (e).

contour3

contour3(X, Y, Z): A 3-D contour plot. Example is shown in figure 2.15 (f).

meshz

meshz(X, Y, Z): Like a mesh plot, but with a curtain around the wireframe mesh. Example is shown in figure 2.15 (g).

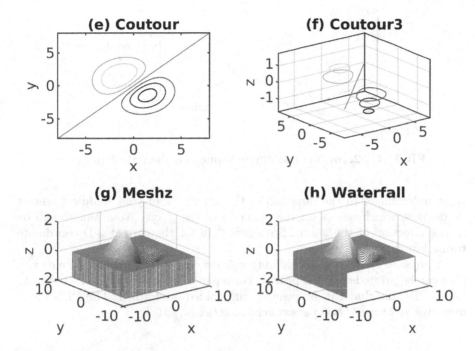

FIGURE 2.15: Three dimensional plots: (e) contour, (f) contour3, (g) meshz, (h) waterfall.

waterfall
waterfall(X, Y, Z): A mesh similar to the meshz function, but it does not generate lines from the columns of the matrices. Example is shown in figure 2.15 (h).

surfl
surfl(X, Y, Z): A shaded surface based on a combination of ambient, diffuse, and specular lighting models. Color is required for this plot to look right and the effect will vary depending on the data. An example plot is not shown.

Tip: Before concluding that you made a mistake if your 3-D plot looks like a 2-D plot, use the rotate tool to move the plot around. The view(3) command ensures that a plot displays in 3-dimensions.

2.3.2 Axis Orientation

If the axes are not labeled, it may be hard to remember which is the x axis and the y axis. Use the right hand rule to help with this. Point your

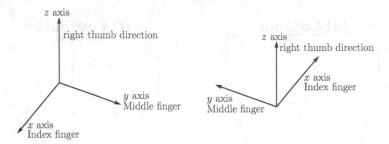

FIGURE 2.16: 3-D coordinate frame and the right hand rule.

right index finger in the direction of the x axis. Hold your middle finger at 90 degrees, which will be in the direction of the y axis. Your thumb will be in the direction of the z axis. See figure 2.16 for the correct 3-D coordinate frame layout.

A physical model is also helpful to visualize the 3-D coordinate frame axes. One can be printed on a 3-D printer. For a paper model, Peter Corke's website has a PDF file that can be printed, cut, folded, and glued. The PDF file is available at http://www.petercorke.com/axes.pdf

2.4 Exercises

Exercise 2.1   Plot 2-D Shapes

Use MATLAB's `plot` function to make plots of the following geometric shapes. Use the `axis equal` command to ensure the axes of the plots use the same scale.

(a) A 4 inch by 3 inch rectangle centered at (0, 0).

(b) A 3 inch by 4 inch diamond centered at (1, 1).

(c) A circle with a 2 inch radius centered at (2, 2). *Hint:* Use the sin and cos functions.

(d) An 5 inch by 3 inch ellipse centered at (0, 0). *Hint:* Start with points on a unit circle and use scalar multipliers to stretch and x and y values.

Exercise 2.2   Plot 3-D Shapes

(a) Use MATLAB's `surf` function to plot a cube centered at (0, 0, 0) with a length of 10, width of 5, and height of 7. *Hint:* Use the `surf` function six times to plot each surface.

Spiral Sphere   Circles Sphere   Surf Sphere

FIGURE 2.17: Three sphere plots.

(b) Use MATLAB's `plot3` function to plot a 3-dimensional spiral in the shape of a sphere of radius 3 inches centered at (0, 0, 0).

(c) Use MATLAB's `plot3` function to plot stacked circles in the shape of a sphere of radius 3 inches centered at (0, 0, 0).

(d) Look up the documentation for the MATLAB `sphere` function and use it along the `surf` function to plot a sphere of radius 3 inches centered at (0, 0, 0).

Examples of the three sphere plots are shown in figure 2.17.

Exercise 2.3   Roof Plot

Make a surface plot of a pitched roof. The roof should be pitched with a slope of 1/3 on all four sides. The size of the roof should be 60 feet by 30 feet. *Hint:* Create a surface covering the whole roof for each side and then use the `min` function to find the final roof surface. To ensure that the three dimensions of the plot use the same scale, use the `axis equal` command. An example plot is shown in figure 2.18.

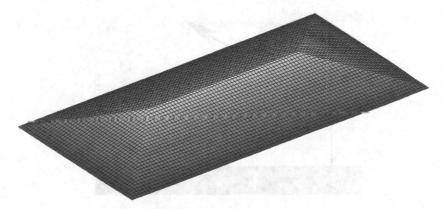

FIGURE 2.18: Pitched roof plot.

Exercise 2.4   The Flight of a Home Run Baseball

It was the bottom of the ninth inning of game 1 of the 2015 World Series. The New York Mets led the Kansas City Royals 4 to 3 with one out in the bottom of the ninth inning and then Alex Gordon tied the game by hitting a home run directly over the wall where it is marked as 410 feet from the home plate. From Major League Baseball's video archive [6], we can watch and make note of two things—how long was the ball in the air (measure with a stopwatch), and estimate how far the ball went before hitting the ground. The marker on the wall gives us a starting point to estimate how far the ball went. Reasonable values are $t_f = 5.9$ seconds and $x_f = 438$ feet.

Write a MATLAB script that will plot the vertical and horizontal path of the ball. An example plot follows. The script should also display the maximum height, initial velocity, and initial projection angle of the ball. The steps listed below will guide you.

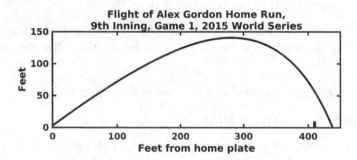

The horizontal and vertical paths of the ball may be calculated independently. The following diagram shows the geometry of how the initial velocity, V_0, and projection angle, θ, relate to the initial horizontal and vertical velocities.

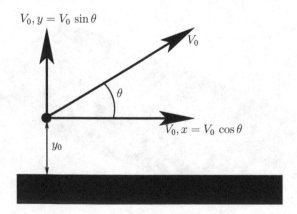

1. First make variables for constant values, gravity: $g = -32$ feet/sec^2, final time of flight: t_f in seconds, initial height: $y_0 = 3$ feet, and the final horizontal displacement: x_f in feet. Also use `linspace` to make a 100 element vector, t for time from 0 to t_f.

2. The vertical path of the ball may be treated the same as for throwing a ball straight up. The only force on the ball worth noting is gravity. The height of the ball is given by

$$y = y_0 + V_{0,y}\, t + 1/2\, g\, t^2$$

Using the t_f value measured with the stopwatch, calculate $V_{0,y}$, which occurs when $y = 0$. The y path of the ball may be calculated now.

3. Next use the final horizontal displacement, x_f, to calculate the initial horizontal velocity. Air resistance will slightly slow the horizontal displacement according to the differential equation [1],

$$\frac{dv_x(t)}{dt} = -0.2\, v_x(t).$$

The solution to the differential equation is

$$v_x(t) = V_{0,x}\, e^{-0.2\, t}.$$

The displacement comes from the definite integral of the velocity.

$$x(t) = V_{0,x} \int_0^t e^{-0.2\tau} d\tau = 5\, V_{0,x}(1 - e^{-0.2\, t})$$

The constant value of 0.2 is an estimated, typical value. The real value is dependent on other factors such as the wind speed and direction that we don't know.

From simple trigonometry, V_0, and θ can be expressed in terms of $V_{0,y}$ and $V_{0,x}$. Hint: Use the `atan2d` function to find θ.

4. Plot the path of the ball with a 'LineWidth' of 2.

   Show the home run wall by plotting a green line with a 'LineWidth' of 3. Make the wall 10 feet tall at 410 feet from home plate.

   Remember to annotate the plot and print in the Command Window the maximum height, initial velocity, and initial projection angle of the ball.

Chapter 3

Statistical Data Analysis

MATLAB contains many functions for statistical data analysis. The coverage here focuses on statistical calculations with MATLAB.

We will also touch on an important statistics topic later when we discuss least squares regression in section 5.10.

3.1 Introduction to Statistics

What does the data tell us? Are our design decisions data-driven? You have probably heard these questions before. Unfortunately, asking these questions is much simpler than answering them. Identifying what data to consider is hard. Distinguishing factual information from the noise and variability is even more complicated. Statistics gives us some tools that can help.

The first step is to find and collect our data, which is drawn from a population. If we are concerned about our company's products, then the set of every product made is the population. Collecting data from every item in the population is likely to be impractical, if not impossible. Collecting data from random samples of the population is more practical and usually gives useful information. Some exceptions allow us to consider an entire population. For example, every hydroelectric power generating facility in a country built during the last 50 years defines a significant but manageable population. Or, we may be able to consider a population derived from established rules of probability, such as the outcomes from a pair of thrown dice. As we define and use statistics, we will indicate with different symbols when we are using a population or sampled data.

A *random variable* is a number whose value is the outcome of an experiment, observation, or measurement influenced by unpredictable randomness. The probability that the random variable can take the given values from its domain defines its probability distribution. We will describe some standard distributions in section 3.4. Random variables can be either discrete or continuous. Discrete random variables take on discrete or countable values. Continuous random variables can be uncountable real numbers.

DOI: 10.1201/9781003271437-3

When we know our data's probability distribution, we can calculate and use statistical parameters to provide the information needed to make inferences, conclusions, and ultimately decisions. The tools that will allow us to determine probabilities are the probability mass function (PMF) for discrete random variables and the probability density function (PDF) for continuous random variables. The PMF defines the chances for the random variable to take each possible value. To calculate a probability, we will find the sum of terms from the PMF. The PDF defines the likelihood of the random variable taking a value within a specified range of numbers. Thus, we use definite integrals of the PDF to find probabilities for continuous random variables. Figure 3.1 shows an example of two PDF plots.

3.2   Common Statistical Functions

A *statistic* is a metric, or measure, taken from the data. That is to say, it is a value resulting from a calculation of the data points that gives useful information about the data. We will refer to the random variable with a capital letter, X, and the individual members of the random variable with a lower case letter, x.

3.2.1   Minimum and Maximum

The `min` and `max` functions take a vector input and can be used with either one or two outputs. The first output is either the minimum or maximum value of the vector. The second output is the index (location) of the value.

```
>> v = [31;12;8;29;36];          >> [mn, idx] = min(v)
                                 mn =
>> mn = min(v)                       8
mn =                             idx =
    8                                3
>> mx = max(v)                   >> [mx, idx] = max(v)
mx =                             mx =
   36                               36
                                 idx =
                                     5
```

3.2.2   Mean, Standard Deviation, Median, and Mode

Mean

The **mean** of a data set is what we also call the *average*, or *expected value* $(E(X))$. The mean is the sum of the values divided by the number of values. MATLAB has a function called **mean** that takes an vector argument and returns the mean of the data. Outliers (a few values significantly different than the rest of the data) can move the mean. So it can be a poor estimator of the center of the data. The median is less affected by outliers. The symbol for the sample mean of a random variable, X, is $\bar{x}$. The symbol for a population mean is μ, which is the expected value, $E(X)$, of the random variable.

$$\bar{x} = \frac{1}{n} \sum_{i=1}^{n} x_i$$

$$\mu = E(X)$$

```
>> v = [31;12;8;29;36];
>> sum(v)/length(v)
ans =
    23.2000
>> mean(v)
ans =
    23.2000
```

Standard Deviation

The **standard deviation**, which is the square root of the **variance**, is a measure of how widely distributed a random variable is about its mean. As shown in figure 3.1, a small standard deviation means that the numbers are close to the mean. A larger standard deviation means that the numbers vary quite a bit. For a normal (Gaussian) distributed variable, 68% of the values are within one standard deviation of the mean, 95% are within

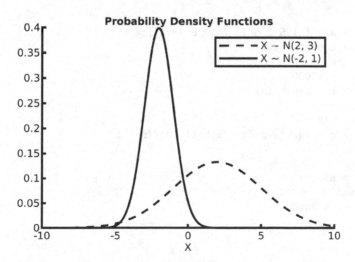

FIGURE 3.1: Probability Density Functions with $\sigma = 3$ and $\sigma = 1$.

two standard deviations and 99.7% are within three standard deviations. The symbol for a sample standard deviation is s and the symbol for a population standard deviation is σ.

We define the difference between a random variable and its mean as $Y = (X - \bar{x})$. Note that owing to the definition of the mean, $\bar{y} = 0$. To account equally for the variability of values less than or greater than the mean, we will use Y^2. The maximum likelihood estimator (MLE) of the variance computes the mean value of Y^2.

$$s^2_{MLE} = \left(\frac{1}{n} \sum_{i=1}^{n} x_i^2 \right) - \bar{x}^2$$

It is usually preferred to use the unbiased estimator of the variance, which divides the sum by $(n - 1)$ instead of by n. The reason is because $\sum_{i=1}^{n} y_i = 0$. If we know the sum of the first $(n - 1)$ terms, then the last term can be determined. Only $(n - 1)$ of the squared deviations can vary freely. The number $(n-1)$ is called the *degrees of freedom* of the variance [58].

The unbiased variance is then calculated as

$$s^2 = \frac{(x_1 - \bar{x})^2 + (x_2 - \bar{x})^2 + \cdots + (x_n - \bar{x})^2}{n - 1} = \frac{1}{n - 1} \sum_{i=1}^{n} (x_i - \bar{x})^2.$$

The standard deviation, s, is the square root of the variance. MATLAB has a function called `std` that takes a vector argument and returns the standard deviation. Similarly, the `var` function returns the variance.

To find the MLE standard deviation or variance, pass a second argument of 1 to the functions (`std(X, 1)`, `var(X, 1)`).

```
>> x = [-1 5 2 6 11 7 10 8 13 11];
>> x_bar = mean(x)
x_bar =
      7.2000
>> n = length(x)
n =
      10
>> s = sqrt(sum((x - x_bar).^2)/(n-1))
s =
      4.3665
>> std(x)
ans =
      4.3665
```

Median

The **median** of a data set is the number which is greater than half of the numbers and less than the other half of the numbers. Although the mean is a more commonly used metric for the center value of a data set, the median is often a better indicator of the center. This is because values that are *outliers* to the main body of data can skew a mean, but will not shift the median value.

When the number of items in the vector is odd, the median is just the center value of the sorted data. When the array has an even number of items, the median is the average of the two center items of the sorted data.

```
>> vs = sort(v)'                    >> vm = sort([v' 17])
vs =                                vm =
     8    12    29    31    36           8    12    17    29    31    36
>> vs(3)                            >> (vm(3) + vm(4))/2
ans =                               ans =
    29                                  23
>> md = median(v)                   >> median(vm)
md =                                ans =
    29                                  23
```

Mode

The **mode** of a data set is the value that appears most frequently. MATLAB has a function called **mode**. If there is a tie for the number of occurrences of values, the smaller value is returned. If no values repeat, then the mode function returns the minimum value of the data set.

3.2.3   Sort, Sum, Cumsum, and Diff

Sort

Use the `sort` function to sort the values of a vector. By default, the values are sorted in ascending order.

```
>> va = sort(v)                     >> vd = sort(v,'descend')
va =                                vd =
     8                                  36
    12                                  31
    29                                  29
    31                                  12
    36                                   8
```

Sum and Cumsum

The sum function find the sum of the values in a vector. The cumsum function finds a cumulative sum as each number is added to the sum.

```
>> s = sum(v)              >> cs = cumsum(v)
s =                        cs =
    116                        31
                               43
                               51
                               80
                              116
```

Diff

The diff function calculates the difference between each successive values of a vector. Note that the length of the returned vector is one less than the original vector.

```
>> diff(v)
ans =
   -19
    -4
    21
     7
```

3.3   Moving Window Statistics

A moving window is a contiguous subset of a vector that is used to compute local statistics. At each step, the window advances to include one new value and leaves off one old value. A statistic is calculated over the values in the window and is placed at the corresponding center position of the returned data. The resulting vector reduces some of the fluctuations that are found in the original vector. MATLAB provides functions for performing several statistical operations on moving windows. The moving window functions are listed in table 3.1. These functions all use the same syntax.

```
>> y = movmean(x, k)
```

Inputs: x – Array of data.
            k – Number of points in the window.
Outputs: y – Result of the k-point moving mean applied to the data in x.

TABLE 3.1: Moving Window Functions

Function	Description
movmin	Moving minimum
movmax	Moving maximum
movsum	Moving sum
movmean	Moving mean
movmedian	Moving median
movstd	Moving standard deviation
movvar	Moving variance

The commands below show how the values are calculated using the movmean function with a window size of 7. Notice that the first three values are calculated with a shortened window size. The full window size is used beginning with the fourth term. The window advances for the first time to not include the first data value at the fifth term. A shortened window size is similarly used at the end of the vector. You can change this behavior by specifying the optional 'Endpoints' argument. A plot of the original data and the filtered data from the movmean function is shown in figure 3.2.

```
>> x = linspace(0, 6, 20);
>> y = 5*x - 0.3*x.^2 + sin(3*x);
>> y1 = movmean(y, 7);
>> y(1:5)
ans =
    0   2.3609   3.9862   4.7626   5.2336
>> y1(1:5)
ans =
    2.7774   3.2687   3.7484   4.3319   5.7502

>> sum(y(1:4))/4
ans =
    2.7774       % y1(1) - startup, short window
>> sum(y(1:5))/5
ans =
    3.2687       % y1(2) - startup, short window
>> sum(y(1:7))/7
ans =
    4.3319       % y1(4) - first full window mean
>> sum(y(2:8))/7
ans =
    5.7502       % y1(5) - window advanced
```

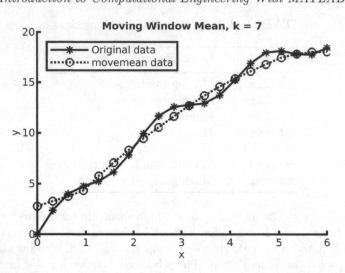

FIGURE 3.2: The moving window mean smooths the data fluctuations.

As with other statistical functions in MATLAB, if **X** is a matrix, the function is applied to the columns of **x** independently. You can change this behavior by specifying an optional dimension argument. You can also provide an optional flag to specify how NaN values are handled.

3.4   Probability Distributions

Recall that discrete random variables take integer values. Whereas, a continuous random variable can be real numbers. Here we will define common discrete and continuous probability distributions with an eye toward computing probabilities.

3.4.1   Discrete Distributions

Here are some examples of discrete random variables.

- The number of times that heads might appear in 100 coin tosses.

- The number of coin tosses needed to get 1 head.

- The number of customers entering a store in 1 hour.

- The number of products with a defect coming out of a machine each day.

The **probability mass**, $p(k) = P(X = x_k)$, defines the set of probabilities for the random variable taking each possible value. The expected value of a discrete random variable represents the long-run average value (mean).

$$E[X] = \sum_{k=1}^{n} x_k \, P(X = x_k)$$

The variance of a discrete random variable is

$$Var[X] = \sum_{k=1}^{n} x_k^2 \, P(X = x_k) - E[X]^2.$$

Bernoulli random variable

A trial that takes one of two outcomes, such as {(success—failure), (head—tail), (true—false), etc. }, has a *Bernoulli distribution*.

$$\begin{aligned} p(0) &= P\{X = 0\} = 1 - p, \quad (0 \le p \le 1) \\ p(1) &= P\{X = 1\} = p \end{aligned}$$

$$E[X] = p$$

Since X has only two values—zero and one, $E[X^2] = E[X] = p$.

$$Var[X] = p - p^2 = p(1 - p)$$

Binomial random variable

A Bernoulli experiment where the variable X represents the number of successes in n independent trials has a *Binomial distribution* with parameters (n, p), denoted $X \sim Bin(n, p)$.

We need to introduce a new operator to describe the probability distribution. The construct $\binom{n}{k}$ is called *n–choose–k* and is the number of different combinations that can be chosen from a set of n objects. For example, there are three ways that three $(n = 3)$ coin tosses can yield two heads $(k = 2)$: {(H H T), (H T H), (T H H)}.

$$\binom{n}{k} = \frac{n!}{(n - k)! \, k!}$$

$$\binom{3}{2} = \frac{3!}{1! \, 2!} = \frac{(3 \times 2 \times 1)}{(1) \times (2 \times 1)} = 3$$

$$p(k) = P(X = k) = \binom{n}{k} p^k (1 - p)^{(n-k)}, \quad k = 0, 1, 2, \ldots n$$

MATLAB has a function called nchoosek.

$$E[X] = \sum_{k=0}^{n} k \binom{n}{k} p^k (1-p)^{(n-k)} = n\,p$$

$$Var[X] = n\,p\,(1-p)$$

For our example of three fair ($p = 0.5$) coin tosses:

```
>> pbin = @(n, k) nchoosek(n, k)*(0.5^k)*(0.5^(n - k));

% Find probabilities for number of heads (0 to 3).
>> p = zeros(1, 4);
>> for k = 0:3
p(k+1) = pbin(3, k);
end
>> p
p =
     0.1250     0.3750     0.3750     0.1250

% Expected value
>> E_x = sum((0:3).*p)
E_x =
     1.5000
```

When p is close to 0.5 the probability mass is symmetric as shown above, but is skewed when p is closer to 0 or 1. The probability mass of binomial variables resembles a discrete version of a normal distribution. We see this in section 3.10 where we use the central limit theory with Bernulli events that sum to give a binomial distribution and ultimately are shifted and scaled to a standard normal distribution.

For help identifying all of the possible permutations of n events, MATLAB has a **perm** function that will list them.

Uniform discrete random variable

Some experiments may have multiple discrete outcomes that are each equally likely. The most obvious example is the throw of a six sided die, which has six possible outcomes each with probability of 1/6.

$$E[X] = \frac{a+b}{2}$$

$$Var[X] = \frac{n^2 - 1}{12}$$

The variables a and b represent minimum and maximum values. The number n represents how many different outcomes are available.

Poisson random variable

The *Poisson distribution* is for random variables that count the number of events that might occur within a fixed unit, such as a quantity, a spatial region, or a time span. A special property of Poisson random variables is that their values do not have a fixed upper limit. Some examples might be the number of defects per 1,000 items produced, the number of customers that a bank teller helps per hour, or the number of telephone calls processed by a switching system per hour. It is denoted as $X \sim Poi(\lambda)$. An example Poisson distribution probability mass plot is show in figure 3.3 for $\lambda = 3$. Note that $0! = 1$, so $k = 0$ is defined in the PMF.

$$p(k) = P(X = k) = \frac{\lambda^k e^{-\lambda}}{k!}, \ k = 0, 1, 2, \ldots, \ \lambda > 0$$

$$E[X] = \lambda$$
$$Var[X] = \lambda$$

```
>> poi = @(l, k) l.^k .* exp(-l) ./ factorial(k);

>> k = 0:10;
>> lambda = 3;
>> p = poi(lambda, k);
>> stem(k, p)
>> title('Poisson Distribution, \lambda = 3')
>> xlabel('k')
>> ylabel('p(k)')
```

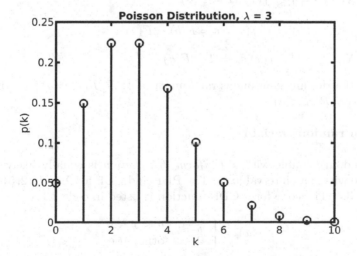

FIGURE 3.3: Poisson distribution probability mass

3.4.2   Continuous Distributions

Since a random variable following a continuous distribution can take on any real number, the probability of the variable being a specific value is zero, $P(X = a) = 0$. Instead a probability is specified only for a range of values ($P(a < X < b)$). Therefore, continuous variables do not have probability mass functions, instead we use a *probability density function* (PDF), $f(x)$. The mean and variance are found from integrals of the PDF.

$$E[X] = \int_{-\infty}^{\infty} x\, f(x)\, dx$$

$$E[X^n] = \int_{-\infty}^{\infty} x^n\, f(x)\, dx$$

$$Var[X] = E\left[(X - E[X])^2\right] = E\left[X^2\right] - [E(X)]^2$$

The area under a region of the PDF, a definite integral, defines a probability. The total area under a PDF must be one. A useful tool for computing probabilities is a *cumulative distribution function* (CDF), which tells us the probability $F(a) = P(X < a)$. The CDF is a function giving the result of integrating the PDF. A CDF plot is always zero on the left side of the plot and one to the right.

$$\int_{-\infty}^{\infty} f(x)\, dx = 1$$

$$F(a) = \int_{-\infty}^{a} f(x)\, dx$$

The CDF simplifies probability computations.

- $P(X < a) = \int_{-\infty}^{a} f(x)\, dx = F(a)$

- $P(a < X < b) = \int_{a}^{b} f(x)\, dx = F(b) - F(a)$

- $P(X > a) = \int_{a}^{\infty} f(x)\, dx = 1 - F(a)$

Note that for any continuous random variable X, $P(X = a) = 0$, therefore $P(X \leq a) = P(X < a)$.

Uniform random variable

A random variable with a *Uniform distribution* is equally likely to take any value within an interval $[a, b]$. The PDF and CDF for $X \sim U(2, 5)$ is show in figure 3.4. The code for a CDF function is listed in code 3.1.

$$f(x) = \begin{cases} \frac{1}{b-a}, & a \leq x \leq b \\ 0, & \text{otherwise} \end{cases}$$

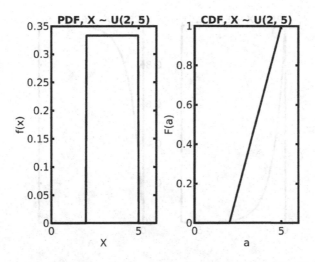

FIGURE 3.4: Uniform distribution PDF and CDF

```
function F = unifcdf(c, varargin)
% UNIFCDF - CDF of continuous uniform distribution U(a, b)
%
%  Inputs: c - variable in question -- may be a scalar or row
vector
%          a - minimum value of distribution
%          b - maximum value of distribution
%       unifcdf(c) is the same as unifcdf(c, 0, 1)

    if length(varargin) == 0
        a = 0;
        b = 1;
    else
        a = varargin{1};
        b = varargin{2};
    end
    F = zeros(1, length(c));
    inRange = (c >= a) & (c <= b);
    F(inRange) = (c(inRange) - a)/(b - a);
    F(c > b) = 1;
end
```

CODE 3.1: CDF of continuous uniform distribution $U(a, b)$.

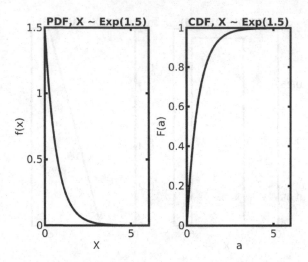

FIGURE 3.5: Exponential distribution

$$F(c) = \begin{cases} 0, & c < a \\ \frac{c-a}{b-a}, & a \leq c \leq b \\ 1, & c > b \end{cases}$$

$$E[X] = \frac{a+b}{2}$$

$$Var[X] = \frac{(b-a)^2}{12}$$

Exponential random variable

Exponential random variables model lengths of time, such as the length of time between events or the duration of events. Example PDF and CDF plots for the exponential distribution are shown in figure 3.5. We might measure the length of time between customers entering a store or how long it takes to help each customer. The *exponential distribution* is often used together with the Poisson distribution to predict capacity requirements. For example, in the telecommunications industry, the Poisson distribution models the expected number of telephone calls to be handled by a switching system per hour while the exponential distribution models the length of each call. The statistical models for the number and length of calls is used to determine the call handling capacity requirements of the equipment. A similar analysis could predict the number of employees needed at a business.

$$f(x) = \begin{cases} \lambda e^{-\lambda x}, & x > 0 \\ 0, & x \leq 0 \end{cases}$$

$$F(a) = \begin{cases} 1 - e^{-\lambda a}, & a > 0 \\ 0, & a \leq 0 \end{cases}$$

$$E[X] = \frac{1}{\lambda}$$

$$Var[X] = \frac{1}{\lambda^2}$$

Normal random variable

The *normal distribution*, also called the *Gaussian distribution*, models random variables that naturally occur in nature and society. Random variables with a normal distribution include measurements (length, width, height, volume, weight, etc.) of plants and animals, scores on standardized tests, income levels, air pollution levels, etc. Its probability density function has the familiar *bell shaped curve*. We denote the distribution as $X \sim N(\mu, \sigma^2)$.

$$f(x) = \frac{1}{\sqrt{2\pi\sigma^2}} e^{\frac{-(x-\mu)^2}{2\sigma^2}}$$

To simplify things, we will map the random variable to a *standard normal distribution* with zero mean and unit variance, $Y \sim N(0, 1)$.

$$y = \frac{x - \mu}{\sigma}$$

$$f(y) = \frac{1}{\sqrt{2\pi}} e^{\frac{-y^2}{2}}$$

Computing the CDF requires numerical integration because there is not a closed form integral solution to the PDF. Fortunately, there is a built-in MATLAB function that we can use to compute the integral. See the documentation for functions `erf` and `erfc`. The *Statistics and Machine Learning Toolbox* includes a function to compute the CDF, but implementing our own function using `erfc` is not difficult.

We desire a function called `normcdf` that behaves as follows for distribution $Y \sim N(0, 1)$. A plot of the needed area calculation is shown in figure 3.6. A plot of the CDF is shown in figure 3.7.

$$\text{normcdf(a)} = P(Y < a) = F(a) = \frac{1}{\sqrt{2\pi}} \int_{-\infty}^{a} e^{\frac{-y^2}{2}} \, dy$$

The `erfc(b)` function gives us the following definite integral.

$$\text{erfc(b)} = \frac{2}{\sqrt{\pi}} \int_{b}^{\infty} e^{-y^2} \, dy.$$

So, if we multiply by 1/2, change the sign of the definite integral boundary value to reflect that we are integrating from the boundary to infinity rather

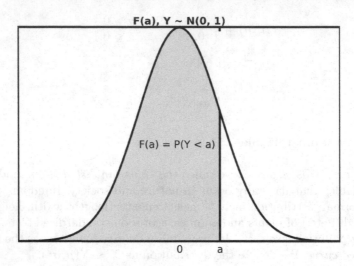

FIGURE 3.6: The shaded area is $P(Y < a)$.

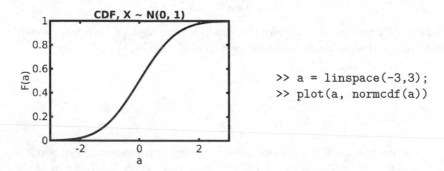

```
>> a = linspace(-3,3);
>> plot(a, normcdf(a))
```

FIGURE 3.7: CDF for $X \sim N(0,1)$

than from negative infinity to the boundary, and divide the boundary-value by $\sqrt{2}$ because the squared integration variable in the `erfc` calculation is not divided by 2 as we need, then the integration will compute an equivalent area to what we need. The code for a normal distribution CDF is listed in code 3.2.

Note that because of the symmetry of the PDF curve about the mean, $P(Y < a)$ is the same as $P(Y > -a)$.

$$\text{normcdf(a)} = F(a) = \frac{1}{2}\left(\frac{2}{\sqrt{\pi}}\int_{-a/\sqrt{2}}^{\infty}e^{-y^2}\,dy\right) = \frac{1}{2}\text{erfc}(-a/\sqrt{2})$$

Probability Example:

You learn that the lot of calves to be auctioned at the local livestock sale have a mean weight of 500 pounds with a standard deviation of 150 pounds. What fraction of the calves likely weigh:

```
function F = normcdf(a, varargin)
% NORMCDF - CDF for normal distribution
%    Inputs: a  - variable in question
%               - may be a scalar or row vector
%            mu - mean of distribution
%            sigma - standard deviation
%       normcdf(a) is the same as normcdf(a, 0, 1)
%
%    Output: Probability P(X < a)

    if length(varargin) ~= 0
        mu = varargin{1};
        sigma = varargin{2};
        a = (a - mu)./sigma;
    end
    F = erfc(-a/sqrt(2))/2;
end
```

CODE 3.2: Code for a normal distribution CDF.

1. Less than 400 pounds?

2. More than 700 pounds?

3. Between 450 and 550 pounds?

```
>> % less than 400 pounds
>> less400 = normcdf(400, 500, 150)
less400 =
    0.2525

>> % more than 700 pounds
>> more700 = 1 - normcdf(700, 500, 150)
more700 =
    0.0912

>> % between 450 and 550 pounds
>> middle100 = normcdf(550, 500, 150) - normcdf(450, 500, 150)
middle100 =
    0.2611
```

3.5   Generating Random Numbers

Random numbers are useful for creating simulations, and to emulate the presence of noise or random chance. MATLAB includes continuous and discrete random number generators for uniform and normal distributions. Random number generators for different probability distributions can be derived from the uniform and normal generators. Section 3.5.1 lists a short function to generate random numbers according to the exponential probability distribution. The *Statistics and Machine Learning Toolbox* has random number generators for other distributions. A search on the Internet will also find the source code for random number generators for other distributions.

rng

All MATLAB random number generating functions can be seeded to generate either a repeatable or unique sequence of pseudo-random numbers with the **rng** function. Seeding the random number generator with a fixed number results in the same sequence of numbers, which is helpful for comparisons of algorithms and is a debugging strategy because runs are reproducible. A strategy to get a unique sequence each time a program runs is to use the current time as the seed number.

rand

The `rand(n)` function creates a $n \times n$ matrix of uniformly distributed random numbers in the interval (0 to 1). `rand(m, n)` creates a $m \times n$ matrix or vector. To generate n random numbers in the interval (a, b) use the formula `r = a + (b-a)*rand(1, n)`. A uniform distribution means that each number is equally likely.

randn

The `randn` function creates random numbers with a normal (Gaussian) distribution. The parameters to `randn` regarding the size of the returned data is the same as for `rand`. The values returned from `randn` have a mean of zero and variance of one. To generate data with a mean of a and standard deviation of b, use the equation `y = a + b*randn(N,1)`.

randi

The `randi(Max, m, n)` function creates a $m \times n$ matrix of uniformly distributed random integers between 1 and a maximum value. Provide the maximum value as the first input, followed by the size of the array.

randperm

The `randperm` function returns a row vector containing a random permutation of the integers from 1 to n without repeating elements. One application of this is to achieve a shuffle operation, like with a deck of playing cards.

Usage code and histograms of each random number generator is shown in code 3.3 and figure 3.8.

```
% File: randGenPlots.m
t = tiledlayout(3, 1);
title(t, 'Random Number Generators')
nexttile;
histogram(20*rand(1, 1000), 40)
title('rand \sim U(0, 20)')
nexttile;
histogram(randi(20, 1, 1000), 20)
title('randi \sim U(1, 20) integers')
nexttile;
histogram(10 + 3*randn(1, 1000), 40)
title('randn \sim N(10, 3)')
```

CODE 3.3: Code for histogram plots of random number generators.

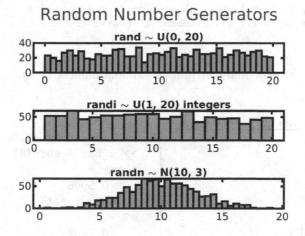

FIGURE 3.8: Histograms of random number generators

3.5.1   Exponential Random Number Generator

Recall from section 3.4.2 that equation (3.1) is the CDF of the exponential probability distribution.

$$F(a) = \begin{cases} 1 - e^{-\lambda a}, & a > 0 \\ 0, & a \le 0 \end{cases} \tag{3.1}$$

Equation (3.2) shows the inverse of the CDF found by letting $y = F(a)$ and solving for a. To make an exponential random number generator we use equation (3.2) and the rand random number generator that has a uniform distribution in the interval $(0, 1)$. Code 3.4 lists an exponential random number generating function. Figure 3.9 shows a histogram of a random variable created with our random number generator.

$$a = F^{-1} = \frac{1}{\lambda} \ln\left(\frac{1}{1-y}\right) \tag{3.2}$$

```
function X = randexp(lambda, m, n)
% RANDEXP - Simple exponential distribution random
%    number generator. If you have the Statistics
%    and Machine Learning Toolbox, use exprnd.
%
% lambda - exponential distribution parameter
%    mean(X) ~ 1/lambda, var(X) ~ 1/lambda^2
%    Output is size m x n
    y = rand(m, n);       % y ~ U(0, 1)
    y(y == 1) = 1 - eps; % protect from divide by 0
    X = log(1./(1-y))./lambda; % inverse of CDF
end
```

CODE 3.4: Exponential distribution random number generating function.

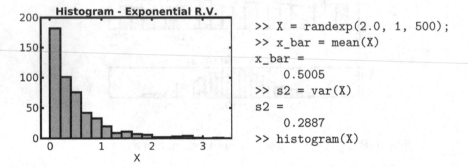

FIGURE 3.9: Histogram from the exponential random number generator

3.5.2   Monte Carlo Simulation

Here is a fun application of a random number generator. The script in code 3.5 performs what is called a Monte Carlo simulation to make an estimate the value of π [30]. This is done by creating a set of uniform distributed random points in a square between -1 and 1 in both the x and y directions. Then we relate the ratio of the number of points that fall inside a circle of radius 1 to the total number of points. The number of points inside the circle should be related to the area of the circle. Since the sides of the square is 2, the area of the square is 4, while the area of the unit circle is π. A plot of the points inside and outside of the circle is shown in figure 3.10.

```
% File: randpi.m
% Script to use a random number generator to estimate the
% value of PI. This is an example of a Monte Carlo simulation.

N = 1000000;

% 2-D points on a plane, uniform distribution in each
% dimension between -1 and 1.
points = 2*rand(2, N) - ones(2, N);
% scatter(samples(1,:), samples(2,:), '.')
radii = vecnorm(points);   % distance of each point from (0, 0).
inCircle = radii <= 1;     % points inside circle

% Area of square is 4. Area of circle is pi. So the ratio
% of points inside the circle to the total should be pi/4.
piEstimate = 4*nnz(inCircle)/N;
error = 100*abs(piEstimate - pi)/pi;
disp(['Estimate of pi is: ', num2str(piEstimate)])
disp(['The error was ', num2str(error), ' percent.'])

% plot the points inside and outside of the circle.
figure, axis equal
hold on
scatter(points(1, inCircle), points(2, inCircle), '.')
scatter(points(1, ~inCircle), points(2, ~inCircle), '.')
hold off
title('Monte Carlo Estimate of \pi')
```

CODE 3.5: Monte Carlo simulation to estimate the value of π.

3.5.3 Random Casino Walk

Another use of random number generators is a random walk simulation. Random walks represent a cumulative series of random numbers that at each step can move the state of the simulation. Random walks can be performed in different dimensions. In the example here, the dimension of the random walk is one.

This is a simulation of playing slot machines at a casino. One hundred players each start with 80 quarters ($20). Each player plays all 80 quarters and counts their winnings at the end. The machines are programmed to return 80% of what they take in. So the expected winnings of each player is $16 after spending $20 (loss of $4). Sometimes the machines give winners 4 quarters and sometimes they payout 8 quarters.

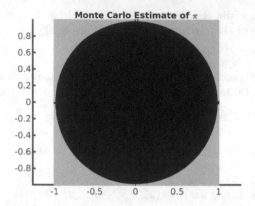

FIGURE 3.10: The ratio of points inside the circle to total points is close to $\pi/4$.

Logical arrays simplify the implementation. We use integer random numbers with a uniform distribution for events with simple percentages of occurrence. The simulation is listed in code 3.6. A histogram of the winnings for the 100 players is shown in figure 3.11. The shape of the histogram also affirms the central limit theorem described in section 3.8.

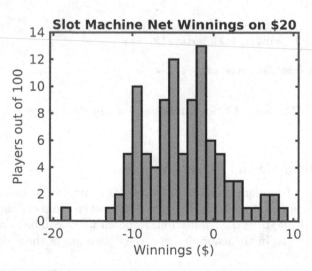

FIGURE 3.11: Histogram of winnings after spending $20 on slot machines.

```
% File: casinoWalk.m
% Random walk simulation of playing slot machines.
% 100 people each start with 80 quarters ($20).
% They play all 80 quarters and see how much money they
% have in winnings. The slot machines payout 80% of
% what they take in. Half of the time they use 1/5 odds
% to pay 4 quarters. The other times they use 1/10 odds
% to pay 8 quarters. You should find the average winning
% to be about -4 dollars.

winning = zeros(1, 100);
for p = 1:100    % for each player
    oneIn10 = randi(2, 1, 80) == 2; % win 1/10 or 1/5
    rand = randi(10, 1, 80);
    payout = zeros(1, 80);
    payout(oneIn10 & rand == 10) = 8;
    payout(~oneIn10 & rand > 8) = 4;
    winning(p) = sum(payout)/4 - 20; % net winnings
end
disp(['Average winning = $', num2str(mean(winning))])
histogram(winning, 25)
xlabel('Winnings ($)')
ylabel('Players out of 100')
title('Slot Machine Net Winnings on $20')
```

CODE 3.6: Random walk simulation of playing casino slot machines.

3.6   Statistics on Matrices

3.6.1   Column Statistics

When statistical functions are passed a matrix, the default is to work on each column of the matrix independently. The result is a row vector, which may in turn be used with other statistical functions if desired.

Some functions return multiple values for each column. For example, the diff function calculates the difference between adjacent elements in each column of the input matrix. The cumsum function also returns a matrix rather than a row vector.

3.6.2   Changing Dimension

Many statistical functions accept an optional dimension argument that specifies whether the operation should be applied to the columns independently (the default) or to the rows.

If matrix **A** has size 10×20, then `mean(A,2)` returns a 10 element column vector; whereas, `mean(A)` returns a 20 element row vector.

```
>> A = randi(25,10,20);
>> mC = mean(A);
>> mr = mean(A,2);
>> whos
Name        Size          Bytes  Class     Attributes

A           10x20          1600  double
mC          1x20            160  double
mr          10x1             80  double
```

Some functions, such as `min`, `max`, and `diff`, use the second argument for other purposes, which makes dimension the third argument. To skip the second argument, use a pair of empty square brackets for an empty vector, `[]`.

```
>> Amin = min(A,[],2);
```

3.6.3   Covariance and Correlation

Covariance shows how distinct variables relate to each other. It is calculated in the same manner that variance is calculated for a single variable. Variance (square of the standard deviation) is the expected value of the squared difference between each value and the mean of the random variable. Note that covariance and correlation can be defined either in terms of populations or samples. We mostly use the notion for sampled data.

$$\sigma_x^2 = E(X - \mu_x)^2$$

$$s^2 = \frac{1}{n-1}\sum_{i=1}^{n}(x_i - \bar{x})^2$$

Similarly, the covariance between two variables is the product of the differences between samples and their respective means.

$$\sigma_{xy} = E[(X - \mu_x)(Y - \mu_y)]$$

$$S_{xy} = \frac{1}{n-1}\sum_{i=1}^{n}(x_i - \bar{x})(y_i - \bar{y})$$

Thus, the covariance between a variable and itself is its variance, $s_{xx} = s_x^2$. Covariance is represented with a symmetric matrix because $s_{xy} = s_{yx}$. The variances of each variable will be on the diagonal of the matrix.

For example, consider taking a sampling of the age, height, and weight of n children. We could construct a covariance matrix as follows.

$$\text{Covariance}(a, h, w) = \begin{bmatrix} s_{aa} & s_{ah} & s_{aw} \\ s_{ha} & s_{hh} & s_{hw} \\ s_{wa} & s_{wh} & s_{ww} \end{bmatrix}$$

The **correlation coefficient** of two variables is a measure of their linear dependence.

$$r_{xy} = \frac{s_{xy}}{s_x s_y}$$

A matrix of correlation coefficients has ones on the diagonal since variables are directly correlated to themselves. Correlation values near zero indicate that the variables are mostly independent of each other, while correlation values near one or negative one indicate positive or negative correlation relationships. Correlation coefficients are generally more useful than covariance values because they are scaled to always have the same range ($-1 \leq r \leq 1$).

$$\mathbf{R}(a, h, w) = \begin{bmatrix} 1 & r_{ah} & r_{aw} \\ r_{ha} & 1 & r_{hw} \\ r_{wa} & r_{wh} & 1 \end{bmatrix}$$

The MATLAB functions to compute the covariance and correlation coefficient matrices are cov and corrcoef. In the following example, matrix **A** has 100 random numbers in each of two columns. Half of the value of the second column come from the first column and half come from another random number generator. The variance of each column is on the diagonal of the covariance matrix. The covariance between the two columns is on the off-diagonal. The off–diagonal of the matrix of correlation coefficients shows that the two columns have a positive correlation.

```
>> A = 10*randn(100, 1);
>> A(:,2) = 0.5*A(:,1) + 5*randn(100, 1);

>> Acov = cov(A)
Acov =
    94.1505    50.0808
    50.0808    50.6121

>> Acorr = corrcoef(A)
Acorr =
     1.0000     0.7255
     0.7255     1.0000
```

3.7   Plots of Statistical Data

Here we generate a data set of 200 random numbers to illustrate two common plots that show the distribution of the data. The data was generated with a normal distribution random number generator.

```
d = 50 + 15*randn(1, 200);    % normal mean=50, std=15
```

3.7.1   Box Plot

A box plot gives us a quick picture of the distribution of the data. The plot makes it easy to see the range of each 25% of the data (called quartiles). A box plot example is shown in figure 3.12. The vertical line at the bottom of the plot shows the lower limit value and extends up to the bottom of a box showing the first quartile, Q_1. The box in the center represents the range of the middle 50% of the data. It goes from the first quartile, Q_1, to the third quartile, Q_3. There is a horizontal line at the second quartile, which is the median of the data. Then there is another vertical line from the third quartile to the upper limit value. The lower and upper limits may be the minimum and maximum values of the data. However, they are often found relative to the center region of the data so that outliers are excluded from the four quartiles. The range of the center region is called IQR, $IQR = Q_3 - Q_1$. The lower and upper limit values (LL and UL) are computed as $LL = Q_1 - 1.5 \times IQR$ and $UL = Q_3 + 1.5 \times IQR$. Any data points less than LL or greater than UL are classified as outliers, and may appear as scatter points in the box plot.

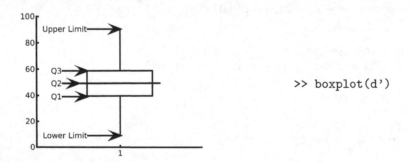

```
>> boxplot(d')
```

FIGURE 3.12: A Box Plot with Quartile's Noted

The box plot function from MathWorks is part of the *Statistics and Machine Learning Toolbox*, which is an extra purchase. However, there are a few free box plot functions available on the *MathWorks File Exchange*. Some of those use functions from extra toolboxes. The free Boxplot function [52] is a simple function that uses only standard MATLAB functions.

The `boxplot` function wants the data to be in a column vector because it can make several box plots in a figure if each data set is a column of a matrix.

3.7.2 Histogram

A histogram plot divides the data into regions (called bins) and shows how many values fall into each region. If the size of the data is large, a histogram plot will begin to take the shape of the PDF.

The histogram function has several possible parameters, but the most common usage is to pass two arguments—the data and the number of bins to use. Another useful pair of options is `'Normalization'`, `'probability'`, which scales the height of each bin to its probability level. This is a good option if you want to overlay another histogram or a PDF plot. An example histogram is shown in figure 3.13.

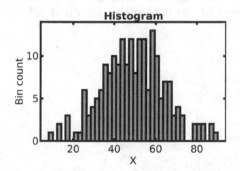

`histogram(d, 40)`

FIGURE 3.13: A Histogram Plot

3.8 Central Limit Theorem

We know that many random variables naturally fit the normal distribution model. It turns out that random variables from other distributions can be mapped to a normal distribution by the *central limit theorem*.

From a population that has mean μ and variance σ^2, draw m sampling sets, each of size n. The central limit theorem says that when n is large, the distribution of the sample means and sample sums is approximately normal regardless of the underlying population distribution. For each sampling of random variable X, let $\bar{X}_i$ be the sample mean, and let Y_i be the sample sum.

$$\bar{X}_i = \frac{1}{n} \sum_{j=1}^{n} x_{i,j}$$

$$Y_i = \sum_{j=1}^{n} x_{i,j}$$

We also define variable Z as follows.

$$Z = \frac{\bar{X} - \mu}{\frac{\sigma}{\sqrt{n}}} = \frac{Y - n\mu}{\frac{\sigma}{\sqrt{n}}}$$

Then we can define normal distributions from $\bar{X}$, Y, and Z.

$$\bar{X} \sim N\left(\mu, \frac{\sigma^2}{n}\right), \text{ and } Z \sim N(0,1)$$

Let's see this in action. We will start with X representing the six sides of a die, so we use a discrete uniform distribution. We will put the data in a 100×100 matrix so that each column will be a sampling. Then we can find the mean and sum of each column to get a new random variable with a normal distribution. Normalized histogram plots of the data along with a standard normal PDF are shown in figure 3.14.

```
>> n = 100;
>> X = randi(6, n);   % 100 x 100
>> X_bar = mean(X);   % 1 x 100
>> mu = mean(X(:))
mu =
       3.4960           % 3.5 expected
>> sigma = std(X(:))
sigma =
       1.7025           % 35/12 = 2.92 expected

% Make Z ~ N(0, 1)
>> Z = (X_bar - mu)/(sigma/sqrt(n));
>> mean(Z)
ans =
    -1.9895e-15
>> var(Z)                % Z ~ N(0,1)
ans =
       0.9791
```

3.9   Sampling and Confidence Intervals

As mentioned in section 3.1, it is difficult to find parameters for a large population. When sampling from a large population, each sample set will likely

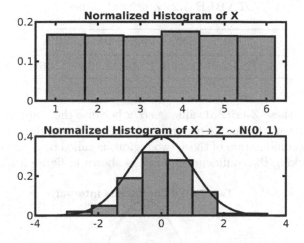

FIGURE 3.14: Histograms of $X \sim U(1,6)$ and $Z \sim N(0,1)$.

have a different mean. However, we can calculate a range of values for which we have confidence that the actual population mean is within. The central limit theorem and what we know about the standard normal distribution give us the tools to do this.

Based on the mapping of the random variable to a standard normal distribution by the central limit theorem, we establish the Z variable.

$$Z = \frac{\bar{x} - \mu}{\left(\frac{\sigma}{\sqrt{n}} \right)}$$

From the CDF of a standard normal distribution, we can establish probabilities and introduce Z-critical values, which are used to find an interval where we have confidence at a level of C that the true population mean is within the interval [37]. The Z-critical values for establishing levels of confidence are shown in table 3.2. A confidence level of 95% is the most commonly used, but other confidence levels, especially 90% and 99% are also useful. We start with the following probabilities for a standard normal distribution.

$$C = Pr\left(-Z_{\alpha/2} < Z < Z_{\alpha/2}\right) = 1 - \alpha$$

Then for the estimated mean of the population, we use the sample mean, population standard deviation, and the size of the sample to find the following probability relationship, which establishes a confidence interval.

$$Pr\left(\bar{x} - Z_{\alpha/2}\frac{\sigma}{\sqrt{n}} < \mu_x < \bar{x} + Z_{\alpha/2}\frac{\sigma}{\sqrt{n}}\right) = C$$

TABLE 3.2: Z-critical values

	80%	85%	90%	95%	99%
$Z_{\alpha/2}$	1.28	1.44	1.645	1.96	2.576

We refer to these Z-critical values as $\alpha/2$ because there are regions greater than and less than the confidence interval where the population mean might be found. The combination of those two regions is called α. A standard normal PDF plot marking the confidence interval is shown in figure 3.15.

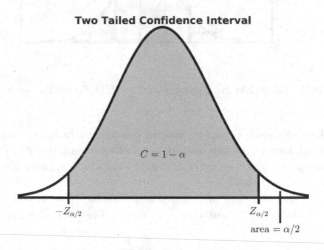

FIGURE 3.15: Two tailed 95% confidence interval, $C = 0.95$, $\alpha = 0.05$, $Z_{\alpha/2} = 1.96$.

In the simulation of a normal random variable listed in code 3.7, we create 100 sampling sets. Using one of the samples, we find a 95% confidence interval. Our population mean is certainly within the confidence interval, but as you can see from the output of the simulation, the means of some samples are outside of our confidence interval.

3.10   Statistical Significance

An observation from an experiment becomes accepted fact when the data proves *statistical significance*. That is, it must be proven that the observation is consistently true and not a random anomaly.

```
% File: testConfidence.m
% Find a 95% confidence interval and see how many
% sample means are inside and outside the interval.
n = 100;
X = 50 + 20*randn(50, n);   % 50 x 100
X_bar = mean(X);
Za2 = 1.96;       % 95% Confidence
mu = mean(X(:))   % population mean
sigma = std(X(:));
% look at one sample
x_bar = X_bar(1);
Zlow = x_bar - Za2*sigma/sqrt(n)
Zhigh = x_bar + Za2*sigma/sqrt(n)
% test all samples
inAlpha = nnz(X_bar < Zlow | X_bar > Zhigh)
inC = nnz(X_bar > Zlow & X_bar < Zhigh)

>> testConfidence
mu =
    49.5914
Zlow =
    45.2956
Zhigh =
    53.5387
inAlpha =
    8
inC =
    92  % About as expected for a 95% confidence interval
```

CODE 3.7: Normal distribution 95% confidence interval test.

Let us consider a specific historical example. In the nineteenth century, lung cancer was a rare disease. Then during the first half of the twentieth century, the smoking of tobacco products, especially cigarettes, grew in popularity. According to Gallup polling data, self-reported adult smoking in the U.S. peaked in 1954 at 45%. Medical doctors noticed that as smoking rates increased, so did occurrences of lung cancer. But observations were not enough to conclude that smoking was causing lung cancer. Some studies in the 1940s and 1950s pointed blame toward smoking, but the public was skeptical because so many were smokers, and the tobacco industry claimed that smoking was safe. Then from 1952 to 1955, E. Cuyler Hammond and Daniel Horn, scientists from the American Cancer Society, conducted an extensive study using over 187,000 male volunteers [33]. They showed that the rate of lung cancer among smokers is outside the range of what is statistically possible for nonsmokers.

The strategy is to show that a *null hypothesis* must be rejected since it is false. A null hypothesis, represented as H_0, claims that the observation in question falls within the range of regular statistical occurrences. It usually advocates for the status quo belief. So a null hypothesis might be: *"Smokers of tobacco products and nonsmokers are equally likely to develop lung cancer."* Whereas, an alternate hypothesis, H_a, might be: *"Smoking tobacco products increases the risk of developing lung cancer."* The alternate hypothesis often promotes belief in a causal effect. It proposes that when some treatment is applied, then those receiving the treatment are changed. Example treatments and changes include: smoking leads to lung cancer, taking a medication heals people from a disease, or exercising reduces cholesterol. It is usually harder to directly show that the alternate hypothesis is true than to show that the null hypothesis is false. If we can show that the null hypothesis is wrong, then the alternate hypothesis must be accepted.

To build a case for rejecting a null hypothesis, we develop the statistics of the population not receiving the treatment in question and then show that the statistic of the treatment population falls outside of the range of possibility for the null hypothesis to be true. So Hammond and Horn established the statistics for nonsmokers developing lung cancer and then showed that smokers develop lung cancer at a rate beyond what is possible for nonsmokers.

The tools that statisticians use to establish confidence intervals for population means and accept or reject null hypotheses with statistical significance require a normal distribution. When the data has a different probability distribution, it is usually possible to frame the problem as an instance of a normal distribution by taking advantage of the central limit theorem.

Hypotheses testing is a matter of testing if two data sets could have come from the same source. We accept the null hypothesis if they likely came from the same source or an equivalent source, and reject the null hypothesis if we conclude that they are from different sources.

If the standard deviation of the null case population is known, use the Z-test with the Z-critical values. When the standard deviation is not known, use the t-test which uses the sample standard deviation and is based on the t-distribution, which is similar to the normal distribution except with more variability.

Counterfeit Competition

As an example, consider the following scenario. Suppose that our company makes a product that is very good at what it does. A new competitor introduced a product that it claims is just as good as your product, but is cheaper. We would like to prove that the competitor's product is not as good as our product. So we draw sample sets of our product and the competitor's product to establish the quality of each, which we will verify with a Z-test and a t-test.

3.10.1 Z-Test

For the Z-test, we only need to calculate the sample mean of the H_a data and then calculate the Z variable that maps the sample mean of the treatment data set to a standard normal distribution.

$$Z = \frac{\bar{x}_a - \mu_0}{\frac{\sigma}{\sqrt{n}}}, \text{where} H_0 : \mu = \mu_0$$

The Z variable is compared to a Z-critical value to determine if the H_a sample mean is consistent with the H_o population. Note that in addition to the two tailed tests, tests based on either the upper or lower tail may also be used (see figure 3.16). Commonly used Z-critical values for one tailed and two tailed tests is shown in table 3.3.

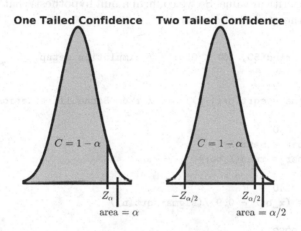

FIGURE 3.16: One and two tailed 95% confidence intervals, $\alpha = 0.05$, $Z_\alpha = 1.645$, $Z_{\alpha/2} = 1.96$.

TABLE 3.3: Z-critical values

	80%	85%	90%	95%	99%
$Z_{\alpha/2}$	1.28	1.44	1.645	1.96	2.576
Z_α	0.84	1.04	1.282	1.645	2.326

Counterfeit Competition: Z-Test

For the sake of our example, let's say that the metric of concern is a Bernoulli event. We obtain 500 each of our competitor's product and our product for testing. We split the 500 products into 10 sampling sets of 50 items.

We can use the Z-test for our product since we have a well established success rate for it. In several tests, our company's product has consistently shown a success rate of at least 90%. We just want to verify that our latest product matches up to our established quality level. We put our successes and failures in a matrix and compute the mean of each column. According to the central limit theory, the means of the sample sets give us a normal distribution that we can use with both Z-tests and t-tests.

You may wonder about the binomial distribution, which comes from the sum of successes of Bernoulli events. A binomial distribution is a discrete facsimile of a normal distribution. The central limit theorem equations can scale and shift a binomial distribution into a standard normal distribution.

In the simulation test of our product, we find a Z-Test score that is well below any Z-critical value. So we confirm a null hypothesis that these samples test match the quality of our product.

```
>> X = rand(50, 10) < 0.9;    % simulation setup
>> n = 10;
>> p = 0.9;
>> sigma = sqrt(p*(1-p))      % from Bernoulli equations
sigma =
    0.3000
>> X_bar = mean(X);
>> x_bar = mean(X_bar)
x_bar =
    0.9240
>> Z = (x_bar - 0.9)/(sigma*sqrt(n))
Z =
    0.0253
```

3.10.2   t-Test

In practice, the t-test is much like the Z-test except the sample's calculated standard deviation (s) is used rather than the population standard deviation (σ). Because each sample from the population will have a slightly different standard deviation, we use the student's t-distribution, which has a slightly wider bell-shaped PDF than the standard normal PDF. The distribution becomes wider for smaller sample sizes (degrees of freedom). As the sample size approaches infinity, the t-distribution approaches the standard normal distribution. The degrees of freedom (df) is one less than the sample size ($df = n - 1$). The t-distribution PDF plot for three df values is shown in figure 3.17.

The T statistic is calculated and compared to critical values to test for acceptance of a null hypothesis as was done with the Z-test. The critical values are a function of both the desired confidence level and the degree of freedom

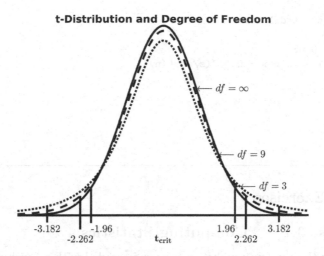

t-Distribution and Degree of Freedom

FIGURE 3.17: t-Distribution PDF plots showing two tailed 95% critical values for infinity, 9, and 3 degrees of freedom.

(sample size). There are several ways to find the critical values. Tables found on websites or in statistics books list t-distribution critical values. Interactive websites as well as several software environments can calculate critical values. Unfortunately, MATLAB requires the *Statistics and Machine Learning Toolbox* to calculate critical values. However, as with the Z-test, the *t*-critical values are constants, so they may be hard-coded into a program to test a null hypothesis.

$$T = \frac{\bar{X} - \mu_0}{\left(\frac{s}{\sqrt{n}}\right)}, \text{where} H_0 : \mu = \mu_0$$

Counterfeit Competition: t-Test

When we apply the t-test to our competitor's product, we use statistics calculated from the data and the established mean success rate of our product. Although the success rate of our competitor's product seems to be not far below that of our own, the calculated T is less than the negative one sided *t*-critical value, which establishes with statistical significance that our competitor's product is inferior to ours.

```
>> X = rand(50, 10) < 0.85;
>> n = 10;
>> X_bar = mean(X);
>> x_bar = mean(X_bar)
x_bar =
    0.8620
```

```
>> s = std(X_bar)
s =
    0.0382
>> T = (x_bar - 0.9)/(s/sqrt(n))
T =
   -3.1425
```

3.11   Exercises

Exercise 3.1   Computing Statistics

When the random number generator is seeded with a fixed number, then the sequence of random numbers will be the same each time. Seed the random number generator and create a data set from the normal distribution random number generator as done below. Calculate to the mean, standard deviation, and median of the data. Use both the MATLAB functions for each and calculate them using lower level functions `sum`, `length`, `sqrt`, and `sort`. Then use the `histogram` function with 20 bins to relate the measured statistics to the distribution of the data.

```
>> rng(2022)
>> v = 20 + 3*randn(1, 200);
```

Exercise 3.2   Probabilities

1. If we take the sum of two thrown six-sided die, the minimum sum is 2 and the maximum sum is 12. Write a MATLAB script to compute the probability of the sum being each number from 2 to 12.

2. Products made at a manufacturing factory occasionally break because they are not aligned properly with the material handling equipment. We can model the number of broken items per day with the Poisson distribution, where the average number broken products per day is λ. If $\lambda = 5$, what is the probability that 10 or more products beak during one day?

3. It takes a work crew a minimum of 43 minutes to complete a task. The addition time required beyond the 43 minutes is modeled by an exponential distribution with $\lambda = 0.271$. What is the probability of the work crew completing the task in less than 50 minutes?

4. If the average amount of paint required to spray paint an object is 8 oz. with a standard deviation $1\frac{1}{3}$ oz., assuming a normal distribution, what is the probability that an object can be painted using less than 10 oz.?

Exercise 3.3   Annual Temperature Plot

This project will demonstrate:

1. Importing data into MATLAB.

2. Fixing out-of-range data.

3. Working with data containing dates.

4. Using statistical moving window functions.

The instructions are given as a tutorial, but depending on the data that you download, some adjustments may be needed. The output of the project will be a plot.

Part 1: Getting the Data

- Search the Internet for hourly temperature data. The National Climatic Data Center of NOAA has data taken hourly from several weather reporting stations. https://www.ncdc.noaa.gov/crn/qcdatasets.html

- The Hourly02 directory has what we want. The folders and files are organized by year and reporting station. Downloaded the data for the location and year of your choice.

- The meaning of fields are given in the documentation. Field 4 is Local Standard Time (LST) date. Field 5 is the Local Standard Time (LST) time of the observation. Field 10 is average air temperature in degrees C for the entire hour.

- Note that data starts based on UTC time, so it includes the last hours of the previous year.

- Use the Import Data tool. Rename and import fields 4, 5, and 10 to Date, Hour, and Temp.

- Check for missing data with the ismissing function. If there is missing data, use the fillmissing function to fill in those values using linear interpolation.

- Check the minimum and maximum temperatures. You will likely find a few clusters of -9999 temps from when no reading was taken. We can replace those values with NaN and again use the fillmissing function to fill in the missing data.

- Use the `writetable` function to save the data.

Part 2: The Plotted x axis

Now for the date and time. The following function returns a `datetime` value from the dates and times given:

```
function date = get_date_time( dateNum, timeNum )
%GET_DATE_TIME Convert numeric date and time to datetime
%    dateNum such as: 20151231
%    timeNum such as: 1900

    year = floor(dateNum/10000);
    month = floor((dateNum - year*10000)/100);
    day = dateNum - year*10000 - month*100;
    hour = timeNum/100;
    date = datetime(year,month,day,hour,0,0);
end
```

For an initial plot:

```
>> dates = get_date_time(tempData.Date, tempData.Hour);
>> plot(dates, temp)
```

Part 3: Statistical Analysis

Write a MATLAB script to convert the temperatures to Fahrenheit (optional), find the daily mean (average), maximum, and minimum temperatures. Experiment with the length of the window size to find a plot that shows any weather pattern lasting for several days, but smooths minor fluctuations.

Exercise 3.4   Confidence Intervals

Your company has designed a new golf driver club. You hire a professional golfer to test your club. The golfer has an average driving distance of 290 yards. He drives 75 balls with your club for an average distance of 296 yards with a standard deviation of 20 yards. Can you conclude with 99 % confidence that your golf club gave the golfer a longer driving distance?

Chapter 4

Using the Symbolic Math Toolbox

The majority of our consideration relates to numerical algorithms. However, we now focus on analytic solutions to mathematical problems. We seek analytical solutions in mathematics courses, and we typically solve them with pencil and paper. Computers are, of course, more suited to numerical algorithms. But MATLAB does have the ability to find analytic solutions when they are available. When a computer solves a mathematics problem analytically, we say that the computer is operating on symbols in addition to numbers. With symbolic operations, the computer would simplify expressions such as 6/3 to 2, but would leave numbers like π, e, and 17/7 as irreducible numbers.

The *Symbolic MATH Toolbox* (SMT) is required to access the features of MATLAB discussed here. The SMT is part of the student MATLAB bundle, but is an additional purchase with other MATLAB licenses.

4.1 Throwing a Ball Up

As a quick tour of some of the features of the SMT, consider throwing a ball straight upward. The scene is depicted in figure 4.1. The algebraic equations describing the force of gravity on a moving ball are readily available on the web or in physics and dynamics textbooks. Let us use the SMT to derive the equations from differential equations and apply them. In this example, we will use functions from the SMT with minimal comments about them. Then we will give more details about the SMT functions beginning in section 4.2.

The vertical position of the ball with respect to time is $y(t)$. The velocity of the ball is the derivative of the position,

$$v(t) = \frac{dy(t)}{dt}. \tag{4.1}$$

The derivative of the velocity is the acceleration, which in this case is a constant from gravity in the negative direction. The acceleration is the second derivative of the position.

$$a(t) = \frac{dv(t)}{dt} = \frac{dy^2(t)}{dt^2} = g \tag{4.2}$$

DOI: 10.1201/9781003271437-4

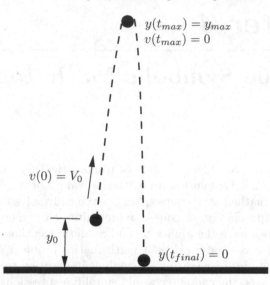

$$y(t_{max}) = y_{max}$$
$$v(t_{max}) = 0$$

$$v(0) = V_0$$

$$y_0$$

$$y(t_{final}) = 0$$

FIGURE 4.1: A ball is thrown upward from height y_0 with an initial velocity of V_0. When the ball reaches its maximum height, y_{max} at time t_{max}, its velocity is zero, $v(t_{max}) = 0$. The ball hits the ground at time t_{final}, $y(t_{final}) = 0$.

$$g = -9.8 \text{ m/sec}^2, \text{ or } g = -32 \text{ feet/sec}^2$$

Our starting point, equation (4.2), is a second order differential equation. In addition to describing the differential equation, we need to describe the initial conditions of the problem. The initial height of the ball is given by $y(0) = y_0$; and the initial velocity from the throw is $v(0) = V_0$.

Symbolic variables and functions are created in MATLAB with the **syms** statement. Another way to create a symbolic variable is to pass a value or expression to the **sym** function, such as z = sym(5). The **sym** function returns a symbolic variable.

The differential equations in equation (4.1) and equation (4.2) are solved in the following MATLAB code. We will declare y(t) (position) as a symbolic function, and y0 (initial position), V0 (initial velocity), and g (gravity) as symbolic variables. Time, t, is implicitly declared from y(t). MATLAB solves symbolic differential equations with the **dsolve** function.

Note in the code below that the equal sign (=) is, as always, used for assignment statements, but the double equal sign (==) is used to declare equality relationships in equations.

```
>> syms y(t) y0 V0 g
>> diffeq = diff(y(t),t,2) == g
diffeq =
```

```
      diff(y(t), t, 2) == g
>> v(t) = diff(y(t),t)
v(t) =
      diff(y(t), t)

>> cond = [y(0) == y0, v(0) == V0]
cond =
      [ y(0) == y0, subs(diff(y(t), t), t, 0) == V0]
>> y(t) = dsolve(diffeq, cond)
y(t) =
      (g*t^2)/2 + V0*t + y0
```

The `diff` function takes the derivative of the position to find the velocity.

```
>> v(t) = diff(y(t),t)
v(t) =
      V0 + g*t
```

The maximum height of the ball occurs when the velocity is zero. The `solve` function finds solutions to algebraic equations.

```
>> t_max = solve(v(t) == 0)
t_max =
      -V0/g
>> y_max = y(t_max)
y_max =
      -V0^2/(2*g) + y0
```

The ball lands on the ground when `y(t)` is zero. Two answers are returned because `y(t)` is a quadratic equation.

```
>> t_final = solve(y(t) == 0, t)
t_final =
      -(V0 + (V0^2 - 2*g*y0)^(1/2))/g
      -(V0 - (V0^2 - 2*g*y0)^(1/2))/g
```

We can give the variables numerical values and substitute (`subs`) the values into the equations, and see a decimal version of the number with vpa.

```
% From 1 meter off ground, throw up at 10 m/s
>> y0 = 1; V0 = 10; g = -9.8;
>> subs(t_max)                % time to max height
ans =
      50/49
>> vpa(subs(t_max))
ans =
```

```
    1.0204081632653061224489795918367 % seconds

>> subs(y_max)
ans =
    299/49
>> vpa(subs(y_max))
ans =
    6.1020408163265306122448979591837 % meters

>> subs(t_final(1)) % time until the ball hits the ground
ans =
    (5^(1/2)*598^(1/2))/49 + 50/49
>> vpa(subs(t_final(1)))
ans =
    2.1363447440401890728050707611418 % seconds
```

4.2   Symbolic Algebra

4.2.1   Collect

The `collect` function combines variables of the same power of the specified variable in an expression.

```
>> syms x y
>> f = (2*x^2 + 3*y) * (x^3 + x^2 + x*y^2 + 2*y);
>> collect(f,x)
ans =
    2*x^5 + 2*x^4 + (2*y^2 + 3*y)*x^3 + 7*y*x^2 + 3*y^3*x
    + 6*y^2

>> collect(f,y)
ans =
    3*x*y^3 + (2*x^3 + 6)*y^2 + (3*x^3 + 7*x^2)*y
    + 2*x^2*(x^3 + x^2)
```

4.2.2   Factor

The `factor` function finds the set of irreducible polynomials whose product form the higher order polynomial argument to `factor`. The factoring operation is useful for finding the roots of the polynomial.

```
>> g = x^4 + 3*x^3 - 19*x^2 - 27*x + 90;
>> factor(g)
```

```
ans =
    [ x - 2, x - 3, x + 5, x + 3]
```

4.2.3 Expand

The expand function multiplies polynomial products to get a single polynomial.

```
>> f = (3*x^2 + 2*x)*(2*x - 5);
>> expand(f)
ans =
    6*x^3 - 11*x^2 - 10*x
>> g
g =
    (x - 2)*(x - 3)*(x + 3)*(x + 5)
>> expand(g)
ans =
    x^4 + 3*x^3 - 19*x^2 - 27*x + 90
```

4.2.4 Simplify

The simplify function generates a simpler form of an expression.

```
>> g
g =
    x^4 + 3*x^3 - 19*x^2 - 27*x + 90
>> h = expand((x-2)*(x+5))
h =
    x^2 + 3*x - 10
>> k = g/h
k =
    (x^4 + 3*x^3 - 19*x^2 - 27*x + 90)/(x^2 + 3*x - 10)
>> simplify(k)
ans =
    x^2 - 9

>> f = cos(x)*cos(y) + sin(x)*sin(y);
>> simplify(f)
ans =
    cos(x - y)

>> f = cos(x)^2 + sin(x)^2;
>> simplify(f)
ans =
    1
```

4.2.5 Solve

Finding the roots of equations, i.e., x, where $f(x) = 0$, is addressed here and also in the context of two other topics.

- The `roots` function described in section 6.3.2 finds the numerical roots of a polynomial using a linear algebra algorithm.

- The `fzero` function described in section 7.1.1 uses numerical methods to find the roots of any equation of real numbers.

The `solve(s,x)` function returns the set of symbolic values of x that satisfy the expression s. The default is to solve for s == 0, but an equality statement may be used as part of the expression. Be certain to use an equality statement (==), not an assignment statement (=).

```
>> g(x) = x^2 - 1;
>> solve(g(x), x)
ans =
    -1
     1

>> solve(g(x) == 4, x)
ans =
      5^(1/2)
     -5^(1/2)

>> h(x) = x + 2;
>> solve(g(x) == h(x), x)
ans =
     1/2 - 13^(1/2)/2
     13^(1/2)/2 + 1/2

>> syms t y(t)
>> y(t) = cos(pi*t)  % non-polynomial
y(t) =
    cos(pi*t)
>> solve(y(t), t)
ans =
    1/2

>> y(t) = 1 + t^2;
>> solve(y(t), t)
ans =                    % even complex roots
    -1i
     1i
```

4.2.6 Subs

The subs(s) function returns a copy of s after replacing the symbolic variables in s with their values from the MATLAB workspace. The subs function was demonstrated in section 4.1.

```
>> syms g(x) a b x
>> g(x) = a*x^2 + b*x;
>> a = 2; b = 6;
>> subs(g)
ans(x) =
    2*x^2 + 6*x
>> subs(g(3))
ans =
    36
>> syms c
>> h = g/c
h(x) =
    (a*x^2 + b*x)/c
>> c = 3;
>> subs(h(3))
ans =
    12
```

4.2.7 Vpa

The *Symbolic Math Toolbox* tries to keep variables as exact symbolic values rather than as decimal approximations. The vpa function converts symbolic expressions to decimal values. The name vpa is an acronym for variable–precision floating-point arithmetic. As the name implies, the vpa(x, d) function evaluates x to at least d digits. The default number of digits is 32. The maximum number of digits is 2^{29}.

```
>> g(x) = x^2 - 1;
>> h = solve(g(x) == 4, x)
h =
    5^(1/2)
    -5^(1/2)
>> vpa(h)
ans =
    2.2360679774997896964091736687313
    -2.2360679774997896964091736687313
```

4.3 Symbolic Calculus

4.3.1 Symbolic Derivatives

The `diff(s, var, n)` function takes the symbolic derivative of an expression with respect to a specified variable. An optional third argument specifies the order of the derivative. As with many SMT functions, the default variable is x. The variable is needed when the equation has more than one variable, but it never hurts to specify it.

```
>> syms x
>> diff(x^2 + 3*x + 2, x)
ans =
    2*x + 3
>> diff(x^2 + 3*x + 2, x, 2)  % 2nd derivative
ans =
    2
>> diff(x*sin(3*x))
ans =
    sin(3*x) + 3*x*cos(3*x)
>> diff(exp(-x/2))
ans =
    -exp(-x/2)/2
```

4.3.2 Symbolic Integration

The `int` function performs integration of an expression. The variable specification is optional if the expression has only one symbolic variable. Thus valid forms for indefinite integrals are `int(s)` and `int(s,var)`.

$$\int x^2 + 3x + 2\,dx = \frac{1}{3}x^3 + \frac{3}{2}x^2 + 2x$$

```
>> syms x t
>> int(x^2 + 3*x + 2, x)
ans =
    (x*(2*x^2 + 9*x + 12))/6
>> expand(int(x^2 + 3*x + 2, x))
ans =
    x^3/3 + (3*x^2)/2 + 2*x
```

For definite integrals, we add the limits of integration to the arguments: `int(s, var, a, b)`.

$$\int_0^5 x^2 + 3x + 2\,dx = \frac{535}{6}$$

```
>> int(x^2 + 3*x + 2, x, 0, 5) % definite integral
ans =
      535/6
```

The limits of integration may be another variable.

$$\int_0^t x^2 + 3x + 2\,dx = \frac{1}{3}t^3 + \frac{3}{2}t^2 + 2t$$

```
>> expand(int(x^2 + 3*x + 2, x, 0, t))
ans =
      t^3/3 + (3*t^2)/2 + 2*t
```

$$\int_0^t e^{-x/2}\,dx = 2 - 2e^{-t/2}$$

```
>> int(exp(-x/2),x,0,t)
ans =
      2 - 2*exp(-t/2)
```

Infinity may also be an integration limit.

$$\int_0^\infty e^{-x/2}\,dx = 2$$

```
>> int(exp(-x/2), x, 0, Inf)  % definite to infinity
ans =
      2
```

When the SMT is not able to find an analytic solution, it may return a result that makes use of numerical integration methods. We also encountered the erf and erfc functions in section 3.4.2.

```
>> p = int(exp(-x^2), -Inf, 2)
p =
      (pi^(1/2)*(erf(2) + 1))/2
```

4.3.3 Symbolic Limits

The limit(s, h, k) function returns the limit of an expression, s as the variable h approaches k.

$$\lim_{x \to 0} \frac{\sin(x)}{x} = 1$$

$$\lim_{x \to \infty} x e^{-x} = 0$$

```
>> limit(sin(x)/x, x, 0)
ans =
    1
>> limit(x*exp(-x), x, Inf)
ans =
    0
```

Another optional argument of 'left' or 'right' is useful when the expression has a discontinuity at $h = k$. The expression $\frac{|x-3|}{x-3}$ has a discontinuity at $x = 3$.

```
>> limit(abs(x-3)/(x-3), x, 3, 'right')
ans =
    1
>> limit(abs(x-3)/(x-3), x, 3, 'left')
ans =
    -1
```

The following computes $\frac{dy}{dx}$ for $y(x) = x^2$ using a limit. Recall the definition of a derivative:

$$\frac{dy}{dx} = \lim_{h \to 0} \frac{y(x+h) - y(x)}{h}.$$

```
>> syms h s x
>> s = ((x + h)^2 - x^2)/h;
>> limit(s, h, 0)
ans =
    2*x
```

4.4   Symbolic Differential Equations

We use the solutions to differential equations in all areas of physics and engineering. Differential equations are just expressions where one or more terms of the expression is the derivative of another variable. Any system that moves or changes in either time or space is defined by a differential equation.

The `dsolve` function returns the solution to differential equations. Note that previous versions of MATLAB used the special variables Dy and D2y for the first and second derivatives of y. That notation is now deprecated. Instead, use the `diff` function to define derivatives in the expression. As an example, consider the simple differential equation in equation (4.3), which is solved by the following MATLAB code.

$$\frac{dy(t)}{dt} = -a\,y(t)$$

$$y(t) = C\,e^{-a\,t} \tag{4.3}$$

```
>> syms t y(t) a
>> y(t) = diff(y(t)) == -a*y(t);
>> Y(t) = dsolve(y(t))
Y(t) =
    C4*exp(-a*t)
```

The example above shows two important points about differential equations. First, the solution contains an exponential function of the irrational number e. This is because $e^{a\,t}$ is the only function for which the derivative is a multiple of the original function. Specifically,

$$\frac{d\,e^{a\,t}}{dt} = a\,e^{a\,t}. \tag{4.4}$$

Thus, exponential functions of e show up as the solution to any differential equation of growth or decay where there is a linear relationship between the rate of change and the function's value.

Secondly, the solution to equation (4.4) contains a constant, C4. This is normal when an initial condition or boundary condition was not specified. With an initial condition, we can either solve for C4 algebraically, or use the condition as an input to `dsolve`. For example, let's say that $Y(0) = 5$. From our solution, $Y(0) = C4$, thus $C4 = 5$. The following code uses the SMT to solve a first order differential equation like equation (4.4) with exponential decay. The solution is plotted in figure 4.2.

Initial or boundary conditions may also be passed to `dsolve` as an extra argument. When there is more than one condition, as is common for second or higher order differential equations, the conditions are passed as an array. See the section 4.1 example for a second order differential equation example with two initial conditions.

```
>> syms y(t) a
>> eqn = diff(y(t)) == -a*y(t);
>> cond = y(0) == 5;
>> Y(t) = dsolve(eqn, cond)
Y(t) =
    5*exp(-a*t)
```

Note: As with function handles, the results computed by the SMT that are functions may be plotted with the `fplot` function (section 2.2.10).

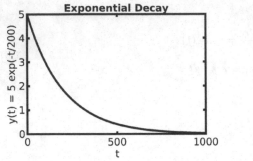

>> a = 0.005;
>> y(t) = subs(Y(t))
y(t) =
        5*exp(-t/200)
>> fplot(y(t), [0 1000])

FIGURE 4.2: The exponential decay solution to a differential of the same form as equation (4.4).

4.5   Exercises

Exercise 4.1   Symbolic Calculations

Use the MATLAB *Symbolic Math Toolbox* functions to complete the following mathematics problems.

1. Use the **expand** function to express this product of polynomials as a single polynomial: $(x^2 + 5x + 6)(x - 4)(4x + 9) = 0$.

2. Factor the following polynomial: $x^3 - 3x^2 - 13x + 15 = 0$.

3. Simplify the equation $(\cos(x) + \sin(x))(\cos(x) + \sin(x))$.

4. Simplify the equation $\cos(x + y) \cdot \sin(x - y)$.

5. Simplify the equation $\frac{\sin(x)}{\cos(2*x)} + \frac{\cos(x)}{\sin(2*x)}$

6. Simplify the equation $\frac{x^4 - 5\,x^3 + 20\,x - 16}{x^2 - 4}$

7. Find the values of x that satisfy $3x^2 - 8x + 5 = 0$.

8. Find the derivative with respect to x of $f(x) = \frac{x^3 + 7x + 1}{x^3 - x}$.

9. Find the derivative with respect to x of $f(x) = x^2 \cos(x)$.

10. Calculate the definite integral $\int_0^4 2\pi x \left(2 - \sqrt{x}\right) dx$

11. Find the integral of $\int \frac{1 + \cos(x)}{\sin^2(x)} dx$.

12. Find the limit $\lim_{x \to 2} \frac{x^2 - 4}{x - 2}$

Exercise 4.2   Symbolic Differential Equation

Find $y(t)$ when $\frac{dy(t)}{dt} = 2\,y(t) - 3\,y^2(t)$, and $y(0) = 1/3$. This is an example of the logistic equation, so you can check your answer in section 7.5.8.

Chapter 5

Introduction to Linear Algebra

Linear algebra is a branch of mathematics that is very useful in science and engineering. However, its frequency of use and prominence in undergraduate science and engineering curriculums has not always been what it is today. Although linear algebra has a long history, the computational requirements limited its adoption until recently. Gottfried Wilhelm Leibnitz, one of the inventors of calculus, used linear algebra to solve systems of equations in 1693 [75]. As with calculus, linear algebra progressed slowly in the eighteenth century but advanced significantly during the nineteenth century [13]. Yet even in the middle of the twentieth century, use of linear algebra by scientists and engineers was limited by the lack of efficient tools. Calculations on vectors and matrices require large numbers of simple multiplications and additions, which are tedious for humans. The arrival of fast computers, more efficient algorithms, and powerful software made linear algebra more accessible. Now that we are better able to leverage the assistance of computers, linear algebra has become a prerequisite to studying several application areas such as control systems, signal and image processing, data analysis, communication systems, and artificial intelligence.

There are at least five important applications of linear algebra:

1. Problems related to spatial vectors and geometry,

2. Solutions to systems of linear equations,

3. Vector projections with application to least square regression,

4. Applications of eigenvalues and eigenvectors to systems of difference and differential equations,

5. Applications of dimensionality reduction from the singular value decomposition and principal component analysis.

It is highly appropriate to learn linear algebra within the context of computational engineering. We will learn the basics of the mathematics, but we will quickly turn to the computer to do the heavy lifting of computing results.

This chapter and the next chapter on the eigenvalues and eigenvectors of matrices relate to linear algebra. The first three sections of this chapter provide an introduction to vectors, matrices, and associated mathematical operations. Then the remainder of the chapter is devoted to finding solutions to systems

DOI: 10.1201/9781003271437-5

of linear equations of the form $\mathbf{A}x = b$. The matrix $\mathbf{A}$ gives us the coefficients of the equations. The x vector represents the unknown variables that we wish to find. The b vector completes the system of equations with the product of $\mathbf{A}$ and x. We will consider three types of linear algebra equations.

- **Square matrix systems**, or critically determined systems[1], have an equal number of equations and unknown variables, so $\mathbf{A}$ is square (equal number of rows, m, and columns, n). Square matrix systems are what first come to mind as a system having a unique, exact answer. Although, some square matrix systems are *singular* and do not have a solution.

- **Under-determined systems** have fewer equations than unknown variables. Matrix $\mathbf{A}$ has more columns than rows ($m < n$). These systems do not have a unique solution, but rather have an infinite number of solutions. Since under-determined systems are common in some application domains, we want to go beyond finding the set of solutions and identify which solutions are considered the best solutions.

- **Over-determined systems** have more equations than unknown variables. Matrix $\mathbf{A}$ has more rows than columns ($m > n$). Over-determined systems may have an exact, unique solution if the equations are all consistent. But the solution is often an approximation. Such is the case for problems requiring least squares regression where we have a large number of equations (rows), but random uncertainty requires an approximation solution.

We conclude the chapter with an introduction to the numerically stable and efficient algorithms used by MATLAB's left-divide (backslash) operator to solve systems of linear equations.

5.1 Working with Vectors

Section 1.7 defines vectors and discusses how to create and use them in MATLAB. Here we define some linear algebra operations on vectors.

Vectors have scalar coefficients defining their displacement in each dimension of the coordinate axis system. We can think of a vector as having a specific length and direction. What we don't always know about a vector is where it begins. Lacking other information about a vector, we can consider that the vector begins at the origin. *Points* are sometimes confused for vectors because they are also defined by a set of coefficients for each dimension. The

---

[1]The term *critically determined* is occasionally used in linear algebra literature to describe square matrix systems of equations. The term is used here to make a clear distinction between under-determined and over-determined systems of equations.

values of points are always relative to the origin. Vectors may be defined as spanning between two points ($v = \mathbf{p}_1 - \mathbf{p}_2$); and one of the points may be the origin. Figure 5.1 shows the relationship of points and vectors relative to the coordinate axes.

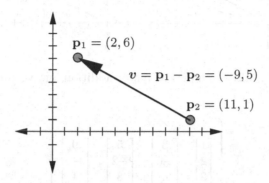

FIGURE 5.1: The elements of points are the coordinates relative to the axes. The elements of vectors tell the length of the vector for each dimension.

Note that vectors are usually stored as column vectors in MATLAB. One might accurately observe that the data of a vector is a set of numbers, which is neither a row vector nor a column vector [48]. However, vectors in MATLAB are stored as either row vectors or column vectors. The need to multiply a vector by a matrix using inner products (defined in section 5.1.4) motivates toward using column vectors. Because column vectors take more space when displayed in written material, they are often displayed in one of two alternate ways—either as the transpose of a row vector $[a, b, c]^T$ or with parentheses (a, b, c).

Note: The word *vector* comes from the Latin word meaning "*carrier*".

5.1.1 Linear Vectors

Vectors are linear, which as shown in figure 5.2 means that they can be added together ($u + v$) and scaled by multiplication by a scalar constant ($k\,v$). Vectors may also be defined as a linear combinations of other vectors ($x = 3\,u + 5\,v$).

5.1.2 Independent Vectors

A vector is independent of other vectors within a set if no linear combination of other vectors defines the vector. Consider the following set of

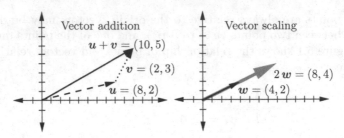

FIGURE 5.2: Examples of vector addition and vector scaling.

four vectors.

$$\left\{ \underbrace{\begin{bmatrix} 1 \\ 2 \\ 2 \end{bmatrix}}_{w}, \underbrace{\begin{bmatrix} 3 \\ -6 \\ 4 \end{bmatrix}}_{x}, \underbrace{\begin{bmatrix} 5 \\ -2 \\ 8 \end{bmatrix}}_{y}, \underbrace{\begin{bmatrix} 9 \\ -2 \\ 1 \end{bmatrix}}_{z} \right\}$$

The first three vectors of the set $\{w, x, \text{and } y\}$ are not independent because $y = 2w + x$. Likewise, $w = \frac{1}{2}(y - x)$ and $x = y - 2w$. However, z is independent of the other three vectors.

Sometimes we can observe dependent relationships, but it is often difficult to see the relationships. MATLAB has a few functions that will help one test for independence and even see what the dependent relationships are. Tests for independence include the `rank` and `det` functions discussed in section 5.2.6. The `rref` function (section 5.5.3) and `cr` function show the relationships between dependent vectors. The `cr` function is not part of MATLAB, but is described in section 5.4.3.

Appendix A.2.2 gives a more mathematically formal definition of linearly independent vectors.

5.1.3   Transpose

The transpose of either a vector or matrix is the reversal of the rows for the columns.

$$\text{Let } a = \begin{bmatrix} a \\ b \\ c \end{bmatrix} \text{ then, } a^T = \begin{bmatrix} a & b & c \end{bmatrix}$$

$$\text{Let } b = \begin{bmatrix} a & b & c \end{bmatrix} \text{ then, } b^T = \begin{bmatrix} a \\ b \\ c \end{bmatrix}$$

$$\text{Let } \mathbf{C} = \begin{bmatrix} a & b & c \\ d & e & f \end{bmatrix} \text{ then, } \mathbf{C}^T = \begin{bmatrix} a & d \\ b & e \\ c & f \end{bmatrix}$$

In MATLAB, the transpose operator is the apostrophe ('):

```
>> A_trans = A';
```

Note: The .' operator performs a simple transpose, and ' performs a complex conjugate transpose. For matrices with only real numbers, the result of the two operators is the same. For matrices with complex numbers, the complex conjugate is usually desired.

5.1.4 Dot Product and Inner Product

The sum of products between the elements of two vectors is called a dot product. The operation yields a scalar.

$$\begin{bmatrix} a \\ b \\ c \end{bmatrix} \cdot \begin{bmatrix} d \\ e \\ f \end{bmatrix} = a\,d + b\,e + c\,f$$

$$\begin{bmatrix} 1 \\ 2 \\ 3 \end{bmatrix} \cdot \begin{bmatrix} 4 \\ 5 \\ 6 \end{bmatrix} = 1 \cdot 4 + 2 \cdot 5 + 3 \cdot 6 = 32$$

Inner Product

An inner product is the sum of products between a row vector and a column vector. The multiply operator (*) in MATLAB performs an inner product. Thus to calculate a dot product, we only need to multiply the transpose of the first vector by the second vector. The resulting scalar is the dot product of the vectors.

For example,

$$a = \begin{bmatrix} 1 \\ 2 \end{bmatrix} \quad b = \begin{bmatrix} 3 \\ 4 \end{bmatrix}$$

$$a \cdot b = a^T b = \begin{bmatrix} 1 & 2 \end{bmatrix} \begin{bmatrix} 3 \\ 4 \end{bmatrix}$$
$$= 1 \cdot 3 + 2 \cdot 4$$
$$= 11$$

The inner product is also used to calculate each value of the product between matrices. Each element of the product is an inner product between a row of the left matrix and a column of the right matrix.

MATLAB has a **dot** function that takes two vector arguments and returns the scalar dot product. However, it is often just as easy to implement a dot product using inner product multiplication.

```
>> a = [3; 5];              >> c = a'*b
>> b = [2; 4];             c =
>> c = dot(a,b)                26
c =
    26
```

Note: If the vectors are both row vectors (nonstandard), then the dot product becomes $a \cdot b = a\, b^T$.

5.1.5   Dot Product Properties

5.1.5.1   Commutative

The dot product $u \cdot v$ equals $v \cdot u$. The order does not matter.

5.1.5.2   Length of Vectors

The *length* of the vector v, $\|v\|$, is the square root of the dot product of v with itself.

$$\|v\| = \sqrt{v \cdot v} = (v_1^2 + v_2^2 + ... + v_n^2)^{1/2}$$

MATLAB has a function called norm that takes a vector as an argument and returns the length. The Euclidean length of a vector is called the l_2-norm, which is the default for the norm function. See appendix A.1 for information on other vector length measurements.

A *normalized* or *unit* vector is a vector of length one. A vector may be normalized by dividing it by its length.

$$\hat{v} = \frac{v}{\|v\|}$$

5.1.5.3   Angle Between Vectors

Consider two unit vectors (length 1) in $\mathbb{R}^2$ as shown in figure 5.3, $u = (1, 0)$ and $v = (\cos\theta, \sin\theta)$. Then consider the same vectors rotated by an angle α, such that $\theta = \beta - \alpha$. Refer to a table of trigonometry identities to verify the final conclusion below.

$$\begin{aligned}
u \cdot v &= 1 \cdot \cos\theta + 0 \cdot \sin\theta = \cos\theta \\
w \cdot z &= \cos\alpha \cdot \cos\beta + \sin\alpha \cdot \sin\beta \\
&= \cos(\beta - \alpha) \\
&= \cos\theta
\end{aligned} \tag{5.1}$$

The result of equation (5.1) can also be found from the *law of cosines*, which tells us that

$$\|z - w\|^2 = \|z\|^2 + \|w\|^2 - 2\|z\|\,\|w\|\cos\theta$$

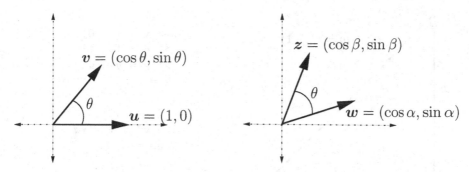

FIGURE 5.3: The dot product between two unit vectors is the cosine of the angle between the vectors.

Since both z and w are unit vectors, several terms become 1.

$$\begin{aligned}
(z - w) \cdot (z - w) &= 2 - 2 \cos \theta \\
z \cdot z - 2\, w \cdot z + w \cdot w &= 2 - 2 \cos \theta \\
-2\, w \cdot z &= -2 \cos \theta \\
w \cdot z &= \cos \theta
\end{aligned}$$

When the vectors are not unit vectors, the vector lengths factor out as constants. A unit vector is obtained by dividing the vector by its length.

$$\left(\frac{v}{\|v\|} \right) \cdot \left(\frac{w}{\|w\|} \right) = \cos \theta$$

$$\boxed{v \cdot w = \|v\|\,\|w\| \cos \theta} \tag{5.2}$$

All angles have $|\cos \theta| \leq 1$. So all vectors have:

$$|v \cdot w| \leq \|v\|\,\|w\|$$

5.1.5.4 Orthogonal Vector Test

When two vectors are perpendicular ($\theta = \pi/2$ or $90°$), then from equation (5.2) their dot product is zero. This property extends to $\mathbb{R}^3$ and beyond, where we say that the vectors in $\mathbb{R}^n$ are *orthogonal* when their dot product is zero.

This is an important result. The geometric properties of orthogonal vectors provide useful strategies for finding optimal solutions for some problems. One example of this is demonstrated in section 5.9.

Perpendicular Vectors

By the *Pythagorean theorem*, if v and w are perpendicular, then as shown in figure 5.4

$$\|v\|^2 + \|w\|^2 = \|v - w\|^2.$$

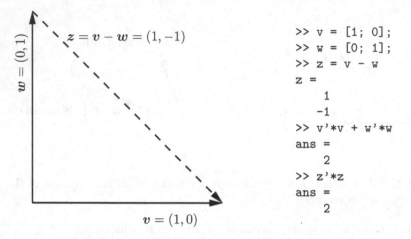

```
>> v = [1; 0];
>> w = [0; 1];
>> z = v - w
z =
    1
   -1
>> v'*v + w'*w
ans =
    2
>> z'*z
ans =
    2
```

FIGURE 5.4: Dot products and the Pythagorean theorem.

5.1.6   Application of Dot Products

5.1.6.1   Perpendicular Rhombus Vectors

As illustrated in figure 5.5, a *rhombus* is any parallelogram whose sides are the same length. Let us use the properties of dot products to show that the diagonals of a rhombus are perpendicular.

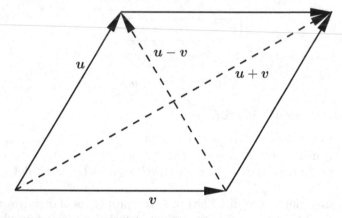

FIGURE 5.5: A rhombus is a parallelogram whose sides are the same length. The dot product can show that the diagonals of a rhombus are perpendicular.

Section 5.2.3.2 lists transpose properties including the transpose with respect to addition, $(a + b)^T = a^T + b^T$. We also need to use the fact that the sides of a rhombus are the same length. We begin by setting the dot product of the diagonals equal to zero.

$$\begin{aligned}
(u + v)^T (u - v) &= 0 \\
(u^T + v^T)(u - v) &= 0 \\
u^T u - u^T v + v^T u - v^T v &= 0 \\
u^T u &= v^T v
\end{aligned} \tag{5.3}$$

Each term of equation (5.3) is a scalar and dot products are commutative, so the middle two terms cancel each other and we are left with the requirement that the lengths of the sides of the rhombus be the same, which is in the definition of a rhombus.

5.1.6.2 Find a Perpendicular Vector

Let us consider a mobile robotics problem. The robot needs to execute a wall–following algorithm. We will use two of the robot's distance sensors — one perpendicular to the robot's direction of travel and one at 45°. As shown in figure 5.6, we establish points $\mathbf{p_1}$ and $\mathbf{p_2}$ from the sensors in the robot's coordinate frame along the wall and find vector $b = \mathbf{p_2} - \mathbf{p_1}$. We need to find a short-term goal location for the robot to drive toward that is perpendicular to the wall at a distance of k from the point $\mathbf{p_2}$. Using vectors a and b, we can find an equation for vector c such that c is perpendicular to b, and c begins at the origin of a. With the coordinates of c, the goal location is a simple calculation.

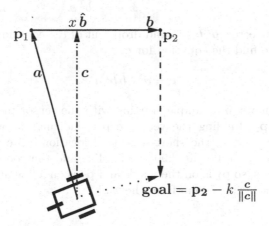

FIGURE 5.6: Finding vector c perpendicular to the wall, b, is needed to find the wall–following goal location.

We want to identify the vector from the terminal point of a to the terminal point of c as $\hat{b}x$ where x is a scalar and $\hat{b}$ is a unit vector in the direction of b.

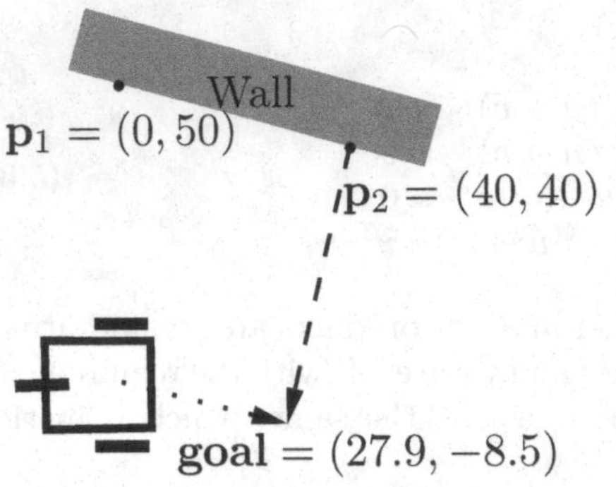

```
>> p1 = [0; 50];
>> p2 = [40; 40];
>> b = p2 - p1;
>> bhat = b/norm(b);
>> a = p1;
>> c = a - bhat*(bhat'*a);
>> k = 50;
>> goal = p2 - k*c/norm(c)
goal =
    27.873
    -8.507
```

FIGURE 5.7: Wall–following example.

$$\hat{b} = \frac{b}{\|b\|}$$

$$c = a + \hat{b}x \qquad (5.4)$$

We find the answer by starting with the orthogonality requirement between vectors c and $\hat{b}$.

$$\hat{b}^T(a + \hat{b}x) = 0$$
$$\hat{b}^T a + \hat{b}^T \hat{b} x = 0$$
$$x = -\hat{b}^T a$$

Since $\hat{b}$ is a unit vector, $\hat{b}^T \hat{b} = 1$ and drops out of the equation. We return to equation (5.4) to find the equation for c.

$$c = a - \hat{b}\hat{b}^T a \qquad (5.5)$$

Figure 5.7 shows an example starting with the sensor measurements for points p_1 and p_2. Finding the vector c perpendicular to wall with equation (5.5) makes finding the short-term goal location quite simple. We are using sensors on the robot at $\pm 90°$ and $\pm 45°$, and the coordinate frame is that of the robot's, so p_1 is on the y axis and p_2 is on a line at $\pm 45°$ from the robot. The robot begins 50 cm from the wall.

5.1.7 Outer Product

Whereas the inner product (dot product) of two vectors is a scalar ($u \cdot v = u^T v$), the *outer product* of two vectors is a matrix. Each element of the outer product is a product of an element from the left column vector and an element from the right row vector. Outer products are not used nearly as often as inner products, but in section 6.8 we will see an important application of

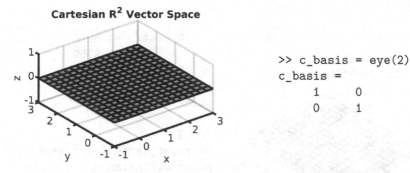

FIGURE 5.8: The 2-dimensional Cartesian vector space is a plane where $z = 0, \forall(x, y)$.

outer products.

$$
\boldsymbol{u} \otimes \boldsymbol{v} = \boldsymbol{u}\,\boldsymbol{v}^T =
\begin{bmatrix} u_1 \\ u_2 \\ \vdots \\ u_m \end{bmatrix}
\begin{bmatrix} v_1 & v_2 & \cdots & v_n \end{bmatrix} =
\begin{bmatrix}
u_1v_1 & u_1v_2 & \cdots & u_1v_n \\
u_2v_1 & u_2v_2 & \cdots & u_2v_n \\
\vdots & \vdots & \ddots & \vdots \\
u_mv_1 & u_mv_2 & \cdots & u_mv_n
\end{bmatrix}
$$

5.1.8 Dimension and Space

Basis vectors provide the coordinate axes used to define the coordinates of points and vectors. The $\mathbb{R}^2$ Cartesian basis vectors are defined by the two columns of the identity matrix, $(1, 0)$ and $(0, 1)$. A plot of a portion of the Cartesian $\mathbb{R}^2$ vector space is shown in figure 5.8.

A **vector space** consists of a set of vectors and a set of scalars that are *closed* under vector addition and scalar multiplication. Saying that they are *closed* just means that we can add any vectors in the vector space together and multiply any vectors in the space by a scalar and the resulting vectors are still in the vector space. The set of all possible vectors in a vector space is called the **span** of the vector space.

For example, vectors in our physical 3-dimensional world are said to be in a vector space called $\mathbb{R}^3$. Vectors in $\mathbb{R}^3$ consist of three real numbers defining their magnitude in the x, y, and z directions. Similarly, vectors on a 2-D plane, like a piece of paper, are said to be in the vector space called $\mathbb{R}^2$.

Let us clarify a subtle matter of definitions. We normally think of the term **dimension** as meaning how many coordinate values are used to define points and vectors, which is accurate for coordinates relative to their basis vectors. However, when points in a custom coordinate system are mapped to Cartesian coordinates via multiplication by the basis vectors, then the number of elements matches the number of elements in the basis vectors rather than

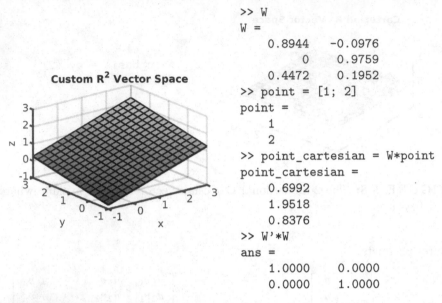

```
>> W
W =
        0.8944    -0.0976
             0     0.9759
        0.4472     0.1952
>> point = [1; 2]
point =
        1
        2
>> point_cartesian = W*point
point_cartesian =
        0.6992
        1.9518
        0.8376
>> W'*W
ans =
        1.0000     0.0000
        0.0000     1.0000
```

FIGURE 5.9: Two custom basis vectors define a 2-dimensional vector space.

the number of basis vectors. In appendix A.2, we give a more precise definition of dimension as the number of basis vectors in the vector space.

In figure 5.9, a $\mathbb{R}^2$ custom vector space has basis vectors as the columns of the $\mathbf{W}$ matrix. Even though the basis vectors are in $\mathbb{R}^3$ Cartesian coordinates, the vector space is 2-dimensional since there are two basis vectors and the span of the vector space is a plane.

Notice that the two basis vectors are orthogonal to each other. Since the basis vectors are in the columns of $\mathbf{W}$, $\mathbf{W}^T \mathbf{W} = \mathbf{I}$ show us that the columns are unitary (length of one and orthogonal to each other).

Other vector spaces may also be used for applications not relating to geometry and may have higher dimension than 3. Generally, we call this $\mathbb{R}^n$. For some applications, the coefficients of the vectors and scalars may also be complex numbers, which is a vector space denoted as $\mathbb{C}^n$.

5.2   Working with Matrices

We introduced both vectors and matrices in section 1.7 and discussed how to create them in MATLAB. Here, we present mathematical operations on matrices and review important properties of matrices.

5.2.1 Matrix Math

5.2.1.1 Addition and Subtraction

Addition and subtraction of matrices is performed element-wise.

$$\begin{bmatrix} a & b \\ c & d \end{bmatrix} - \begin{bmatrix} e & f \\ g & h \end{bmatrix} = \begin{bmatrix} a-e & b-f \\ c-g & d-h \end{bmatrix}$$

Scalar and element-wise operations between matrices work the same as with vectors. Of course, the size of both matrices must be the same (section 1.7.3).

5.2.1.2 Matrix Multiplication

Multiplication of a vector by a matrix and multiplication of two matrices are common linear algebra operations. The former is used in the expression of a system of linear equations of the form $\mathbf{A}x = b$. Points and vectors can be moved, scaled, rotated, or skewed when multiplied by a matrix. This has application to physical geometry and also to pixels in an image. Two primary uses of matrix–to–matrix multiplication are explored in later sections. First, we discuss geometric transformation matrices in section 5.3. The transformation matrices are multiplied with moveable points to determine new coordinate locations. Multiplication of the transformation matrices is also used to combine transformations. Secondly, we will cover several matrix factorings later in this chapter and in the next chapter, such as $\mathbf{A} = \mathbf{L}\mathbf{U}$, $\mathbf{A} = \mathbf{X}\mathbf{\Lambda}\mathbf{X}^{-1}$, $\mathbf{A} = \mathbf{Q}\mathbf{R}$, and $\mathbf{A} = \mathbf{U}\mathbf{\Sigma}\mathbf{V}^T$. Each matrix factoring facilitates solving particular problems. It is matrix multiplication that verifies that the product of the factors yields the original matrix.

Here is an example showing both the product of matrix factors and how multiplying a vector by a matrix scales and rotates vectors and points. We begin with a vector on the x axis and a matrix that rotates points and vectors $\pi/4$ radians (45°). Rotation matrices are discussed in section 5.3.1. We multiply the rotation matrix by two so that it also scales the vector. Then we use the singular value decomposition[2] (SVD), which is later described in section 6.8. With matrix multiplication, we see that the product of the factors is the same as our original matrix. Then figure 5.10 shows that successive multiplications by the factors rotate, scale, and again rotate the vector to yield the same result as multiplication by the original scaling and rotating matrix $(\mathbf{p}_3 = 2\,\mathbf{R}\,\mathbf{p} = \mathbf{U}\,\mathbf{\Sigma}\,\mathbf{V}^T\mathbf{p})$.[3]

$$2\,\mathbf{R} = \underbrace{\begin{bmatrix} 1.4142 & 1.4142 \\ 1.4142 & 1.4142 \end{bmatrix}}_{} = \underbrace{\begin{bmatrix} -0.7071 & -0.7071 \\ -0.7071 & 0.7071 \end{bmatrix}}_{\mathbf{U}} \underbrace{\begin{bmatrix} 2 & 0 \\ 0 & 2 \end{bmatrix}}_{\mathbf{\Sigma}} \underbrace{\begin{bmatrix} -1 & 0 \\ 0 & 1 \end{bmatrix}}_{\mathbf{V}^T}$$

---

[2]The meaning of the SVD factors is not our concern at this point. In this application, it is only an example factoring. However, the SVD was chosen because the three factors have the interesting property of rotating, scaling, and then rotating points or vectors again.

[3]The factors of the SVD are $\mathbf{A} = \mathbf{U}\,\mathbf{\Sigma}\,\mathbf{V}^T$. In this example, $\mathbf{V}$ is symmetric, so $\mathbf{V} = \mathbf{V}^T$.

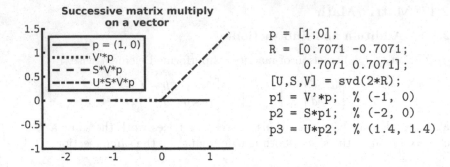

```
p = [1;0];
R = [0.7071 -0.7071;
     0.7071 0.7071];
[U,S,V] = svd(2*R);
p1 = V'*p;   % (-1, 0)
p2 = S*p1;   % (-2, 0)
p3 = U*p2;   % (1.4, 1.4)
```

FIGURE 5.10: The first matrix multiplication rotates the vector. The second multiplication scales the vector. The third multiplication rotates the vector again. The three successive multiplications achieve the same result as multiplying the vector by $2\mathbf{R}$.

The elements of a matrix product are found by inner products. To complete the inner products, the number of columns of the left matrix must match the number of rows in the right matrix. We say that the inner dimensions of the two matrices must agree. As shown in table 5.1, A $m \times p$ matrix may be multiplied by a $p \times n$ matrix to yield a $m \times n$ matrix. The size of the product matrix is the outer dimensions of the matrices.

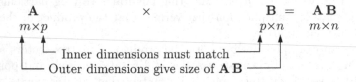

With regard to matrix multiplication, we will view vectors as a matrix with only one column (size $m \times 1$). Multiplication of a matrix by a vector is defined as either the linear combination of the columns of the matrix with the vector elements, or as the inner product between the rows of the left matrix

TABLE 5.1: Size of Matrix Multiplications

Left Matrix	Right Matrix	Output Matrix
$1 \times n$ row vector	$n \times 1$ column vector	1×1 scalar
$1 \times p$ row vector	$p \times n$ matrix	$1 \times n$ row vector
$m \times 1$ column vector	$1 \times n$ row vector	$m \times n$ matrix
$m \times p$ matrix	$p \times 1$ column vector	$m \times 1$ column vector
$m \times p$ matrix	$p \times n$ matrix	$m \times n$ matrix

and the column vector.

$$
\mathbf{A}\,x \;=\;
\begin{bmatrix} 1 & 2 & 3 \\ 2 & 5 & 2 \\ 6 & -3 & 1 \end{bmatrix}
\begin{bmatrix} 2 \\ 1 \\ 1 \end{bmatrix}
$$

$$
= 2\begin{bmatrix} 1 \\ 2 \\ 6 \end{bmatrix}
+ 1\begin{bmatrix} 2 \\ 5 \\ -3 \end{bmatrix}
+ 1\begin{bmatrix} 3 \\ 2 \\ 1 \end{bmatrix}
$$

$$
=
\begin{bmatrix} [1 \;\; 2 \;\; 3] \\ [2 \;\; 5 \;\; 2] \\ [6 \;\; -3 \;\; 1] \end{bmatrix}
\begin{bmatrix} 2 \\ 1 \\ 1 \end{bmatrix}
$$

$$
=
\begin{bmatrix} 7 \\ 11 \\ 10 \end{bmatrix}
$$

The individual values of a matrix product $\mathbf{C}_{i,j}$ are calculated as inner products between the rows of the left matrix ($\mathbf{A}$) and the columns of the right matrix ($\mathbf{B}$).

$$
\mathbf{C}_{i,j} = \sum_{n=1}^{p} \mathbf{A}_{i,n}\,\mathbf{B}_{n,j}
$$

For 2×2 matrices:

$$
\begin{bmatrix} a & b \\ c & d \end{bmatrix}
\begin{bmatrix} e & f \\ g & h \end{bmatrix}
=
\begin{bmatrix} (ae + bg) & (af + bh) \\ (ce + dg) & (cf + dh) \end{bmatrix}
$$

$$
\begin{bmatrix} 1 & 2 \\ 3 & 4 \end{bmatrix}
\begin{bmatrix} 5 & 6 \\ 7 & 8 \end{bmatrix}
=
\begin{bmatrix} (1\cdot5 + 2\cdot7) & (1\cdot6 + 2\cdot8) \\ (3\cdot5 + 4\cdot7) & (3\cdot6 + 4\cdot8) \end{bmatrix}
=
\begin{bmatrix} 19 & 22 \\ 43 & 50 \end{bmatrix}
$$

5.2.1.3 Matrix Multiplication Properties

1. Matrix multiplication is *associative*: $(\mathbf{A}\,\mathbf{B})\,\mathbf{C} = \mathbf{A}\,(\mathbf{B}\,\mathbf{C})$.

2. In general, matrix multiplication is **NOT** *commutative* even for square matrices: $(\mathbf{A}\mathbf{B} \neq \mathbf{B}\mathbf{A})$. The more obvious exceptions to this are multiplication by the identity matrix and an inverse matrix. Note below that the two products of the same matrices are not the same.

$$
\begin{bmatrix} 3 & 0 \\ 4 & 3 \end{bmatrix}
\begin{bmatrix} 2 & -2 \\ 3 & 4 \end{bmatrix}
=
\begin{bmatrix} 6 & -6 \\ 3 & 4 \end{bmatrix}
$$

$$
\begin{bmatrix} 2 & -2 \\ 3 & 4 \end{bmatrix}
\begin{bmatrix} 3 & 0 \\ 4 & 3 \end{bmatrix}
=
\begin{bmatrix} -2 & -6 \\ 25 & 12 \end{bmatrix}
$$

3. $\mathbf{A}\,\mathbf{B} = \mathbf{A}\,\mathbf{C}$ does **not** imply that $\mathbf{B} = \mathbf{C}$. For example, consider matrices:

$$\mathbf{A} = \begin{bmatrix} 1 & 2 \\ 2 & 4 \end{bmatrix}, \mathbf{B} = \begin{bmatrix} 2 & 1 \\ 1 & 3 \end{bmatrix}, \mathbf{C} = \begin{bmatrix} 4 & 3 \\ 0 & 2 \end{bmatrix}$$

$$\mathbf{A}\,\mathbf{B} = \mathbf{A}\,\mathbf{C} = \begin{bmatrix} 4 & 7 \\ 8 & 14 \end{bmatrix}$$

The Outer Product View of Matrix Multiplication

We normally think of matrix multiplication as finding each term of the matrix product from the sum of products between the rows of the left matrix and the columns of the right matrix. This is the *inner product* view of matrix multiplication. But there is also an outer product view that is columns times rows.

$$\mathbf{A}\,\mathbf{B} = \begin{bmatrix} | & | & & | \\ a_1 & a_2 & \cdots & a_p \\ | & | & & | \end{bmatrix} \begin{bmatrix} - & b_1 & - \\ - & b_2 & - \\ & \vdots & \\ - & b_p & - \end{bmatrix} = a_1\,b_1 + \cdots + a_p\,b_p$$

Each a_k column of a $m \times p$ matrix multiplies b_k, the kth row of a $p \times n$ matrix. The product $a_k b_k$ is an $m \times n$ matrix of *rank one* formed from the outer product of the two vectors. The final product $\mathbf{A}\mathbf{B}$ is the sum of the rank-one matrices.

Outer product matrix multiplication is not how we usually multiply two matrices. Rather, it gives an alternative perspective of what matrix multiplication accomplishes. This perspective will be useful later when we learn about the singular value decomposition.

5.2.1.4   Matrix Division

Matrix division, in the sense of scalar division, is not defined for matrices. Multiplication by the inverse of a matrix accomplishes the same objective as division does in scalar arithmetic. MATLAB also has two alternative operators to the matrix inverse that act as one might expect from division operators, left-divide ($\backslash$) and right-divide ($/$); however, they are more complicated than simple division. As shown in table 5.2, they are used to solve systems of linear equations as described in section 5.4.

TABLE 5.2: MATLAB's Matrix Division Operators

Operator	Name	Need	Usage	Replaces
\	left-divide	A*x = b	x = A\b	x = inv(A)*b
/	right-divide	x*A = b	x = b/A	x = b*inv(A)

5.2.2 Special Matrices

Zero Matrix

A *zero* vector or matrix contains all zero elements and is denoted as **0**.

Diagonal Matrix

A *diagonal* matrix has zeros everywhere except on the main diagonal, which is the elements where the row index and column index are the same. Diagonal matrices are usually square (same number of rows and columns), but they may be rectangular with extra rows or columns of zeros.

$$\begin{bmatrix} d_1 & 0 & 0 \\ 0 & d_2 & 0 \\ 0 & 0 & d_3 \end{bmatrix}$$

The *main* diagonal, also called the *forward* diagonal, or the *major diagonal*, is the set of diagonal matrix elements from upper left to lower right. The set of diagonal elements from lower left to upper right is of significantly less interest to us, but has several names including the *antidiagonal*, *back diagonal*, *secondary diagonal*, or the *minor diagonal*.

Identity Matrix

An *identity* matrix (**I**) is a square, diagonal matrix where all of the elements on the main diagonal are one. Identity matrices are like a one in scalar math. That is, the product of any matrix with the identity matrix yields itself.

$$\mathbf{A\,I} = \mathbf{A} = \mathbf{I\,A}$$

$$\mathbf{I}_{2\times 2} = \begin{bmatrix} 1 & 0 \\ 0 & 1 \end{bmatrix}$$

$$\mathbf{I}_{3\times 3} = \begin{bmatrix} 1 & 0 & 0 \\ 0 & 1 & 0 \\ 0 & 0 & 1 \end{bmatrix}$$

$$\mathbf{I}_{4\times 4} = \begin{bmatrix} 1 & 0 & 0 & 0 \\ 0 & 1 & 0 & 0 \\ 0 & 0 & 1 & 0 \\ 0 & 0 & 0 & 1 \end{bmatrix}$$

MATLAB has a function called eye that takes one argument for the matrix size and returns an identity matrix.

```
>> I2 = eye(2)              >> I3 = eye(3)
I2 =                       I3 =
     1    0                     1    0    0
     0    1                     0    1    0
                                0    0    1
```

Upper Triangular Matrix

Upper triangular matrices have all zero values below the main diagonal. Any nonzero elements are on or above the main diagonal.

$$\mathbf{U} = \begin{bmatrix} a & b & c \\ 0 & d & e \\ 0 & 0 & f \end{bmatrix}$$

Lower Triangular Matrix

Lower triangular matrices have all zero values above the main diagonal. Any nonzero elements are on or below the main diagonal.

$$\mathbf{L} = \begin{bmatrix} a & 0 & 0 \\ b & c & 0 \\ d & e & f \end{bmatrix}$$

Symmetric Matrix

A *symmetric* matrix ($\mathbf{S}$) has symmetry relative to the main diagonal. If the matrix where written on a piece of paper and you folded the paper along the main diagonal then the off-diagonal elements with the same value would lie on top of each other. For symmetric matrices $\mathbf{S}^T = \mathbf{S}$. Note that for any matrix, $\mathbf{A}$, the products $\mathbf{A}^T\mathbf{A}$ and $\mathbf{A}\mathbf{A}^T$ both yield symmetric matrices. Here are a couple examples of symmetric matrices.

$$\begin{bmatrix} 1 & 2 \\ 2 & 3 \end{bmatrix} \qquad \begin{bmatrix} 2 & 3 & 6 \\ 3 & 1 & 5 \\ 6 & 5 & 2 \end{bmatrix}$$

Orthonormal Matrix

A matrix, $\mathbf{Q}$, is called *orthonormal* if the columns are unit vectors (length of 1) and the dot product between the columns is zero ($\cos(\theta) = 0$). That is to say, the columns are all orthogonal to each other.

Orthogonal or Unitary Matrix

A matrix is *orthogonal* and *unitary* if it is both orthonormal and square. The product of these matrices with its transpose is the identity matrix, $\mathbf{Q}\mathbf{Q}^T = \mathbf{I}$, and $\mathbf{Q}^T\mathbf{Q} = \mathbf{I}$, which means that its inverse is its transpose, $\mathbf{Q}^{-1} = \mathbf{Q}^T$. Here is an example of multiplication of an orthogonal matrix by its transpose.

$$\begin{bmatrix} 0.5 & -0.866 \\ 0.866 & 0.5 \end{bmatrix} \begin{bmatrix} 0.5 & 0.866 \\ -0.866 & 0.5 \end{bmatrix} =$$

$$\begin{bmatrix} (0.5^2 & + & 0.866^2) & (0.5 \cdot 0.866 & - & 0.866 \cdot 0.5) \\ (0.866 \cdot 0.5 & - & 0.5 \cdot 0.866) & (0.866^2 & + & 0.5^2) \end{bmatrix} =$$

$$\begin{bmatrix} 1 & 0 \\ 0 & 1 \end{bmatrix}$$

Note: Orthogonal matrices may seem a bit abstract at this point because we have not yet encountered an application for them. We need an orthogonal matrix to provide the basis vectors for vector projections, which is described in section 5.9.3.2. Rotation matrices discussed in section 5.3.1 are orthogonal. We can obtain orthogonal columns from an existing matrix by using the modified Gram–Schmidt algorithm (appendix A.4.3), QR factorization (appendix A.5), or from the SVD (appendix A.4 and section 6.8.5.2). Orthogonal matrices are also the result from finding the eigenvectors of symmetric matrices, which is proven in appendix A.9.

5.2.3 Special Matrix Relationships

5.2.3.1 Matrix Inverse

The inverse of a square matrix, $\mathbf{A}$, is another matrix, $\mathbf{A}^{-1}$, that multiplies with the original matrix to yield the identity matrix.

$$\mathbf{A}^{-1}\mathbf{A} = \mathbf{A}\mathbf{A}^{-1} = \mathbf{I}$$

Not all square matrices have an inverse and calculating the inverse for larger matrices is slow, requiring significant calculations.

In MATLAB, the function inv(A) returns the inverse of matrix $\mathbf{A}$. MATLAB's left-divide operator provides a more efficient and numerically stable alternative to using a matrix inverse to solve a system of equations. So while matrix inverses frequently appear in written documentation, we seldom calculate a matrix inverse.

Here are a couple properties for matrix inverses.

- $(\mathbf{A}\mathbf{B})^{-1} = \mathbf{B}^{-1}\mathbf{A}^{-1}$

$$(\mathbf{A}\mathbf{B})^{-1}(\mathbf{A}\mathbf{B}) = \mathbf{I}$$
$$(\mathbf{A}\mathbf{B})^{-1}\mathbf{A}\mathbf{B}\mathbf{B}^{-1}\mathbf{A}^{-1} = \mathbf{I}\mathbf{B}^{-1}\mathbf{A}^{-1}$$
$$(\mathbf{A}\mathbf{B})^{-1} = \mathbf{B}^{-1}\mathbf{A}^{-1}$$

- $\left(\mathbf{A}^T\right)^{-1} = \left(\mathbf{A}^{-1}\right)^T$.

  Start with the simple observation that $\mathbf{I}^T = \mathbf{I}$ and $(\mathbf{A}\mathbf{B})^T = \mathbf{B}^T\mathbf{A}^T$ (section 5.2.3.2).

$$\left(\mathbf{A}^{-1}\mathbf{A}\right)^T = \mathbf{I}^T = \mathbf{I}$$
$$\mathbf{A}^T\left(\mathbf{A}^{-1}\right)^T = \mathbf{I}$$
$$\left(\mathbf{A}^T\right)^{-1}\mathbf{A}^T\left(\mathbf{A}^{-1}\right)^T = \left(\mathbf{A}^T\right)^{-1}$$
$$\left(\mathbf{A}^{-1}\right)^T = \left(\mathbf{A}^T\right)^{-1}$$

5.2.3.2 Matrix Transpose Properties

We described the transpose for vectors and matrices in section 5.1.3.

- $\left(\mathbf{A}^T\right)^T = \mathbf{A}$

- The transpose with respect to addition, $(\mathbf{A} + \mathbf{B})^T = \mathbf{A}^T + \mathbf{B}^T$.

- $(\mathbf{A}\mathbf{B})^T = \mathbf{B}^T\mathbf{A}^T$. Notice that the order is reversed.

$$
\begin{aligned}
(i,j) \text{ value of } (\mathbf{A}\mathbf{B})^T &= (j,i) \text{ value of } \mathbf{A}\mathbf{B} \\
&= j \text{ row of } \mathbf{A} \times i \text{ column of } \mathbf{B} \\
&= i \text{ row of } \mathbf{B}^T \times j \text{ column of } \mathbf{A}^T \\
&= (i,j) \text{ value of } \mathbf{B}^T\mathbf{A}^T
\end{aligned}
$$

- We often pre-multiply vectors and matrices by their transpose $(\mathbf{A}^T\mathbf{A})$. The result is a scalar for a column vector, and a square, symmetric matrix for a row vector, rectangular matrix, and square matrix. If $\mathbf{A}$ is a $m{\times}n$ matrix, $\mathbf{A}^T\mathbf{A}$ is a $n{\times}n$ square, symmetric matrix.

  Since $\mathbf{S} = \mathbf{A}^T\mathbf{A}$ is symmetric, then $\mathbf{S}^T = \mathbf{S}$.

$$\mathbf{S}^T = \left(\mathbf{A}^T\mathbf{A}\right)^T = \mathbf{A}^T\left(\mathbf{A}^T\right)^T = \mathbf{A}^T\mathbf{A} = \mathbf{S}$$

Another interesting result is that the trace of $\mathbf{S} = \mathbf{A}^T\mathbf{A}$, denoted as $Tr(\mathbf{S})$, is the sum of the squares of all the elements of $\mathbf{A}$. The trace of a matrix is the sum of its diagonal elements.

$$
\mathbf{A}^T\mathbf{A} = \begin{bmatrix} a & b & c \\ d & e & f \end{bmatrix} \begin{bmatrix} a & d \\ b & e \\ c & f \end{bmatrix} = \begin{bmatrix} a^2 + b^2 + c^2 & ad + be + cf \\ ad + be + cf & d^2 + e^2 + f^2 \end{bmatrix}
$$

- For an orthogonal matrix, $\mathbf{Q}^{-1} = \mathbf{Q}^T$. This is called the *Spectral Theorem*. Note that the columns of the matrix must be unit vectors for this property to be true.

Since each of the columns are unit vectors, it follows that the diagonal of $\mathbf{Q}^T\mathbf{Q}$ is ones. The off-diagonal values are zeros because the dot product between the columns is zero. Thus,

$$\mathbf{Q}^T\mathbf{Q} = \mathbf{Q}\,\mathbf{Q}^T = \mathbf{I}$$

$$\mathbf{Q}^T = \mathbf{Q}^{-1}$$

Here is an example of an orthogonal rotation matrix.

```
>> Q
Q =
     0.5000    -0.8660
     0.8660     0.5000

>> Q'
ans =
     0.5000     0.8660
    -0.8660     0.5000

>> Q(:,1)'*Q(:,2)    % column dot products => orthogonal
ans =
     0

>> Q'*Q              % orthogonal columns
ans =
     1.0000    -0.0000
    -0.0000     1.0000
>> Q*Q'              % orthogonal rows
ans =
     1.0000     0.0000
     0.0000     1.0000
```

Another interesting result about orthogonal matrices is that the product of two orthogonal matrices is also orthogonal. This is because both the inverse and transpose of matrices reverse the matrix order. Proof of orthogonality only requires that $\mathbf{Q}^T = \mathbf{Q}^{-1}$.

Consider two orthogonal matrices $\mathbf{Q}_1$ and $\mathbf{Q}_2$.

$$(\mathbf{Q}_1\mathbf{Q}_2)^{-1} = \mathbf{Q}_2^{-1}\mathbf{Q}_1^{-1} = \mathbf{Q}_2^T\mathbf{Q}_1^T = (\mathbf{Q}_1\mathbf{Q}_2)^T$$

5.2.4 Determinant

The determinant of a square matrix is a number. At one time, determinants were a significant component of linear algebra. But that is not so much the case today [69]. The importance of determinants is now more for analytical than computational purposes. Thus, while it is important to know what a determinant is, we will avoid using them when possible.

- Either of two notations are used to indicate the determinant of a matrix, det(**A**) or |**A**|.

- The classic technique for computing a determinant is called the *Laplace expansion*. The complexity of computing a determinant by the classic Laplace expansion method has run-time complexity on the order of $n!$ (n–factorial), which is fine for small matrices, but is too slow for large ($n > 3$) matrices.

- In practice, the determinant for matrices is computed using a faster method that is described in section 5.6.5 and improves the performance to $\mathcal{O}(n^3)$.[4] The faster method uses the result that det(**A B**) = det(**A**) det(**B**). The LU decomposition factors are computed as described in section 5.6, then the determinant is the product of the determinants of the LU decomposition factors.

- In some applications, such as machine learning, very large matrices are often used, so computing determinants is impractical. Fortunately, we now have alternate algorithms that do not require determinants.

The determinant of a 2×2 matrix is simple, but it gets more complicated as the matrix dimension increases.

$$\begin{vmatrix} a & b \\ c & d \end{vmatrix} = a\,d - c\,b$$

The $n{\times}n$ determinant makes use of n $(n-1){\times}(n-1)$ determinants, called cofactors. In each stage of Laplace expansion the elements in a row or column is multiplied by a cofactor determinant from the other rows and columns excluding the row and column of the multiplication element. The determinant is then the sum of the expansion along the chosen row or column. But then the $(n-1){\times}(n-1)$ determinants need to be calculated. Likewise for the following smaller determinants until the needed determinant is a 2×2 that is solved directly. It works as follows for a 3×3 determinant.

---

[4]The symbol $\mathcal{O}(\ )$ is called *Big–O*. It is a measure of the order of magnitude of the number of calculations needed relative to the size of the data operated on, n. If we say that an algorithm has run time complexity of $\mathcal{O}(n^3)$, we are not saying that n^3 is the exact number of calculations needed to run the algorithm. Rather, we are saying that the order of magnitude of the number of calculations needed has a cubic relationship to n. So if the number of data elements is doubled, then the number of calculations needed grows by a factor of 8, $\mathcal{O}((2n)^3)$.

$$\begin{vmatrix} a & b & c \\ d & e & f \\ g & h & i \end{vmatrix} = a\begin{vmatrix} e & f \\ h & i \end{vmatrix} - b\begin{vmatrix} d & f \\ g & i \end{vmatrix} + c\begin{vmatrix} d & e \\ g & h \end{vmatrix}$$

Another approach to remembering the sum of products needed to compute a 3×3 determinant is to compute diagonal product terms. The terms going from left to right are added while the right to left terms are subtracted. Note: This only works for 3×3 determinants.

$$\begin{vmatrix} a & b & c \\ d & e & f \\ g & h & i \end{vmatrix} = \begin{bmatrix} a & b & c & a & b \\ d & e & f & d & e \\ g & h & i & g & h \end{bmatrix}$$

$$- \begin{bmatrix} a & b & c & a & b \\ d & e & f & d & e \\ g & h & i & g & h \end{bmatrix}$$

$$\begin{vmatrix} a & b & c \\ d & e & f \\ g & h & i \end{vmatrix} = aei + bfg + cdh - ceg - afh - bdi$$

For a 4×4 determinant, the pattern continues with each cofactor term now being a 3×3 determinant. The pattern likewise continues for higher order matrices. The number of computations needed with this method of computing a determinant is on the order of $n!$, which is prohibitive for large matrices.

Notice that the sign of the cofactor additions alternate according to the pattern:

$$sign_{i,j} = (-1)^{i+j} = \begin{bmatrix} + & - & + & - \\ - & + & - & + \\ + & - & + & - \\ - & + & - & + \end{bmatrix}$$

It is not necessary to always use the top row for Laplace expansion. Any row or column will work, just make note of the signs of the cofactors. The best strategy is to apply Laplace expansion along the row or column with the most zeros.

Here is bigger determinant problem with lots of zeros so it is not too bad. At each stage of the expansion there is a row or column with only one nonzero element, so it is not necessary to find the sum of cofactor determinants. Trace the Laplace expansion as we find the determinant.

$$\begin{vmatrix} 8 & 5 & 4 & 3 & 0 \\ 7 & 0 & 6 & 1 & 0 \\ 8 & -5 & 4 & 3 & -5 \\ -3 & 0 & 0 & 0 & 0 \\ 4 & 0 & 2 & 2 & 0 \end{vmatrix} = 3 \times \begin{vmatrix} 5 & 4 & 3 & 0 \\ 0 & 6 & 1 & 0 \\ -5 & 4 & 3 & -5 \\ 0 & 2 & 2 & 0 \end{vmatrix}$$

$$= 3 \times 5 \times \begin{vmatrix} 5 & 4 & 3 \\ 0 & 6 & 1 \\ 0 & 2 & 2 \end{vmatrix} = 3 \times 5 \times 5 \times \begin{vmatrix} 6 & 1 \\ 2 & 2 \end{vmatrix} = 750$$

MATLAB has a function called det that takes a square matrix as input and returns the determinant.

5.2.5   Calculating a Matrix Inverse

The inverse of a square matrix, $\mathbf{A}$, is another matrix, $\mathbf{A}^{-1}$ that multiplies with the original matrix to yield the identity matrix.

$$\mathbf{A}^{-1}\mathbf{A} = \mathbf{A}\,\mathbf{A}^{-1} = \mathbf{I}$$

Unfortunately, calculating the inverse of a matrix is computationally slow. A formula known as *Cramer's Rule* provides a neat and tidy equation for the inverse, but for matrices beyond a 2×2, it is prohibitively slow. Cramer's rule requires the calculation of the determinant and the full set of cofactors for the matrix. Cofactors are determinants of matrices one dimension less than the original matrix. To find the inverse of a 3×3 matrix requires finding one 3×3 determinant and 9 2×2 determinants. The *run time complexity* of Cramer's rule is $\mathcal{O}(n \cdot n!)$.

For a 2×2 matrix, Cramer's rule can be found by algebraic substitution.

$$\mathbf{A}\,\mathbf{A}^{-1} = \mathbf{I}$$
$$\begin{bmatrix} a & b \\ c & d \end{bmatrix} \begin{bmatrix} x_1 & x_2 \\ y_1 & y_2 \end{bmatrix} = \begin{bmatrix} 1 & 0 \\ 0 & 1 \end{bmatrix}$$

One can perform the above matrix multiplication and find four simple equations from which the terms of the matrix inverse may be derived.

$$\mathbf{A} = \begin{bmatrix} a & b \\ c & d \end{bmatrix} \text{ has inverse } \mathbf{A}^{-1} = \frac{\begin{bmatrix} d & -b \\ -c & a \end{bmatrix}}{det(\mathbf{A})} = \frac{1}{(a\,d - c\,d)} \begin{bmatrix} d & -b \\ -c & a \end{bmatrix}$$

Because of the computational complexity, Cramer's rule is not used by MATLAB or similar software. Another technique known as elimination is used. The elimination algorithm is descried in section 5.5. The *run time complexity* of calculating a matrix inverse by elimination is $\mathcal{O}(n^3)$. The good news is that we do not need to manually calculate the inverse of matrices because MATLAB has a matrix inverse function called inv.

```
>> A = [2 -3 0;4 -5 1;2 -1 -3];
>> A_inv = inv(A)
A_inv =
    -1.6000      0.9000      0.3000
    -1.4000      0.6000      0.2000
    -0.6000      0.4000     -0.2000
```

Note: For square matrices, the left-divide operator uses an efficient elimination technique called LU decomposition to solve a system of equations. Thus, it is seldom necessary to compute the full inverse of the
matrix.

5.2.6   Invertible Test

Not all matrices can be inverted. A matrix that is not invertible is said to
be *singular*. A matrix is singular if some of its rows or columns are dependent
on each other as described in section 5.1.2. An invertible matrix must derive
from the same number of independent equations as the number of unknown
variables. A square matrix from independent equations has *rank* equal to the
dimension of the matrix, which means that it is *full rank*, has a nonzero determinant, and is invertible. We will define *rank* more completely after discussing
elimination. For now, just think of *rank* as the number of independent rows
and columns in the matrix. See also section 5.4 that describes how linear equations are mapped to matrix equations. A function described in section 5.4.3
shows which columns are independent and how the columns are combined to
yield any dependent columns. Appendix A.2.2 gives more formal definitions
of independent and dependent vectors.

The following matrix is singular because column 3 is just column 1 multiplied by 2.

$$\begin{bmatrix} 1 & 0 & 2 \\ 2 & 1 & 4 \\ 1 & 2 & 2 \end{bmatrix}$$

Sometimes it is difficult to observe that the rows or columns are not independent. When a matrix is singular, its rank is less than the dimension of the
matrix, and its determinant also evaluates to zero.

```
>> A = [1 0 2;2 1 4;1 2 2];
>> rank(A)
ans =
     2
>> det(A)
ans =
     0
>> det(A')  % the transpose is also singular
ans =
     0
```

If you work much with MATLAB, you will occasionally run across a warning message saying that a matrix is close to singular. The message may also reference a rcondition number. The condition number and rcondition number are metrics indicating if a matrix is invertible, which we will describe in section 5.5.4.3. For our discussion here, we will focus on the rank as our invertible test.

5.2.7   Cross Product

The cross product is a special operation using two vectors in $\mathbb{R}^3$ with application to geometry and physical systems. Although cross product is an operation for vectors, it is included here because matrix operations (determinant or multiplication) is used to compute it.

The cross product of two nonparallel 3-D vectors is a vector perpendicular to both vectors and the plane which they span and is called the normal vector to the plane. The direction of the normal vector follows the right hand rule as shown in figure 5.11. Finding the normal vector to a plane is needed to write the equation of a plane. Section 5.9.3 shows an example of using a cross product to find the equation of a plane.

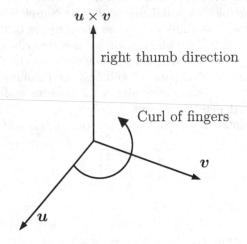

FIGURE 5.11: *Right Hand Rule:* $u \times v$ points in the direction of your right thumb when the fingers curl from u to v.

- $u \times v$ is spoken as "u cross v".

- $u \times v$ is perpendicular to both u and v.

- $v \times u$ is $-(u \times v)$.

- The length $\|v \times u\| = \|u\| \, \|v\| \, |\sin\theta|$, where θ is the angle between u and v.

- The MATLAB function `cross(u, v)` returns $u \times v$.

Cross Product by Determinant

The cross product of $u = (u_1, u_2, u_3)$ and $v = (v_1, v_2, v_3)$ is a vector.

$$u \times v = \begin{vmatrix} \vec{i} & \vec{j} & \vec{k} \\ u_1 & u_2 & u_3 \\ v_1 & v_2 & v_3 \end{vmatrix}$$

$$= \vec{i}(u_2 v_3 - u_3 v_2) - \vec{j}(u_1 v_3 - u_3 v_1) + \vec{k}(u_1 v_2 - u_2 v_1)$$

Vectors $\vec{i}$, $\vec{j}$, and $\vec{k}$ are the unit vectors in the directions of x, y, and z of the 3-D axis.

Cross Product by Skew-Symmetric Multiplication

An alternative way to compute $u \times v$ is by multiplication of a *skew-symmetric*, or *anti-symmetric* matrix.

- The skew-symmetric matrix of u is given the math symbol, $[u]_\times$. Such a matrix has a zero diagonal and is always singular. The transpose of a skew-symmetric matrix is equal to its negative.

$$[u]_\times^T = -[u]_\times$$

- MATLAB function `skew(u)` returns $[u]_\times$.

- For vectors in $\mathbb{R}^3$:

$$[u]_\times = \begin{bmatrix} 0 & -u_3 & u_2 \\ u_3 & 0 & -u_1 \\ -u_2 & u_1 & 0 \end{bmatrix}$$

$$u \times v = [u]_\times v$$

```
>> u = [1 2 3]';              >> s = skew(u)
>> v = [1 0 1]';              s =
>> cross(u,v)                       0    -3     2
ans =                               3     0    -1
     2                             -2     1     0
     2                        >> s*v
    -2                        ans =
                                    2
                                    2
                                   -2
```

5.3   Geometric Transforms

We now discuss an application of matrix multiplication related to robotics, machine design, image processing, animation, and any other field of study that is concerned with the geometry of items that move in space. The discussion here is directed toward machines, especially robots, but the concepts can be directly applied to other domains. Geometric transformations are used to determine the coordinate location and orientation of items in space. At the heart of geometric transformations is matrix multiplication.

A detailed discussion of geometric transformations is found in chapters 2 and 7 of Peter Corke's text on robotics control [15].

A coordinate frame in three-dimensional space consists of three perpendicular (orthogonal) axes. Points have a (x, y, z) position relative to the origin of the coordinate frame. We often use multiple coordinate frames to simplify the expression of point locations. For example, we usually have a world coordinate frame, which for stationary robots uses the base of the robot as its origin. There might also be coordinates frames for the workpiece that the robot is working on, the robot's end effector, or a camera if we are using a vision system. The location and orientation of each coordinate frame is called the *pose* of the coordinate frame, which is expressed by a matrix. This allows us to use matrix multiplication to express the pose of the coordinate frame and associated points in other coordinate frames.

We will discuss rotation and translation of points and coordinate frames in two dimensions. The concepts described in two dimensions can then be extended to three dimensions.

5.3.1   Rotation of a Point

Rotation of points is an important topic to both machine vision and robotics. Pixels in an image might be rotated to align objects with a model. After describing rotation of a point, we can extend the concept of a rotation matrix to transformations consisting of both rotation and translation. We then consider transformations of coordinate frames that are used to describe the pose of robots and moving parts.

Here, we only consider rotating points about the origin. Rotation about other points is an extension of rotation about the origin.

As shown in figure 5.12, point $\mathbf{p} = (a, b)$ is rotated by an angle θ about the origin to point $\mathbf{p}' = (a', b')$. To facilitate the discussion, the point $\mathbf{p}$ is defined in terms of unit vectors $\hat{\boldsymbol{x}} = (1, 0)$ and $\hat{\boldsymbol{y}} = (0, 1)$. As shown in figure 5.13, the new location, $\mathbf{p}'$ is then defined by unit vectors $\hat{\boldsymbol{x}}'$ and $\hat{\boldsymbol{y}}'$ formed by rotating $\hat{\boldsymbol{x}}$ and $\hat{\boldsymbol{y}}$ by the angle θ.

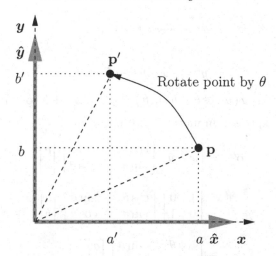

FIGURE 5.12: Point $\mathbf{p} = (a, b)$ is rotated by θ to $\mathbf{p}' = (a', b')$.

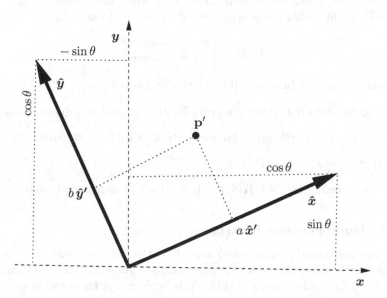

FIGURE 5.13: Rotating the axis allows us to see the coordinates of the rotated point as scalar multiples of the rotated unit vectors $\hat{\boldsymbol{x}}$ and $\hat{\boldsymbol{y}}$.

$$\mathbf{p} = a\,\hat{\boldsymbol{x}} + b\,\hat{\boldsymbol{y}}$$

$$\mathbf{p}' = a\,\hat{\boldsymbol{x}}' + b\,\hat{\boldsymbol{y}}'$$

$$\hat{x}' = \hat{x} \cos\theta + \hat{y} \sin\theta$$

$$\hat{y}' = -\hat{x} \sin\theta + \hat{y} \cos\theta$$

$$\mathbf{p}' = \hat{x}(a \cos\theta - b \sin\theta) + \hat{y}(a \sin\theta + b \cos\theta) \qquad (5.6)$$

Equation (5.6) expressed in matrix notation becomes:

$$\mathbf{p}' = \begin{bmatrix} \hat{x} & \hat{y} \end{bmatrix} \begin{bmatrix} \cos\theta & -\sin\theta \\ \sin\theta & \cos\theta \end{bmatrix} \begin{bmatrix} a \\ b \end{bmatrix}$$

$$= \begin{bmatrix} 1 & 0 \\ 0 & 1 \end{bmatrix} \begin{bmatrix} \cos\theta & -\sin\theta \\ \sin\theta & \cos\theta \end{bmatrix} \begin{bmatrix} a \\ b \end{bmatrix}$$

$$= \begin{bmatrix} \cos\theta & -\sin\theta \\ \sin\theta & \cos\theta \end{bmatrix} \begin{bmatrix} a \\ b \end{bmatrix}$$

Thus, we have a 2×2 rotation matrix, which when multiplied by the coordinates of a point yields the coordinates of the rotated point.

$$\mathbf{R}(\theta) = \begin{bmatrix} \cos\theta & -\sin\theta \\ \sin\theta & \cos\theta \end{bmatrix}$$

The rotation matrix has the following special properties.

1. The columns define basis vectors for the rotated coordinate frame.

2. The matrix is orthogonal (square with unit length, orthogonal columns).

3. $\mathbf{R}(\theta)^{-1} = \mathbf{R}(-\theta) = \mathbf{R}^T(\theta)$.

4. The determinant, $det(\mathbf{R}(\theta)) = 1$, $\forall\theta$, thus $\mathbf{R}$ is never singular.

5.3.2   Homogeneous Matrix

Geometric translation is often added to the rotation matrix to make a matrix that is called the *homogeneous transformation matrix*. The translation coordinates (x_t and y_t) are added in a third column. A third row is also added so that the resulting matrix is square.

$$\mathbf{T}(\theta, x_t, y_t) = \begin{bmatrix} \cos\theta & -\sin\theta & x_t \\ \sin\theta & \cos\theta & y_t \\ 0 & 0 & 1 \end{bmatrix}$$

Homogeneous rotation alone is given by the following matrix.

$$\mathbf{R}(\theta) = \begin{bmatrix} \cos\theta & -\sin\theta & 0 \\ \sin\theta & \cos\theta & 0 \\ 0 & 0 & 1 \end{bmatrix}$$

Homogeneous translation alone is given by the following matrix.

$$\mathbf{T}_t(x_t, y_t) = \begin{bmatrix} 1 & 0 & x_t \\ 0 & 1 & y_t \\ 0 & 0 & 1 \end{bmatrix}$$

The combined rotation and translation transformation matrix can be found by matrix multiplication.

$$\mathbf{T}(\theta, x_t, y_t) = \mathbf{T}_t(x_t, y_t)\,\mathbf{R}(\theta)$$

5.3.3 Applying Transformations to a Point

When applied to a point, the homogeneous transformation matrix uses rotation followed by translation in the original coordinate frame. It is not translation followed by rotation.

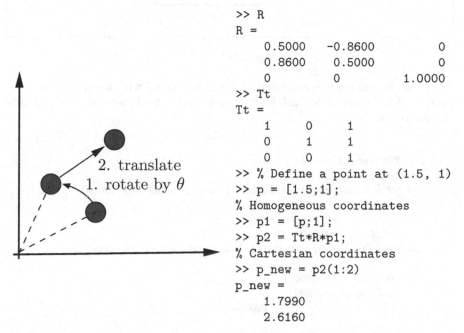

```
>> R
R =
        0.5000   -0.8600         0
        0.8600    0.5000         0
             0         0    1.0000
>> Tt
Tt =
        1         0         1
        0         1         1
        0         0         1
>> % Define a point at (1.5, 1)
>> p = [1.5;1];
% Homogeneous coordinates
>> p1 = [p;1];
>> p2 = Tt*R*p1;
% Cartesian coordinates
>> p_new = p2(1:2)
p_new =
        1.7990
        2.6160
```

FIGURE 5.14: A transformation applied to a point, $\mathbf{p}_2 - \mathbf{T}_t\,\mathbf{R}_\theta\,\mathbf{p}_1$, rotates and translates the point relative to the coordinate frame. The point in Cartesian coordinates is the first two values of the point in homogeneous coordinates.

In the following MATLAB session, we define matrix and translation matrices using homogeneous coordinates. We compound the two with matrix multiplication and apply the compound transformation matrix to a point. Here

---

Why do we say that rotation comes before translation?

The rotated and translated point is given by the product of two matrices and the original point location.

$$\mathbf{p}' = \mathbf{T}_t(x_t, y_t)\,\mathbf{R}(\theta)\,\mathbf{p} \tag{5.7}$$

The left to right reading of equation (5.7) might lead one to say that the translation occurs before the rotation. We say that rotation occurs first because the rotation matrix is immediately left of the point location. So it is the multiplication of the point with the rotation matrix that moves the point first. Since multiplication of matrices is not commutative, the rotation matrix must be immediately left of the point location to get the desired result.

$$\mathbf{p}' = \underbrace{\mathbf{T}_t(x_t, y_t)\,\overbrace{\mathbf{R}(\theta)\,\mathbf{p}}^{\text{first product}}}_{\text{second product}}$$

---

the rotation and translation matrices are constructed with a sequence of commands. Peter Corke's *Spatial Math Toolbox* provides functions that expedite the effort [16].[5]

```
>> % Rotation matrix
>> theta = pi/3;
>> R2 = [cos(theta) -sin(theta); sin(theta) cos(theta)]
R2 =
     0.5000    -0.8660
     0.8660     0.5000
>> R = [R2 [0; 0]; [0 0 1]] % homogeneous coordinates
R =
     0.5000    -0.8660         0
     0.8660     0.5000         0
          0          0    1.0000

>> % translation matrix
>> xt = 2; yt = 1;
>> tr = eye(3); tr(1:2,3) = [xt; yt]
tr =
     1     0     2
     0     1     1
     0     0     1
```

---

[5]You might also be interested in his toolboxes for robotics and machine vision.

```
>> % transformation matrix: pi/3 rotation, (2, 1) translation
>> T = tr*R
T =
    0.5000   -0.8660    2.0000
    0.8660    0.5000    1.0000
         0         0    1.0000

>> % Define a point at (1, 1)
>> p = [1;1]; % Cartesian coordinates
>> p1 = [p;1] % Homogeneous coordinates
p1 =
     1
     1
     1

>> % transformation of p
>> p2 = T*p1
p2 =
    1.6340
    2.3660
    1.0000
>> p_new = p2(1:2) % Cartesian coordinates
p_new =
    1.6340
    2.3660
```

We can also apply the transformation in Cartesian coordinates, but it is
not a simple matrix multiplication as it is in homogeneous coordinates. We
rotate the point and then add the translation to the coordinates.

```
>> R2*p + [xt; yt]
ans =
    1.6340
    2.3660
```

5.3.4 Coordinate Transformations in 2-D

Transformations of coordinate frames work different than transformations
of points. As shown in figure 5.15, each coordinate frame transformation starts
with rotation of the frame and then translation relative to the rotated frame.
To build a compound coordinate frame describing one frame relative to another frame as is shown in figures 5.16 and 5.17 requires a sequence of matrix
multiplications. We start with the previous coordinate frame and then multiply each sequential transform matrix until we reach the desired coordinate
frame position. We write the order of transformation multiplications from left
to right.

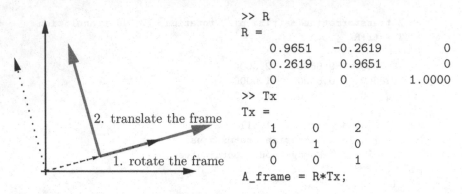

```
>> R
R =
      0.9651    -0.2619         0
      0.2619     0.9651         0
           0          0    1.0000
>> Tx
Tx =
      1     0     2
      0     1     0
      0     0     1
A_frame = R*Tx;
```

FIGURE 5.15: A transformation that traverses a straight rigid body to a new coordinate frame at the other end of the rigid body. It rotates the frame and then translates the frame along the x axis.

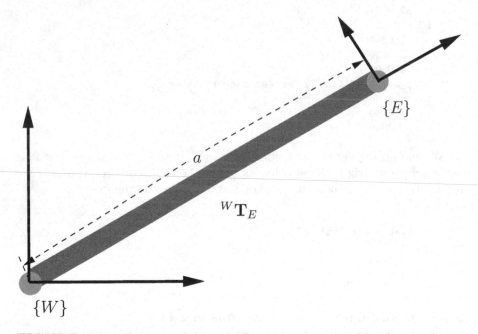

FIGURE 5.16: The pose of the end effector is represented by the coordinate frame $\{E\}$, which is defined relative to the world frame by the homogeneous coordinate transformation $^{W}\mathbf{T}_E = \mathbf{R}(\theta)\mathbf{Tx}(a)$.

Note: The base or world coordinate frame matrix is an identity matrix, so it is not necessary to specify it as the starting coordinate frame.

Consider the pose at the end of a single robotic arm as shown in figure 5.16. Since the rotation changes the orientation of the coordinate frame, the translation is only along the x axis. If a rigid body also has a displacement in the direction of the y axis, we can either use a second translation matrix, or specify the translation matrix as having both x and y components.

$$
{}^{W}T_E = \begin{bmatrix} \cos\theta & -\sin\theta & 0 \\ \sin\theta & \cos\theta & 0 \\ 0 & 0 & 1 \end{bmatrix} \begin{bmatrix} 1 & 0 & a \\ 0 & 1 & 0 \\ 0 & 0 & 1 \end{bmatrix} = \begin{bmatrix} \cos\theta & -\sin\theta & a\cos\theta \\ \sin\theta & \cos\theta & a\sin\theta \\ 0 & 0 & 1 \end{bmatrix}
$$

We read the nomenclature of the coordinate frame transformation ${}^{W}T_E$ as *the transformation with respect to the world frame, $\{W\}$, yielding frame $\{E\}$.* The reference frame is a prescript to the transformation symbol $\mathbf{T}$, while the label of the resulting frame is a subscript.

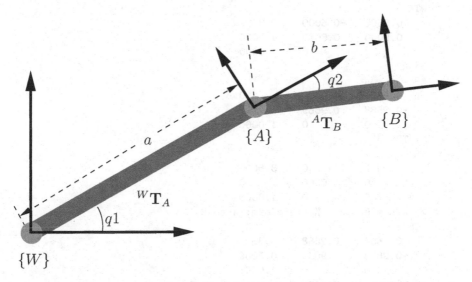

FIGURE 5.17: A coordinate frame for each joint is created by multiplication of transform matrices.

For serial-link robot arms, we define a coordinate frame for each joint of the robot. As shown in figure 5.17, the composite coordinate frames are products of the transformation matrices from the world coordinate frame to each joint.

If we know the position of a point relative to a transformed coordinate frame, then the point's location in the world coordinate frame is the product of the transformation matrices and the point location in the transformed coordinate frame. However, if we know a point location in the world coordinate frame and want its location in another coordinate frame, we multiply the point location by the inverse of the transformation.

```
>> q1 = 30; q2 = -15; % degrees
>> a = 8; b = 3.5;
>> RA
RA =
    0.8660   -0.5000        0
    0.5000    0.8660        0
         0         0   1.0000
>> TA
TA =
    1    0    8
    0    1    0
    0    0    1
>> WTA = RA*TA    % transform: world to A
WTA =
    0.8660   -0.5000   6.9282
    0.5000    0.8660   4.0000
         0         0   1.0000
>> RB
RB =
    0.9659    0.2588        0
   -0.2588    0.9659        0
         0         0   1.0000
>> TB
TB =
    1.0000         0   3.5000
         0    1.0000        0
         0         0   1.0000
>> ATB = RB*TB    % transform: A to B
ATB =
    0.9659    0.2588   3.3807
   -0.2588    0.9659  -0.9059
         0         0   1.0000
>> B_frame = WTA*ATB % transform: world to B
B_frame =
    0.9659   -0.2588  10.3089
    0.2588    0.9659   4.9059
         0         0   1.0000

% point in B to World
>> p1_B = [1;1;1]; % homogeneous
>> p1_W = B_frame*p1_B
p1_W =
   11.0161
    6.1306
    1.0000

% point in World to B
>> p2_W = [12;7;1];
>> p2_B = inv(B_frame)*p2_W
```

```
p2_B =
    2.1754
    1.5851
    1.0000
```

5.4   Systems of Linear Equations

Systems of linear equations are everywhere in the systems that engineers design. Many (probably most) systems are linear. Differential equations sometimes describe systems, but we can change differential equations to linear equations by applying the Laplace transform. The main point, though, is that in actual design and analysis applications, we have *systems of equations* with multiple unknown variables, not just one equation. In section 5.7, we will consider some application examples of systems of linear equations. Initially, we will focus on how to solve them.

The variables in linear equations are only multiplied by numbers. We do not have products of variables such as xy or x^2. We also do not have exponentials such as e^x or functions of variables.

5.4.1   An Example

The following system of equations has three unknown variables.

$$\begin{cases} 2x_1 - 3x_2 & = 3 \\ 4x_1 - 5x_2 + x_3 = 7 \\ 2x_1 - x_2 - 3x_3 = 5 \end{cases}$$

The first step is to represent the equations as a matrix equation.

$$\begin{bmatrix} 2 & -3 & 0 \\ 4 & -5 & 1 \\ 2 & -1 & -3 \end{bmatrix} \begin{bmatrix} x_1 \\ x_2 \\ x_3 \end{bmatrix} = \begin{bmatrix} 3 \\ 7 \\ 5 \end{bmatrix}$$

The common notation for describing this equation is $\mathbf{A}x = b$, where the vector x represents the unknowns. In linear algebra notation, we describe the solution to the problem in terms of the inverse of the $\mathbf{A}$ matrix, which is discussed in sections 5.2.5 and 5.5.2. Note that because matrix multiplication is not commutative, we have to be careful about the order of the terms.

$$\mathbf{A}^{-1}\mathbf{A}x = x = \mathbf{A}^{-1}b \qquad (5.8)$$

We can solve equation (5.8) with the `inv` function, but MATLAB provides a more efficient way to solve it.

5.4.2   Jumping Ahead to MATLAB

The left-divide operator (\) solves systems of linear equations more efficiently than multiplying by the inverse of a matrix.

```
>> A = [2 -3 0; 4 -5 1; 2 -1 -3];
>> b = [3 7 5]';
>> x = A\b
x =
    3.0000
    1.0000
    0.0000
```

Thus, the values of the three unknown variables are: $x_1 = 3$, $x_2 = 1$, and $x_3 = 0$.

5.4.3   The Column and Row Factorization

A matrix factoring converts a matrix into two or more sub-matrices whose product yields the original matrix. Matrix factorization is quite common in linear algebra and is used when the sub-matrices are needed for implementing algorithms. Here we present a factorization that is not used in any algorithms. Its purpose is solely to provide information and, more importantly, to facilitate learning about systems of equations. It is not part of MATLAB, but code 5.1 lists the code. The *CR* (for column–row) factoring of matrix **A** produces matrix **C** that consists of the independent columns of **A** and another matrix, **R**, showing how combinations of the columns of **C** are used in **A** [57]. The CR factoring is a recent proposal from Dr. Gilbert Strang[6] of the Massachusetts Institute of Technology. He is now proposing a new vision for the teaching of linear algebra that presents the CR factorization early in an introductory course [72]. The objective of the CR factorization is to illustrate concepts related to independent and dependent columns and how dependent columns degrade systems of equations.

The `cr` function uses the `rref` function to perform Gaussian elimination as described in section 5.5.3. For now, just consider the output of `cr` rather than its implementation.

A critically determined system of equations has the same number of equations as unknown variables. In **Ax** = **b** matrix form, the **A** matrix is square. For such a system to have a unique solution, all rows and columns must be independent vectors from each other, which is discussed in section 5.1.2 and formally defined in appendix A.2.2. This means that no row or column can be a linear combination of other rows or columns.

---

[6]Dr. Gilbert Strang is widely considered to be a leading authority and advocate for teaching science and engineering applications of linear algebra.

```
function [C,R] = cr(A)
% [C,R] = cr(A) is Strang's column-row factorization, A = C*R.
% If A is m-by-n with rank r, then C is m-by-r and R is r-by-n.

    [R,j] = rref(A);
    r = length(j);      % r = rank.
    R = R(1:r,:);
    C = A(:,j);
end
```

CODE 5.1: The CR factoring of matrix **A** produces matrix **C** that consists of the independent columns of **A** and another matrix, **R**, showing how combinations of the columns of **C** are used in both the independent and dependent columns of **A**.

The matrix in the following example has independent columns. So in the CR factoring, the **C** matrix is the same as **A**. The **R** matrix is an identity matrix indicating that each column of **C** is found unchanged in **A**. Matrices with independent rows and columns are said to be *full rank*.

```
>> A = randi(10, 3);          >> [C, R] = cr(A)
A =                           C =
      3    10     8                 3    10    10
      6     2     7                 6     2     5
     10    10    15                10    10     9
                              R =
                                    1     0     0
                                    0     1     0
                                    0     0     1
```

Rank

The rank, r, of a matrix is the number of independent rows and columns of the matrix. Rank is never more than the smaller dimension of the matrix.

Sometimes we can look at a matrix and observe if some rows or columns are linear combinations of other rows or columns. The CR factoring helps us to see those relationships. The **rank** function tells us the rank for comparison to the size of the matrix. A square matrix is *full rank* when the rank is equal to the matrix size. Code 5.1 shows one way to compute the rank of a matrix, which is the number of nonzero pivot variables from the elimination algorithm, which are the diagonal elements of **R**. See appendix A.2.3 and section 6.8.5.3 for more information on the calculation of a matrix's rank.

Now we change the **A** matrix making the third column a linear combination of the first two columns. The rank of the matrix is then 2. We can see the rank from the CR factorization as the number of columns in **C** and the number of rows in **R**. The last column of **R** shows how the first two columns of **C** combine to make the third column of **A**.

When a square matrix is not full rank, we say that the matrix is *singular*. Singular matrices do not have a matrix inverse, and the left-divide operator will not find a solution to the system of equations.

```
>> A(:,3) = A(:,1) + A(:,2)/2    >> [C, R] = cr(A)
A =                              C =
     3    10     8                    3    10
     6     2     7                    6     2
    10    10    15                   10    10
                                 R =
                                     1.0000         0    1.0000
                                          0    1.0000    0.5000
```

When a matrix has fewer rows than columns, we say that the matrix is under-determined. The rank of under-determined matrices is less than or equal to the number of rows, $r \le m$. The CR factorization will show the columns beyond the rank as linear combinations of the first r columns.

```
>> A = randi(10, 3, 4)           >> [C, R] = cr(A)
A =                              C =
     9    10     3    10              9    10     3
    10     7     6     2             10     7     6
     2     1    10    10              2     1    10


R =
     1.0000         0         0   -2.6456
          0    1.0000         0    3.0127
          0         0    1.0000    1.2278
```

When a matrix has fewer columns than rows, we say that the matrix is over-determined. If the columns of **A** are independent, the CR factorization looks the same as that of the full rank, square matrix—all of the columns of **A** are in **C**.

```
                                      5    10     1
>> A = randi(10, 4, 3)                9     8     9
A =                                   2    10    10
    10     5     7
```

```
                                        9    8     9
                                        2   10    10
  >> [C, R] = cr(A)            R =
  C =                                   1    0     0
        10     5     7                  0    1     0
         5    10     1                  0    0     1
```

5.4.4   The Row and Column View

A system of linear equations may be viewed from either the perspective of its rows or its columns. Plots of row line equations and column vectors give us a geometric understanding of systems of linear equations. The row view shows each row as a line where the solution is the point where the lines intersect. The column view shows each column as a vector and presents the solution as a linear combination of the column vectors. The column view yields a perspective that will be particularly useful when we consider over-determined systems.

Let's illustrate the two views with a simple example.

$$\begin{cases} 2x + y = 4 \\ -x + y = 1 \end{cases}$$

$$\begin{bmatrix} 2 & 1 \\ -1 & 1 \end{bmatrix} \begin{bmatrix} x \\ y \end{bmatrix} = \begin{bmatrix} 4 \\ 1 \end{bmatrix}$$

To find the solution, we can either use the left-divide operator or elimination, which is discussed in section 5.5.

$$\begin{bmatrix} 2 & 1 \\ -1 & 1 \end{bmatrix} \begin{bmatrix} 1 \\ 2 \end{bmatrix} = \begin{bmatrix} 4 \\ 1 \end{bmatrix}$$

Row View Plot

We rearrange our equations into line equations.

$$\begin{cases} y = -2x + 4 \\ y = x + 1 \end{cases}$$

As shown in figure 5.18, the point where the lines intersect is the solution to the system of equations. Two parallel lines represent a singular system that does not have a solution. If we have only one equation, we have an under-determined system with no unique solution.

If another row (equation) is added to the system making it over-determined, there can still be an exact solution if the new row is consistent with the first two rows. For example, we could add the equation $3x + 2y = 7$ to the system and the line plot for the new equation will intersect the other lines at the same point, so $x = 1; y = 2$ is still valid. We could still say that b

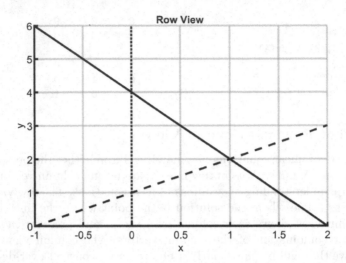

FIGURE 5.18: In the row view of a system of equations, the solution is where the line equations intersect.

is in the column space of **A**. Of course, most rows that might be added will not be consistent with the existing rows, so then we can only approximate a solution, which we will cover in section 5.9.

Column View Plot

Let's view the system of equations as a linear combination of column vectors.

$$1 \times \begin{bmatrix} 2 \\ -1 \end{bmatrix} + 2 \times \begin{bmatrix} 1 \\ 1 \end{bmatrix} = \begin{bmatrix} 4 \\ 1 \end{bmatrix}$$

As depicted in figure 5.19, there is a nonzero linear combination of the column vectors of a full rank matrix that can reach any point in the vector space except the origin. When **A** is full rank and vector **b** is nonzero, then **b** is always in the column space of **A**. The all-zero multipliers provide the only way back to the origin for a full rank system. But a singular system with parallel vectors can have a nonzero vector leading back to the origin, which is called the *null solution* (appendix A.3.2).

5.4.5 When Does a Solution Exist?

Not every matrix equation of the form $\mathbf{A}x = b$ has a unique solution. A few tests can help to determine if a unique solution exists. The matrix in question here has m rows and n columns ($m \times n$).

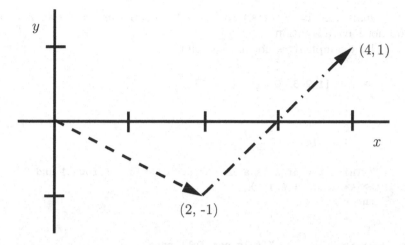

FIGURE 5.19: The column view of a system of equations shows a linear combination of the column vectors. The solution is the scaling of each vector such that the sum of the vectors is the same as the vector on the right side of the equation.

In the Column Space of A

Consider the columns of a matrix **A** as a set of vectors. The set of vectors that can be formed from linear combinations of this set of vectors is called the *span* of the columns of **A**. A vector that is part of the span of the columns of **A** is said to be *in the column space* of **A**. This phrase has particular relevance to over-determined systems.

Solutions to systems of equations of the form $\mathbf{A}x = b$, may:

1. Be an exact, unique solution.

2. Not exit.

3. Be an infinite set of vectors.

4. Not exist, but can be approximated.

Critically determined case:

A square matrix system has a unique solution when:

1. The matrix is full rank, $\text{rank}(\mathbf{A}) = m = n$.

2. The determinant of **A** is not equal to zero. This is a well-known test for **A** being invertible (not singular), but it is slow for larger matrices. The previous rank test is sufficient.

Square matrix systems that are not full rank are called *singular* and do not have a solution.

This example does not have a solution.

```
>> A = [1 2 3; 0 3 1; 1 14 7]
A =
    1    2    3
    0    3    1
    1   14    7

% Third row of A is a linear combination of rows 1 and 2
>> 4*A(2,:) + A(1,:)
ans =
    1   14    7

>> rank(A)      % A is not full rank
ans =
    2

>> det(A)       % Not invertible
ans =
    0

>> b = [1; 2; 3];

>> % This won't go well ...
>> A \ b
Warning: Matrix is singular to working precision.
ans =
     NaN
    -Inf
     Inf
```

The next example will go better. Notice that it is not necessary to calculate the determinant of **A**. Sufficient condition is determined by the rank: $\text{rank}(\mathbf{A}) = m = n$.

```
>> A = randi(10, 3, 3) - 4      >> rank(A)    % full rank
A =                             ans =
    5    6   -1                     3
    6    3    2                 >> x = A \ b
   -2   -3    6                 x =
>> b = randi(10, 3, 1) - 2         -2.8889
b =                                 4.1481
    8                               2.4444
    0
    8
```

Under-determined case:

When $m < n$, there are not enough equations, and no unique solution exists. In this case, the **A** matrix is called *under-determined*. Although a unique solution does not exist, it is possible to find an infinite set of solutions, such as all points on a line or a plane.

Over-determined case:

When $m > n$, there are more equations than unknowns. In this case, the **A** matrix is called *over-determined*. It is required that rank(**A**) = n to find a solution.

When **b** is in the column space of **A**, then an exact solution exists. That is, when **b** can be written as a linear combination of the columns of **A**. We can test for this by comparing the rank of the augmented matrix of both **A** and **b** to the rank of **A**, rank([**A** **b**]) = rank(**A**). If the rank of the augmented matrix and **A** are the same, then **b** is in the column space of **A**.

When **b** is *not* in the column space of **A**, the only solution available is an approximation. We will discuss the over-determined case in more detail in section 5.9.

In the following example, we use the rank test and see that **b** is in the column space of **A**. Then after a change to the **A** matrix, we see from the rank test that **b** is no longer in the column space of **A**

```
>> A                    >> b                % Rank test:
A =                     b =                 % b in column space
     5     3                  7             >> rank([A b])
     6    -3                 15             ans =
    -2    -1                 -3                  2
     6     2                 10             >> rank(A)
                                            ans =
                                                 2
```

```
% Change A               % Rank test:
>> A(4,:) = [5 1];        % b not in column space
A =                       >> rank([A b])
     5     3              ans =
     6    -3                   3
    -2    -1              >> rank(A)
     5     1              ans =
                               2
```

5.5   Elimination

The elimination procedure is used to find solutions to matrix equations. This section reviews the elimination procedure as it is usually performed with pencil and paper. A basic understanding of how to manually implement the elimination procedure is a prerequisite to understanding and applying the functions available in MATLAB for solving systems of equations. In section 5.5.4, we consider round-off errors and poorly conditioned matrices that complicate finding accurate solutions with elimination based algorithms on a computer. Two elimination based MATLAB functions are introduced. The rref function performs elimination as it is described here and is a useful educational and analysis tool. The lu function implements a variation of elimination known as LU decomposition (section 5.6) that is the preferred method for solving square matrix systems of equations.

5.5.1   The Gaussian Elimination Procedure

Gaussian elimination and Gauss–Jordan elimination[7] are procedures for changing a matrix to a form where the solution is simple to calculate. A sequence of linear operations is applied to the matrix's rows to produce an upper triangular matrix. Such a matrix has all zeros below the main diagonal.

Three operations are allowed in elimination.

1. Reorder the rows, which is called row exchanges or partial pivoting. This step is critical to finding an accurate solution to a system of equations. In section 5.5.4, we will address row exchanges in the context of numerical accuracy. Initially, we will use systems of equations where row exchanges are not needed.

2. Add a multiple of one row to another row, replacing that row.

3. Multiply a row by a nonzero constant.

When a row has all zeros to the left of the main diagonal, that nonzero element on the diagonal is called the *pivot*. The pivot is used to determine a multiple of the row to add to another row to produce a needed zero.

We begin with our classic matrix equation $\mathbf{A}x = b$. Any operations done on the $\mathbf{A}$ matrix must also be applied to the b vector. After elimination, we

---

[7]Before row operations are performed using a pivot variable, the Gauss–Jordan elimination algorithm divides the pivot row by the pivot value. Thus the pivot values used are always one. This step is needed when the objective is RREF. The Gaussian elimination algorithm, which is used in the examples presented here, does not take this step.

have $\mathbf{U}x = c$, where $\mathbf{U}$ is an upper triangular matrix.

$$\mathbf{U} = \begin{bmatrix} u_{1,1} & u_{1,2} & u_{1,3} \\ 0 & u_{2,2} & u_{2,3} \\ 0 & 0 & u_{3,3} \end{bmatrix}$$

Note that one could continue the elimination steps until $\mathbf{U}$ is an identity matrix, but it is faster to stop at an upper triangular matrix and use back substitution as illustrated in the example below.

To help carry forward the operations to the right side of the equation, we form an augmented matrix. Using the numbers from our previous example, we have the following. The pivots in each step are underlined.

$$\left[\begin{array}{ccc|c} \underline{2} & -3 & 0 & 3 \\ 4 & -5 & 1 & 7 \\ 2 & -1 & -3 & 5 \end{array}\right]$$

Add -1 of *row 1* to *row 3*.

$$\left[\begin{array}{ccc|c} \underline{2} & -3 & 0 & 3 \\ 4 & -5 & 1 & 7 \\ 0 & 2 & -3 & 2 \end{array}\right]$$

Add -2 of *row 1* to *row 2*. The pivot then moves to *row 2*.

$$\left[\begin{array}{ccc|c} 2 & -3 & 0 & 3 \\ 0 & \underline{1} & 1 & 1 \\ 0 & 2 & -3 & 2 \end{array}\right]$$

Add -2 of *row 2* to *row 3* to finish the row operations.

$$\left[\begin{array}{ccc|c} 2 & -3 & 0 & 3 \\ 0 & 1 & 1 & 1 \\ 0 & 0 & -5 & 0 \end{array}\right]$$

Our matrix equation is now in the upper triangular form.

$$\begin{bmatrix} 2 & -3 & 0 \\ 0 & 1 & 1 \\ 0 & 0 & -5 \end{bmatrix} \begin{bmatrix} x_1 \\ x_2 \\ x_3 \end{bmatrix} = \begin{bmatrix} 3 \\ 1 \\ 0 \end{bmatrix}$$

The final step is now called back substitution where we start at the last row and work up to determine the values of the variables.

$$\begin{cases} 2\,x_1 - 3 + 0 = 3 \\ \qquad x_2 + 0 = 1 \\ \qquad\qquad -5\,x_3 = 0 \end{cases}$$

$$x_1 = 3, x_2 = 1, x_3 = 0$$

Practice Problem

Here is another system of equations with an integer solution. Use elimination to solve for variables x_1, x_2, and x_3. Then use MATLAB to verify your answer.

$$\begin{cases} -3x_1 + 2x_2 - x_3 = -1 \\ 6x_1 - 6x_2 + 7x_3 = -7 \\ 3x_1 - 4x_2 + 4x_3 = -6 \end{cases}$$

5.5.2   Elimination to Find the Matrix Inverse

Gaussian elimination can calculate the inverse of a matrix when we start with an augmented matrix of the form $[\mathbf{A}|\mathbf{I}]$ and do row operations until we have $[\mathbf{I}|\mathbf{A}^{-1}]$.

$$\left[\begin{array}{ccc|ccc} \underline{2} & -3 & 0 & 1 & 0 & 0 \\ 4 & -5 & 1 & 0 & 1 & 0 \\ 2 & -1 & -3 & 0 & 0 & 1 \end{array}\right]$$

Add -1 of *row 1* to *row 3*.

$$\left[\begin{array}{ccc|ccc} \underline{2} & -3 & 0 & 1 & 0 & 0 \\ 4 & -5 & 1 & 0 & 1 & 0 \\ 0 & 2 & -3 & -1 & 0 & 1 \end{array}\right]$$

Add -2 of *row 1* to *row 2*. The pivot then moves to *row 2*.

$$\left[\begin{array}{ccc|ccc} 2 & -3 & 0 & 1 & 0 & 0 \\ 0 & \underline{1} & 1 & -2 & 1 & 0 \\ 0 & 2 & -3 & -1 & 0 & 1 \end{array}\right]$$

Add -2 of *row 2* to *row 3*.

$$\left[\begin{array}{ccc|ccc} 2 & -3 & 0 & 1 & 0 & 0 \\ 0 & \underline{1} & 1 & -2 & 1 & 0 \\ 0 & 0 & -5 & 3 & -2 & 1 \end{array}\right]$$

Add 3 of *row 2* to *row 1*. The pivot then moves to *row 3*.

$$\left[\begin{array}{ccc|ccc} 2 & 0 & 3 & -5 & 3 & 0 \\ 0 & 1 & 1 & -2 & 1 & 0 \\ 0 & 0 & \underline{-5} & 3 & -2 & 1 \end{array}\right]$$

Add 3/5 of *row 3* to *row 1*; and add 1/5 of *row 3* to *row 2*. Then a pivot is no longer needed.

$$\begin{bmatrix} 2 & 0 & 0 & -16/5 & 9/5 & 3/5 \\ 0 & 1 & 0 & -7/5 & 3/5 & 1/5 \\ 0 & 0 & -5 & 3 & -2 & 1 \end{bmatrix}$$

Divide *row 1* by 2; and divide *row 3* by −5.

$$\begin{bmatrix} 1 & 0 & 0 & -16/10 & 9/10 & 3/10 \\ 0 & 1 & 0 & -7/5 & 3/5 & 1/5 \\ 0 & 0 & 1 & -3/5 & 2/5 & -1/5 \end{bmatrix}$$

$$\mathbf{A}^{-1} = \begin{bmatrix} -1.6 & 0.9 & 0.3 \\ -1.4 & 0.6 & 0.2 \\ -0.6 & 0.4 & -0.2 \end{bmatrix}$$

Note: This result matches what MATLAB found in section 5.2.5. All of the tedious row operations certainly makes one appreciate that MATLAB can perform the calculations for us.

5.5.3 Reduced Row Echelon Form

The word *echelon* just means resembling stair steps. A matrix is in row echelon form when:

- All nonzero values in each column are above all zeros.

- The leading coefficient of each row is strictly to the right of the leading coefficient in the row above it.

When Gaussian elimination is used on an augmented matrix to make an upper-triangular matrix, the matrix is put into row echelon form.

Reduced row echelon form (RREF) requires that the elimination procedure continues to address elements above and on the diagonal. The additional requirements for reduced row echelon form are:

- Every leading coefficient must be 1.

- The leading coefficients must be the only nonzero value in its column. In RREF, the first m (number of rows) columns of the matrix should look as close as possible to an identity matrix.

A full rank matrix system in augmented form has the identity matrix in the first m columns when in RREF. The last column is the solution to the system of equations.

```
>> Ab = [A b]                    >> rref(Ab)
Ab =                             ans =
   8     1     6    27              1     0     0     2
   3     5     7    38              0     1     0     5
   4     9     2    55              0     0     1     1
```

As we saw from the CR factoring in section 5.4.3, the RREF of a singular matrix shows how dependent columns are weighted sums of the independent columns. With $r = \text{rank}(\mathbf{A})$, the last $(n - r)$ rows are all zeros.

```
>> A =                           >> rref(A)
   1    -8    -7                  ans =
   4    -5    -1                     1     0     1
   6     2     8                     0     1     1
                                     0     0     0
```

5.5.4   Accuracy of Elimination

We now turn our attention to the accuracy of elimination based functions in MATLAB. MATLAB has two functions that use elimination—rref and lu. Both functions use partial pivoting (also called row exchanges) to improve the accuracy of the results. We demonstrate in this section why partial pivoting improves the accuracy, or numerical stability, of the results from elimination. The underlying problem that can degrade the accuracy of results from elimination when performed by a computer is round-off errors caused by the finite storage space available to hold variables in the computer's memory. We first encountered round-off errors in section 1.6.3.

The rref function has a bad reputation for not being accurate. Actually, it is quite accurate for most applications, but the other elimination based function, lu (for LU decomposition), is even more accurate. The LU decomposition algorithm is presented in section 5.6. The LU decomposition algorithm is also faster. One does not need to look long in the linear algebra literature before finding admonishments telling users to always use LU decomposition and to never use RREF to find the solution to a system of equations. This is good advice because there are some examples where lu gives the correct answer while rref does not. MATLAB's left-divide operator uses LU decomposition for square matrix systems. An example of rref failing to find the correct result is shown in section 5.5.4.2. However, such cases require huge value differences between matrix elements and are unusual systems of equations. There are two strategies to guard against problems resulting from round-off errors. The first, and most important, is partial pivoting, which both algorithms use. The second strategy that LU decomposition uses is an altered elimination algorithm that is less likely to incur round-off errors. Unfortunately, the structure of the LU elimination algorithm is specific to LU decomposition and can not be

Why does addition and subtraction cause round-off errors?

As we saw in the example presented in section 1.6.3, MATLAB has a constant value called `eps` for the smallest value that can be added to one, which is approximately 2.22×10^{-16}. The source of the limitation is the number of binary bits used to hold the mantissa (known digits) of floating point numbers. A double precision, 64-bit floating point number is specified by the *IEEE Standard for Binary Floating-Point Arithmetic* to store 52 binary digits to the right of the binary point [3]. That means that the variable N in the interpretation of the stored data holds 52 bits, while the *sign* is one bit, and the *exponent* takes 11 bits.

$$(-1)^{sign} \times 1.N \times 2^{exponent}$$

A larger exponent increases the value of the smallest number that can be added to that number. A value of one can only be added to numbers smaller than 2^{53}.

```
>> x = 2^53
x =
   9.0072e+15
>> isequal(x, x+1)  % x+1 is the same as x
ans =
  logical
   1
>> y = 2^53 - 1;
>> isequal(y, y+1)
ans =
  logical
   0
```

used to change a matrix into RREF. The LU elimination algorithm is listed in section 5.6.4.

Although we do not use `rref` to solve systems of equations, it does have value as an educational and analysis tool for the majority of systems where partial pivoting is enough to prevent problematic round-off errors. We use it with the column–row factorization in section 5.4.3 to learn about independent and dependent columns of a matrix. It also shows us that the solution to an under-determined system of equations is an infinite set of vectors (section 5.8).

5.5.4.1 The Need for Partial Pivoting

For numerical stability, very small numbers should not be used as pivot variables in the elimination procedure. Each time the pivot is advanced, the remaining rows are sorted to maximize the absolute value of the pivot variable. The new row order is documented with a permutation matrix to associate results with their unknown variable. Reordering rows, or rows exchanges, is also called partial pivoting. Full pivoting is when columns are also reordered.

Here is an example showing why partial pivoting is needed for numerical stability. This example uses a 2×2 matrix where we could see some round-off error. The ϵ value used in the example is taken to be a very small number, such as the machine epsilon value (**eps**) that we used in section 1.6.3. Keep in mind that round-off errors will propagate and build when solving a larger system of equations. Without pivoting, we end up dividing potentially large numbers by ϵ producing even larger numbers. In the algebra, we can avoid the large numbers, but get very small numbers in places. In either case, during the row operation phase of elimination, we add or subtract numbers of very different value resulting in round-off errors. The accumulated round-off errors become especially problematic in the substitution phase where we might divide two numbers that are both close to zero.

$$\begin{cases} \epsilon x_1 + 2x_2 = 4 \\ 3x_1 - x_2 = 7 \end{cases}$$

$$\begin{bmatrix} \epsilon & 2 \\ 3 & -1 \end{bmatrix} \begin{bmatrix} x_1 \\ x_2 \end{bmatrix} = \begin{bmatrix} 4 \\ 7 \end{bmatrix}$$

Regarding ϵ as nearly zero relative to the other coefficients, we can use substitution to quickly see the approximate values of x_1 and x_2.

$$x_2 \approx \frac{4}{2} = 2$$

$$x_1 \approx \frac{7+2}{3} = 3$$

Elimination Without Partial Pivoting

If we don't first exchange rows, we get the approximate value of x_2 with some round-off error, but fail to find a correct value for x_1.

$$\begin{bmatrix} \epsilon & 2 & | & 4 \\ 3 & -1 & | & 7 \end{bmatrix}$$

Only one elimination step is needed.

$$\begin{bmatrix} \epsilon & 2 & | & 4 \\ 0 & -(1 + \frac{6}{\epsilon}) & | & (7 - \frac{12}{\epsilon}) \end{bmatrix}$$

$$\left(1 + \frac{6}{\epsilon}\right) x_2 = \frac{12}{\epsilon} - 7$$

We could multiply both sides of the equation by ϵ to simplify it or calculate the fractions. In either case, the 1 and 7 terms will be largely lost to round off error.

$$x_2 \approx \frac{12}{6} = 2$$

Our calculation of x_1 is completely wrong. To get the correct answer would require division of two numbers both close to zero and the numerator likely suffers from significant round-off error.

$$\epsilon x_1 + 2x_2 = 4$$

$$\epsilon x_1 = 4 - 2x_2 = 0$$

$$x_1 = \frac{4 - 2x_2}{\epsilon} \approx \frac{0}{\epsilon} = 0$$

Elimination With Partial Pivoting

Now we begin with the rows exchanged.

$$\begin{bmatrix} 3 & -1 & | & 7 \\ \epsilon & 2 & | & 4 \end{bmatrix}$$

Again, only one elimination step is needed.

$$\begin{bmatrix} 3 & -1 & | & 7 \\ 0 & (2 + \frac{\epsilon}{3}) & | & (4 - \frac{7\epsilon}{3}) \end{bmatrix}$$

We will again lose some accuracy to round-off errors, but our small ϵ values are paired with larger values by addition or subtraction and have minimal impact on the back substitution equations.

$$x_2 \approx \frac{4}{2} = 2$$

$$3x_1 - x_2 = 7$$

$$x_1 \approx \frac{7+2}{3} = 3$$

MATLAB uses partial pivoting in both RREF and LU decomposition to minimize the impact of the round-off error. It will use the small values in the calculation that we rounded to zero, so the results will be more accurate than our approximations.

```
>> x = A\b
x =
    3.0000
    2.0000
>> isequal(x, [3;2])  % not completely rounded
ans =
    logical
      0
```

5.5.4.2   RREF and Round-off Errors

Here is an example showing an instance of the MATLAB function `rref` failing to find a correct answer while LU decomposition with the left-divide operator succeeds. As the example shows, it takes quite large differences between elements of **A** before RREF fails. Even so, one should use the left-divide operator or LU decomposition directly instead of `rref` to find the solution to an equation.

```
% Large and small numbers together is asking for trouble.
>> A = [0.0001 0.0002; 8e10 2e4];

% x = [5; 2] is what we hope to find.
>> b = A*[5; 2];

% Left-divide gets it right with LU decomposition.
>> A\b
ans =
     5
     2

% RREF fails
>> rref([A b])
ans =
    1.0000    0.0000         0
         0         0    1.0000
```

In the MATLAB documentation for `rref`, there is mention of a tolerance value to prevent errors such as division by a near zero value. The tolerance value may be specified as an argument to the function or it will calculate a default tolerance. The equation for the default tolerance is `max(size(A))*eps*norm(A,inf)`. In this example, the default tolerance is 3.55×10^{-5}, which is less than any value in **A**. Here is what the documentation says about how the tolerance is used.

> If the largest element (by absolute value) in a pivot column is below the tolerance, then the column is zeroed out. This prevents division and multiplication with nonzero pivot elements smaller than the tolerance.

5.5.4.3   Poorly Conditioned Matrices

Another problem besides round-off errors also degrades the accuracy of solutions to systems of equations. The symptom of the problem is an over sensitivity to perturbations of the elements of **A** or **b** when the system is nearly singular, which is also called poorly conditioned, or ill-conditioned. If the coefficients are taken from experiments, then they likely contain a random component, resulting in large variations in the solution if the system is nearly singular.

The *condition number* is a metric to measure how well or poorly conditioned a matrix is. The condition number is related to the size of the elements in $\mathbf{A}$ and the size of the elements of its inverse. Matrices that either contain a mixture of large and small numbers or are close to singular have large condition numbers. The common definition of the condition number is

$$\text{cond}(\mathbf{A}) = \|\mathbf{A}\| \, \|\mathbf{A}^{-1}\|.$$

The cond function in MATLAB uses the $\|\cdot\|_2$ matrix norm by default, but may also compute the condition number using other matrix norms defined in appendix A.1.3. However, to avoid calculating a matrix inverse, MATLAB calculates the condition number using the singular values of the matrix as described in section 6.8.5.5. Using the singular values, there is a concern about division by zero, so MATLAB calculates the reciprocal, which it calls RCOND. MATLAB's left-divide operator checks the condition of a matrix before attempting to find a solution. When using the left-divide operator, one may occasionally see a warning message such as follows.

Warning: Matrix is close to singular or badly scaled. Results may be inaccurate. RCOND = 3.700743e-18.

Any errors in $\mathbf{A}$ or $\mathbf{b}$ caused by either round-off errors or measurement errors ($\Delta\mathbf{A}$ and $\Delta\mathbf{b}$) are conveyed to the solution $\mathbf{x}$. The matrix equation $\mathbf{A}\mathbf{x} = \mathbf{b}$ becomes $(\mathbf{A} + \Delta\mathbf{A})(\mathbf{x} + \Delta\mathbf{x}) = (\mathbf{b} + \Delta\mathbf{b})$. If cond($\mathbf{A}$) is small, then small errors in $\mathbf{A}$ or $\mathbf{b}$ result in only small errors in $\mathbf{x}$. Whereas if cond($\mathbf{A}$) is large, then even small perturbations in $\mathbf{A}$ or $\mathbf{b}$ can result in a large error in $\mathbf{x}$. The condition number of an identity matrix is 1, but nearly singular matrices can have condition numbers of several thousand to even over a million.

A handy rule of thumb for interpreting the condition number is as follows.

If cond($\mathbf{A}$) $\approx 0.d \times 10^k$ then the solution $\mathbf{x}$ can be expected to have k fewer digits of accuracy than the elements of $\mathbf{A}$ [77].

Thus, if cond($\mathbf{A}$) $= 10 = 0.1 \times 10^1$ and the elements of $\mathbf{A}$ are given with 3 digits, then the solution in $\mathbf{x}$ can be trusted to 2 digits of accuracy.

Here is an example to illustrate the problem. To be confident of the correct answer, we begin with a well-conditioned system.

$$\begin{cases} 17\,x_1 - 2\,x_2 = -3 \\ -13\,x_1 + 29\,x_2 = 29 \end{cases} \tag{5.9}$$

It is easily confirmed with MATLAB that the solution is $x_1 = -0.0621, x_2 = 0.9722$.

We make the assertion that the system of equations in equation (5.9) is *well-conditioned* based on two pieces of information. First, its *condition number* is a relatively small (2.3678). Secondly, minor changes to the equation have a minor impact on the solution. For example, if we add 10% of A(1, 1) to itself, we get a new solution of $x_1 = -0.0562, x_2 = 0.9748$, which is a change of 9.5% to the solution to x_1 and 0.286% to the x_2 solution.

Now, to illustrate a poorly conditioned system, we will alter the first equation of equation (5.9) to be the sum of the second equation and only $1/10,000$ of itself.

$$\begin{cases} -12.9983\,x_1 + 28.998\,x_2 = 28.9997 \\ -13\,x_1 + \quad 29\,x_2 = \quad 29 \end{cases} \tag{5.10}$$

Perhaps surprisingly, both LU decomposition and RREF are able to find the correct solution to equation (5.10). One can see that equation (5.10) is nearly singular because the two equations are close to parallel. The condition number is quite large (43,254). If again we add 10% of A(1, 1) to itself, we get a new solution of $x_1 = 0.0001, x_2 = 1.0$, which is a change to the x_1 solution of 100.16% and the x_2 solution changed by 2.86% to the original solution.

5.6   LU Decomposition

We present here a variant of Gaussian elimination called *LU decomposition* (for Lower–Upper). It is used internally by MATLAB for computing inverses, the left- and right-divide operators, and determinants. MATLAB has two functions that implement a form of the elimination procedure to solve systems of equations—lu and rref. The lu function is the faster and more numerically accurate of the two.

Two features of the algorithm lend to its numerical accuracy. First, the LU algorithm uses partial pivoting. As we saw in section 5.5.4.1, round-off errors are less likely when the pivot variable is the largest element in its column. Secondly, LU decomposition uses a different elimination algorithm that is less likely to incur round-off errors than the traditional Gauss–Jordan algorithm used by rref.

Recall that Gaussian elimination uses an augmented matrix so that changes made to matrix **A** are also applied to vector **b**. A sequence of row operations changes the matrix equation.

$$\mathbf{A}x = b \quad \mapsto \quad \mathbf{U}x = c$$

In the modified equation, **U** is an upper triangular matrix for which simple back substitution may be used to solve for the unknown vector **x**. The LU decomposition method differs in that it operates only on the **A** matrix to find a matrix factoring, $\mathbf{A} = \mathbf{L}\,\mathbf{U}$, where **L** is lower triangular and **U** is upper triangular. Thus the substitution phase for solving $\mathbf{A}x = b$ has two steps—solving $\mathbf{L}y = b$ and then solving $\mathbf{U}x = y$.

Note: LU decomposition is also called LU factorization because it is one of the ways that a matrix can be factored into a product of sub-matrices.

LU decomposition changes the equations used to solve systems of equations as follows. In addition to $\mathbf{L}$ and $\mathbf{U}$, we have an elimination matrix, $\mathbf{E}$. Its inverse is $\mathbf{L}$. We also introduce the $\mathbf{P}$ matrix, which is the permutation matrix that orders the rows to match the row exchanges.[8]

$$
\begin{aligned}
\mathbf{A} &\mapsto \mathbf{E}\,\mathbf{P}\,\mathbf{A} = \mathbf{U} \\
\mathbf{P}\,\mathbf{A} = \mathbf{E}^{-1}\mathbf{U} &\mapsto \mathbf{P}\,\mathbf{A} = \mathbf{L}\,\mathbf{U} \\
\mathbf{A}x = b &\mapsto \mathbf{P}^{T}\mathbf{L}\,\mathbf{U}x = b \\
x = \mathbf{A}^{-1}b &\mapsto x = \mathbf{U}^{-1}\mathbf{L}^{-1}\mathbf{P}\,b
\end{aligned}
$$

$$\boxed{x = \mathbf{U}^{-1}\mathbf{L}^{-1}\mathbf{P}\,b} \tag{5.11}$$

5.6.1   LU Example

Before demonstrating how to use the lu function in MATLAB, we will first work part of a problem manually and then turn to MATLAB to help with the computation. We will use the same example that was solved in section 5.5. For purposes of understanding how the $\mathbf{U}$ and $\mathbf{L}$ matrices are factors of the $\mathbf{A}$ matrix, we will follow a traditional elimination strategy to find $\mathbf{U}$ and will form $\mathbf{L}$ from a record of the row operations needed to convert $\mathbf{A}$ into $\mathbf{U}$. We will focus on how the $\mathbf{U}$ and $\mathbf{L}$ matrices are found, and will not make row exchanges in this example, but will later work an example with row exchanges.

$$
\begin{cases}
2x - 3y = 3 \\
4x - 5y + z = 7 \\
2x - y - 3z = 5
\end{cases}
$$

The needed row operations are the same as noted in section 5.5. Each row operation is represented by a matrix $\mathbf{E}_{i,j}$ that can multiply with $\mathbf{A}$ to affect the change. These matrices start with an identity matrix and then change one element to represent each operation. We call these matrices *Elementary Matrices*. Then the product of the elementary matrices is called the *elimination matrix*.

1. Add -1 of *row 1* to *row 3*, called $\mathbf{E}_1$.

2. Add -2 of *row 1* to *row 2*, called $\mathbf{E}_2$.

3. Add -2 of *row 2* to *row 3*, called $\mathbf{E}_3$.

---

[8]Equation (5.11) is expressed with a pure mathematics notation using the inverse of matrices. The solution expression that has better value is the MATLAB expression that directly uses back substitution on triangular matrices (x = U\(L\(P*b))).

Which algorithm is faster?

There are some applications where there is a need to solve several systems of equations with the same **A** matrix but with different **b** vectors. It is faster to find the LU factors once and then use forward and backward substitution with each **b** vector than to do full elimination each time. The combined forward and backward substitution for LU decomposition has run time of $\mathcal{O}(n^2)$.

Elimination with an augmented matrix and the LU decomposition algorithm have the same order of magnitude run time performance, which depending on the exact implementation is approximately $\mathcal{O}(n^3)$. But more precisely, the cost of computation for LU decomposition is $(2/3)n^3 + n^2$ floating point operations (flops) [36], which explains why experimentation shows that LU decomposition is faster than the rref function, especially for larger matrices. Here is a quick experiment performed on a personal computer.

```
>> A = rand(100); b = rand(100,1);

>> tic; [L,U,P] = lu(A);  x = U\(L\(P*b)); toc;
Elapsed time is 0.000605 seconds.

>> tic; x = rref([A b]); toc;
Elapsed time is 0.069932 seconds.
```

```
>> E1            >> E2            >> E3
E1 =             E2 =             E3 =
   1    0    0      1    0    0      1    0    0
   0    1    0     -2    1    0      0    1    0
  -1    0    1      0    0    1      0   -2    1
```

The combined row operations can be found by multiplying these together. The order of matrices must be such that the first operation applied is next to **A** in the equation $\mathbf{U} = \mathbf{E}\,\mathbf{A}$.

```
>> E = E3*E2*E1        >> U = E*A
E =                    U =
   1    0    0            2   -3    0
  -2    1    0            0    1    1
  -1   -2    1            0    0   -5
```

Another way to find E and U is with elimination of an augmented matrix. Start with $[\,A\,|\,I\,]$ and perform row operations until the matrix in the position of A is upper triangular. The resulting augmented matrix is then $[\,U\,|\,E\,]$ [49].

$$\left[\begin{array}{ccc|ccc} 2 & -3 & 0 & 1 & 0 & 0 \\ 4 & -5 & 1 & 0 & 1 & 0 \\ 2 & -1 & -3 & 0 & 0 & 1 \end{array}\right]$$

$$\left[\begin{array}{ccc|ccc} 2 & -3 & 0 & 1 & 0 & 0 \\ 0 & 1 & 1 & -2 & 1 & 0 \\ 0 & 0 & -5 & -1 & -2 & 1 \end{array}\right]$$

The L matrix is the inverse of E. Recall that to take the inverse of a product of matrices, we reverse the order of the matrices—$(A\,B)^{-1} = B^{-1}A^{-1}$.

Note that each elementary matrix only differs from the identity matrix at one element. The inverse of each elementary matrix is found by just changing the sign of the additional nonzero value to the identity matrix.

```
>> inv(E1)                  >> L = inv(E1)*inv(E2)*inv(E3)
ans =                       L =
     1     0     0              1       0       0
     0     1     0              2       1       0
     1     0     1              1       2       1

>> inv(E2)                  >> inv(E)
ans =                       ans =
     1     0     0              1.0000        0        0
     2     1     0              2.0000   1.0000        0
     0     0     1              1.0000   2.0000   1.0000

>> inv(E3)
ans =
     1     0     0
     0     1     0
     0     2     1
```

Also note that the L matrix does the reverse of E, so L could be found directly from the list of row operations.

Now, with L and U found, we can find the solution to $L\,U\,x = b$. Since these are lower and upper triangular matrices, the left-divide operator will quickly solve them. If several solutions are needed, the one time calculations of L and U will substantially expedite the calculations.

```
                        3
>> b                    7
b =                     5
```

```
                                        3.0000
                                        1.0000
   >> x = U\(L\b)                      -0.0000
   x =
```

Note: The substitution phase has two steps—solving $\mathbf{L}y = b$ and $\mathbf{U}x = y$. When presented with upper or lower triangular matrices, the left-divide operator will skip directly to substitution. So either another vector, y is needed, or parenthesis may be used to dictate the ordering of the operations.

5.6.2   Example with Row Exchanges

In this example find the LU decomposition with partial pivoting (row exchanges). We will show the row exchanges with a permutation matrix. A permutation matrix has the rows of an identity matrix, but the order of the rows shows what row exchanges were performed.

$$\mathbf{A} = \begin{bmatrix} 0 & 12 & -3 \\ 8 & -4 & -6 \\ -4 & -2 & 12 \end{bmatrix}$$

Looking at the first column, we will use the largest value, 8, as our first pivot, so we exchange rows one and two and define a permutation matrix. The $\mathbf{U}$ matrix will reflect the row exchanges. The $\mathbf{L}$ matrix begins as an identity matrix and the ones on the diagonal will remain. We will document our elimination row operations directly in the $\mathbf{L}$ matrix as the negative values of what we would add to the elementary matrices.

$$\mathbf{P} = \begin{bmatrix} 0 & 1 & 0 \\ 1 & 0 & 0 \\ 0 & 0 & 1 \end{bmatrix}$$

$$\mathbf{L} = \begin{bmatrix} 1 & 0 & 0 \\ 0 & 1 & 0 \\ 0 & 0 & 1 \end{bmatrix} \quad \mathbf{U} = \begin{bmatrix} 8 & -4 & -6 \\ 0 & 12 & -3 \\ -4 & -2 & 12 \end{bmatrix}$$

Add 1/2 of *row 1* to *row 3*.

$$\mathbf{L} = \begin{bmatrix} 1 & 0 & 0 \\ 0 & 1 & 0 \\ -1/2 & 0 & 1 \end{bmatrix} \quad \mathbf{U} = \begin{bmatrix} 8 & -4 & -6 \\ 0 & 12 & -3 \\ 0 & -4 & 9 \end{bmatrix}$$

We will keep the pivot of 12 in *row 2*, so no change to the permutation matrix is needed.
Add 1/3 of *row 2* to *row 3*.

$$\mathbf{L} = \begin{bmatrix} 1 & 0 & 0 \\ 0 & 1 & 0 \\ -1/2 & -1/3 & 1 \end{bmatrix} \quad \mathbf{U} = \begin{bmatrix} 8 & -4 & -6 \\ 0 & 12 & -3 \\ 0 & 0 & 8 \end{bmatrix}$$

5.6.3 MATLAB Examples of LU

First, we ask MATLAB to just compute **L** and **U**. We can see that **L** is not lower triangular as we think it should be. That is because we did not accept a permutation matrix in the output.

```
>> [L, U] = lu(A)              U =
L =                            4.0000    -5.0000     1.0000
     0.5000   -0.3333   1.0000      0      1.5000    -3.5000
     1.0000         0        0      0           0    -1.6667
     0.5000    1.0000        0
```

Now we will accept a permutation matrix as an output matrix. The relationship of the matrices is $\mathbf{P}\,\mathbf{A} = \mathbf{L}\,\mathbf{U}$. Note that the permutation matrix merely swaps the rows of whatever vector or matrix it is multiplied by. Like the identity matrix, permutation matrices are orthogonal, thus their inverse is just their transpose, so $\mathbf{A} = \mathbf{P}^T\mathbf{L}\,\mathbf{U}$. Now, **L** is lower triangular as expected, but **L** and **U** are different than they were with the manual solution because of the permutation matrix. So the solution comes from equation (5.11), or in MATLAB, x = U\(L\(P*b)). Note here that the *parenthesis are important*. The left-divide operators will not use the correct matrices without the parenthesis.

```
% Returning to              P =
% our previous example ...       0     1     0
>> [L, U, P] = lu(A)             0     0     1
L =                              1     0     0
     1.0000        0        0   >> x = U\(L\(P*b))
     0.5000   1.0000        0   x =
     0.5000  -0.3333   1.0000       3.0000
U =                                 1.0000
     4.0000  -5.0000   1.0000      -0.0000
          0   1.5000  -3.5000
          0        0  -1.6667
```

Another option is to have MATLAB return the row exchange information as an ordering of the rows rather than as a permutation matrix.

```
>> [L, U, P] = ...          >> x = U\(L\(b(P)))
   lu(A, 'vector');         x =
>> P                            3.0000
P =                             1.0000
     2     3     1              0.0000
```

5.6.4   LU's Variant Elimination Algorithm

The MATLAB documentation for lu states that "The LU factorization is computed using a variant of Gaussian elimination". The original algorithm[9] was developed and reported in a 1948 paper by British mathematician Alan Turing who is considered to be a founding father of computer science [76, 19]. The algorithm has two noticeable differences to the elimination algorithm used by rref. First the LU algorithm uses Gaussian elimination instead of Gauss-Jordan. That is, the pivot variables are not normalized to one and the pivot rows are not divided by the pivot variable. Secondly, the steps of the elimination procedure are changed so that round-off errors are less likely. Additionally, the **L** and **U** matrices are taken directly from the modified **A** matrix after the algorithm is completed. After the elimination steps are finished, the **L** matrix is in the lower triangular of **A** and the **U** matrix is in the upper triangular of **A**. Turing's algorithm has good performance and produces accurate results. Unfortunately, the algorithm is specific to LU decomposition and not usable with traditional elimination.

A simple function is presented in code 5.2 that implements Turing's algorithm with a few added lines of code for partial pivoting. It is a little difficult to see from reading the code how it achieves the same result as found from normal Gaussian elimination. With elimination, row operations zero out the lower triangular part of the matrix and then **L** is found from the action of the row operations. Whereas, in Turing's algorithm the row operations directly yield both **L** an **U**. The variable k references the pivot variable. Partial pivoting is performed each time the pivot is advanced. There are only two statements that alter the values of the matrix. First, the elements in the column below the pivot are divided by the pivot. Secondly, row operations are performed between the pivot row and the rows below the pivot row.

Experimentation with this function found that it reliably returns the same results as MATLAB's lu function. Please experiment with it yourself. Set breakpoints for the debugging tool or add printing statements to learn how the algorithm works. But for applications other than education, use the lu function that comes with MATLAB, which runs faster, has more flexible usage options, and will be more reliable. Note that this function could be implemented with vectorized array indices, but uses nested for loops just for clarity of reading the code.

5.6.5   Determinant Shortcut

Another application of LU decomposition is a faster algorithm for computing the determinant of a matrix. The product of the determinants of a matrix

---

[9]Modern numerical linear algebra literature use the name *"kij" version of outer product LU* for the algorithm presented in code 5.2. Golub and Van Loan give a proof showing that potential round-off errors are minimized by this algorithm [28].

```
function [L, U, P] = turingLU(A)
% [L,U,P] = TURINGLU(A)
% A simplified LU factorization of a matrix using Alan Turing's
% variant of Gaussian elimination algorithm.
%
% Input:  A - square matrix
% Output: L - lower triangular matrix
%         U - upper triangular matrix
%         P - permutation matrix
% P*A = L*U, A = P'*L*U
%
% This code is for educational purposes only.
% For any other purpose, use MATLAB's lu function.

[m, n] = size(A);
if m ~= n
    error('Matrix is not square')
end
P = eye(n);
% Turing's nested elimination loop
for k = 1:(n - 1)
    [A(k:n,:), idx] = sortrows(A(k:n,:), k, 'descend', ...
        'ComparisonMethod','abs');
    I = P(k:n,:);
    P(k:n,:) = I(idx,:); % Permutation matrix
    for i = k + 1:n
        A(i, k) = A(i, k)/A(k, k);
        for j = (k + 1):n
            A(i, j) = A(i, j) - A(i, k) * A(k, j);
        end
    end
end
% L is now in the lower triangular of A.
% U is now in the upper triangular of A.
L = tril(A, -1) + eye(n); % extract lower
U = triu(A);              % and upper triangular
```

CODE 5.2: Implementation of LU decomposition for educational purposes.

factorization is equal to the determinant of the original matrix. Thus,

$$|\mathbf{A}| = |\mathbf{P}^T|\,|\mathbf{L}|\,|\mathbf{U}|.$$

Because of the zeros in all three of these matrices, their determinants are quick to compute. It is not necessary to take the transpose of $\mathbf{P}$ before finding its determinant because the determinant of a square matrix is the same as the determinant of its transpose, det(P) = det(P'). The determinant of $\mathbf{P}$ is either 1 or -1. Since each row has only a single one, the determinant is quick to calculate. The determinant of **L** is always 1, which is the product of the ones along the diagonal. The determinant of **U** is the product of the values on its diagonal.

Refer to the normal method for calculating the determinant in section 5.2.4 and take a moment to see why the determinant of an upper or lower triangular matrix is the product of the diagonal values.

The MATLAB det function computes the determinant of a matrix as follows.

```
>> [L, U, P] = lu(A)
>> determinant = det(P)*prod(diag(U))
```

5.7 Linear System Applications

Systems of linear equations are routinely solved using matrices in science and engineering. Here, two applications are reviewed.

5.7.1 DC Electric Circuit

A direct current (DC) electrical circuit with only a battery and resistors provides a system of linear equations. We want to determine the current and voltage drop across each resistor in the circuit. We only need *Ohm's Law* and either *Kirchhoff's Voltage Law* (KVL) or *Kirchhoff's Current Law* (KCL) to find a solution. Both the KVL and KCL methods lead directly to matrix equations.

Note: An AC circuit with capacitors and inductors in addition to resistors is described by a system of differential equations that can also be solved with linear algebra. MATLAB and linear algebra make determining such solutions straight forward; however, additional mathematical abstractions are needed, so we'll stick with the simpler DC circuit with only resistors.

Ohm's Law
When a current of I amperes flows through a resistor of R ohms, the voltage drop, V, in volts across the resistor is $V = I\,R$.

Kirchhoff's Voltage Law (KVL)
Each closed loop of a circuit is assigned a loop current. The current through resistors that are part of two loops is a combination of the loop currents. Then, the sum of the voltages around each loop is zero.

Kirchhoff's Current Law (KCL)

The sum of the currents entering and leaving each node of the circuit is zero. The voltages at each node of the circuit may be quantified in terms of any known voltages, voltage drops across resistors, and other node voltages.

5.7.1.1 KVL Method

The circuit is shown in figure 5.20 with the loop currents and their directions identified. The constant values used in the MATLAB program are: R1 = 1000; R2 = 1500; R3 = 2000; R4 = 1000; R5 = 1500; V = 12. The three loop

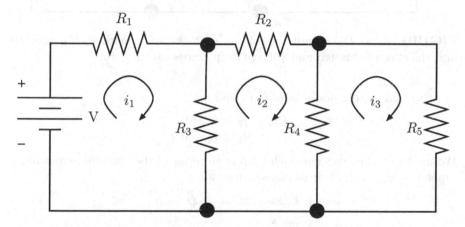

FIGURE 5.20: DC circuit for solving with the KVL method. We need to find the current through and voltage drop across each resistor.

equations are given in terms of the loop currents.

$$\begin{cases} i_1 R_1 + (i_1 - i_2)R_3 = V \\ i_2 R_2 + (i_2 - i_3)R_4 + (i_2 - i_1)R_3 = 0 \\ i_3 R_5 + (i_3 - i_2)R_4 = 0 \end{cases}$$

$$\begin{bmatrix} (R_1 + R_3) & -R3 & 0 \\ -R3 & (R2 + R3 + R4) & -R4 \\ 0 & -R4 & (R4 + R5) \end{bmatrix} \begin{bmatrix} i_1 \\ i_2 \\ i_3 \end{bmatrix} = \begin{bmatrix} V \\ 0 \\ 0 \end{bmatrix}$$

5.7.1.2 KCL Method

With the KCL method, the sum of the currents entering or exiting each node must equal zero. We only need to concern ourselves with the nodes labeled v_1 and v_2 in figure 5.21. Our system of equations is the currents

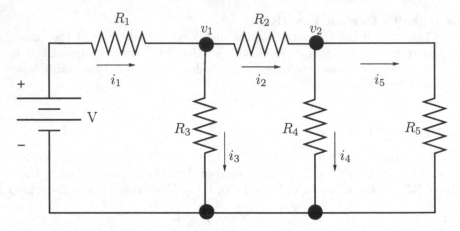

FIGURE 5.21: DC circuit for solving with the KCL method. We need to find the current through and voltage drop across each resistor.

entering and exiting nodes labeled v_1 and v_2.

$$\begin{cases} i_1 - i_2 - i_3 = 0 \\ i_2 - i_4 - i_5 = 0 \end{cases}$$

We use Ohm's law to write each current in terms of the voltages across resistance values, which gives us sums of fractions.

$$\begin{cases} \frac{V-v_1}{R_1} - \frac{v_1-v_2}{R_2} - \frac{v_1}{R_3} = 0 \\ \frac{v_1-v_2}{R_2} - \frac{v_2}{R_4} - \frac{v_2}{R_5} = 0 \end{cases}$$

To remove the fractions, multiply each equation by the products of the denominators. Thus we multiply the terms of the first equation by $-R_1\,R_2\,R_3$. The second equation is multiplied by $R_2\,R_4\,R_5$. Then we combine terms by the unknown node voltages to make the matrix equations.

$$\begin{bmatrix} (R_2\,R_3 & + & R_1\,R_3 & + & R_1\,R_2) & -R_1\,R_3 \\ R_4\,R_5 & (-R_4\,R_5 & - & R_2\,R_5 & - & R_2\,R_4) \end{bmatrix} \begin{bmatrix} v_1 \\ v_2 \end{bmatrix} = \begin{bmatrix} R_2\,R_3\,V \\ 0 \end{bmatrix}$$

5.7.1.3 MATLAB Code

After doing the KVL calculation, the code in code 5.3 computes some node voltages to help verify the correctness of the voltages from the KVL and KCL solutions.

The program displayed the following output. One can see that the KVL and KCL methods give the same results. Remember that the currents reported for the KVL method are loop currents, not the currents through all of the resistors. The current through R_3, and R_4 are combinations of the loop currents.

```
        KVL Solution:              KCL Solution:
   Current:                     Current:
     i1 = 0.005928                i1 = 0.005928
     i2 = 0.002892                i2 = 0.002892
     i3 = 0.001157                i3 = 0.003036
   Voltages:                      i4 = 0.001735
     V1 = 5.927711                i5 = 0.001157
     V2 = 4.337349              Voltages:
     V3 = 6.072289                V1 = 5.927711
     V4 = 1.734940                V2 = 4.337349
     V5 = 1.734940                V3 = 6.072289
   Verify:                        V4 = 1.734940
     1.734940 == 1.734940 ?       V5 = 1.734940
     6.072289 == 6.072289 ?
     12.000000 == 12.000000 ?
```

5.7.2   The Statics of Trusses

The field of statics is concerned with determining the forces acting on members of stationary rigid bodies. We need to know the tension or compression forces on each member of the structure to ensure that it is strong enough. Here we consider a truss structure made of four members. Statics textbooks such as [54] explain how to solve such systems.

Statics problems such as shown in figure 5.22 are solved at two levels. First, we consider the external forces on the overall structure, which are later applied to the members of the structure. Three equations are used for this step.

1. Sum of the moments is zero. Here we define a focal point and then calculate the moment of each external force as the product of the force value perpendicular to the structure times the distance from the focal point. This ensures that the structure does not rotate. Note that in the problem that we solve here, the tension force of the CD member is not perpendicular to the structure, so the cosine function is used to find a perpendicular force.

$$\sum_{\forall i} F_i \, d_i = 0$$

2. Sum of forces in the x direction is zero.

$$\sum_{\forall i} F_{i,x} = 0$$

3. Sum of forces in the y direction is zero.

$$\sum_{\forall i} F_{i,y} = 0$$

```
% File: circuitSys.m
%% Constants
R1 = 1000; R2 = 1500; R3 = 2000; R4 = 1000; R5 = 1500;
v = 12;


%% KVL Solution
R = [(R1+R3) -R3 0;
     -R3 (R2+R3+R4) -R4;
     0 -R4 (R4+R5)];
V = [v; 0; 0];
I = R \ V;        % V = RI
i1 = I(1); i2 = I(2); i3 = I(3);
V1 = R1*i1; V2 = R2*i2; V3 = R3*(i1-i2); V4 = R4*(i2-i3); V5 = R5*i3;
fprintf('\n\tKVL Solution:\n')
fprintf('Current:\n i1 = %f\n i2 = %f\n i3 = %f\n',i1,i2,i3);
fprintf(Voltages:\n)
fprintf(...
   V1 = %f\n V2 = %f\n V3 = %f\n V4 = %f\n V5 = %f\n', ...
            V1,V2,V3,V4,V5);
disp('Verify: ')
fprintf(' %f == %f ?\n',V4,V5);
fprintf(' %f == %f ?\n',V3,(V2+V4));
fprintf(' %f == %f ?\n',v,(V3+V1));

%% KCL Solution

A = [(R2*R3 + R1*R3 + R1*R2) -R1*R3;
      R4*R5 (-R4*R5 - R2*R5 - R2*R4)];
I = [R2*R3*v 0]';
V = A \ I; % (1/R)*V = I, A = 1/R -> AV = I
v1 = V(1);
v2 = V(2);
fprintf('\n\tKCL Solution:\n')
fprintf('Current:\n i1 = %f\n i2 = %f\n',(v-v1)/R1, (v1-v2)/R2);
fprintf(' i3 = %f\n i4 = %f\n i5 = %f\n',v1/R3, v2/R4, v2/R5);
fprintf('Voltages:\n V1 = %f\n V2 = %f\n ', v-v1, v1-v2);
fprintf('V3 = %f\n V4 = %f\n V5 = %f\n', v1, v2, v2);
```

CODE 5.3: Script: `circuitSys`, DC circuit by KVL and KCL.

Secondly, the *method of joints* is used to determine the force on each member. We examine each joint and write equations for the forces in the x and the y directions. These forces must also sum to zero for each joint.

As you can see from figure 5.22, this is a fairly simple problem—4 members between 3 joints and the external supports. As a full system of equations, there are 9 equations—3 external force equations and 2 equations each for the 3 joints: A, B, and C. When taken in the desired order, each algebra

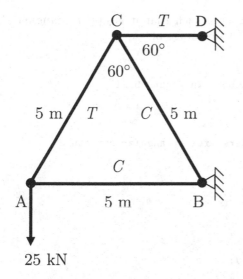

% Forces on each·member
>> truss
AB: 14.4338 kN C
AC: 28.8675 kN T
BC: 28.8675 kN C
CD: 28.8675 kN T

25 kN

FIGURE 5.22: A static truss based support. We determine the tension or compression force on each member of the truss in code 5.4 and report the results above. Each member is marked with a T or C as either bearing a tension or compression force.

equation reduces to having only one unknown variable. We'll take a middle path between an over-determined system of 9 equations with 4 unknowns and a carefully ordered sequence of algebra problems. We list the 9 equations and then pick 4 equations for a 4×4 matrix equation. Three of the picked equations have only one unknown variable. The fourth equation has two unknown variables. The 9 equations are listed below. The numbered equations are used in the matrix solution.

1 – Sum of moments: $\sum M_B = 0$: $(5\text{m})(25\text{kN}) - (5\text{m})\, CD \cos(30°) = 0$

External forces: $\sum F_x = 0$: $B_x = CD$

External forces: $\sum F_x = 0$: $B_y = 25kN$

2 – A: $\sum F_x = 0$: $AB - AC \cos(60°) = 0$

3 – A: $\sum F_y - 0$: $AC \cos(60°) - 25$

B: $\sum F_x = 0$: $AB + BC \cos(60°) = B_x$

4 – B: $\sum F_y = 0$: $BC \cos(60°) = B_y = 25$

C: $\sum F_x = 0$: $AC \cos(60°) + BC \cos(60°) = CD$

C: $\sum F_y = 0$: $BC \cos(30°) + BC \cos(30°) = 0$

The rows of the **A** matrix are ordered such that it is upper triangular, which the left-divide operator will take advantage of.

```
% File: truss.m
% Find the forces on truss members in figure 5.21.
c60 = cosd(60);
c30 = cosd(30);
s30 = sind(30);
A = [1 -c60 0   0; % rows ordered for triangular structure
     0  c30 0   0;
     0  0   c30 0;
     0  0   0   c30];
b = [0; 25; 25; 25];
x = A\b;
fprintf('AB: %g kN C\n', x(1));
fprintf('AC: %g kN T\n', x(2));
fprintf('BC: %g kN C\n', x(3));
fprintf('CD: %g kN T\n', x(4));
```

CODE 5.4: Code to solve the system of equations for the forces on a truss based support. The diagram and results are in figure 5.22.

5.8   Under-determined Systems

As discussed in section 5.4.5, under-determined systems of linear equations have an infinite set of solutions rather than a unique solution. If additional equations can not be found to make the system full rank, then one might find an acceptable solution from the set of vectors that satisfy the system of equations. Under-determined systems are found in linear programming problems, especially in the context of machine learning and artificial intelligence [14, 12].

5.8.1   RREF and Under-determined Systems

Using elimination, we can put a matrix into what is called *reduced row echelon form* (RREF) as described in section 5.5.3, which will reveal the set of solution vectors where the system of equations is satisfied. In MATLAB, the **rref** function performs elimination with partial pivoting to put a matrix into RREF. However, as described in section 5.5.4, **rref** gives inaccurate results for some matrix equations. Still, it is a useful elimination based tool that will help us understand the available solutions to under-determined systems. It gives a set of solutions but does not tell us the single *best* solution. When we need a working solution, MATLAB's left-divide operator usually provides a solution that will meet our needs.

When the first m columns of an under-determined system are independent of each other then the augmented matrix in RREF takes the form of the following example that shows an identity matrix in the first m columns.

```
>> A = randi(10, 3, 4) - randi(5, 3, 4)
A =
     6     0     4     0
     1     1     9     8
     5     9     5     5
>> rank(A)
ans =
     3
>> x = randi(5, 4, 1)
x =
     4
     2
     5
     1
>> b = A*x
b =
    44
    59
    68
>> C = rref([A b])
C =
    1.0000         0         0   -0.6091    3.3909
         0    1.0000         0    0.3864    2.3864
         0         0    1.0000    0.9136    5.9136
```

Notice that each of the first three columns have only a single one. But the fourth column, for x_4, has values in each row. MATLAB found the fourth column to be a linear combination of the first three columns by the weights indicated. So the values of the first three variables, which are pivot columns, are fixed by the equation. We call the variables associated with the pivot columns *basic variables*. Any variables, such as, x_4, that are not associated with pivot columns are called *free variables*, meaning that they can take any value. We don't have an equation for x_4 since it is a free variable, so we can just say that $x_4 = x_4$. We can also replace it with an independent scalar variable, a.

The new system of equations is:

$$\begin{aligned}
x_1 - 0.6091\,x_4 &= 3.3909 \\
x_2 + 0.3864\,x_4 &= 2.3864 \\
x_3 + 0.9136\,x_4 &= 5.9136 \\
x_4 &= x_4
\end{aligned}$$

The solution is a set of lines. We can see this if we write the line equations in what is called *parametric form*.

$$\begin{bmatrix} x_1 \\ x_2 \\ x_3 \\ x_4 \end{bmatrix} = \begin{bmatrix} 3.3909 \\ 2.3864 \\ 5.9136 \\ 0 \end{bmatrix} + a \begin{bmatrix} 0.6091 \\ -0.3864 \\ -0.9136 \\ 1 \end{bmatrix} = u + a\,v$$

Particular and General Solution

Part of the solution that we found by elimination with the RREF is called the *particular* solution, which is any vector that solves $Au = b$ when $a = 0$. The u vector is the last column of the augmented matrix in RREF. We also found a *general* solution, which are the vectors that can be multiplied by any scalar, a. In addition to finding the general solution from RREF, it is also found from the null space solution of A, $a\,v = \text{Null}(A)$. If $x = u + a\,v$, then

$$Ax = A\,(u + a\,v) = b + 0 = b.$$

Note that the general solutions found from the `rref` and `null` MATLAB functions are likely not the same, but may be scalar multiples of each other. In the previous example, if $a = 0.6516$ then $a\,v = \text{Null}(A)$. However, the general solutions of `rref` and `null` will likely not have such a relationship if $m - n > 1$ and the general solution has more than one column. If the general solution has two columns that we call v and w, then we still have the same basic relationship.

$$Ax = A\,(u + a\,v + b\,w) = b + 0 + 0 = b$$

See appendix A.3.2 for more information about the null space of a matrix.

Here is a quick example showing a general solution with two columns. Notice that as we put the equation into parametric form that we change the sign of the general solution columns to move them from the left to right side of the equal sign. We also add ones in the appropriate rows on v and w so that $x_3 = a$ and $x_4 = b$.

```
>> A
A =
       8     -6      4     -3
      10     10     -7      2
>> b
b =
      14
      58
>> z = rref([[A b];zeros(2,5)])
z =
```

```
    1.0000          0   -0.0143   -0.1286    3.4857
         0     1.0000   -0.6857    0.3286    2.3143
         0          0         0         0         0
         0          0         0         0         0
% general solutions
>> v = -z(:,3);
>> v(3) = 1;
>> w = -z(:,4);
>> w(4) = 1;
>> A*(v + w)    % a = 1; b = 1;
ans =
     0
     0
>> A*(v + 2*w)  % a = 1; b = 2;
ans =
     0
     0
% particular solution
>> A*z(:,5)
ans =
    14.0000
    58.0000
```

5.8.2 The Preferred Under-determined Solution

The best solution to an under-determined system of equations is usually the smallest x that satisfies the equations. The length of x is found with the l_2–norm, but the l_1–norm or other vector norms described in appendix A.1.1 may be used.

One possibility is to find a scalar multiplier, a, that minimizes $\|x = u + a\,v\|_2$. We could use an optimization algorithm as described in section 7.1.2 to minimize $\|x\|_2$ with respect to the multipliers $(a, b, \ldots)$. MATLAB's lsqminnorm combines the particular and general solutions to find a solution with a smaller l_2–norm than what the left-divide operators finds using just the particular solution. In this example, lsqminnorm uses the same A and b as the previous examples.

```
>> x = lsqminnorm(A, b)        >> x = A\b
x =                            x =
     3.4774                         3.4857
     1.5379                         2.3143
    -1.1043                              0
     0.0582                              0
>> norm(x)                     >> norm(x)
ans =                          ans =
     3.9599                         4.1840
```

Another approach is to let $a = 0$ and focus on the particular solution. Cline and Plemmons showed that the particular solution that minimizes $\|x = u\|_2$ is found with the Moore–Penrose pseudo-inverse of under-determined matrices [14]. Solutions that minimize the l_2–norm of x are also referred to as the *least squares* solution.

$$x = \mathbf{A}^+ b$$

Under-determined Pseudo-inverse

The Moore–Penrose pseudo-inverse, denoted $\mathbf{A}^+$, fills the role of a matrix inverse for rectangular matrices. We can find a pseudo-inverse for an over-determined matrix $(\mathbf{A}_o^+)$ and an under-determined matrix $(\mathbf{A}_u^+)$. We show in section 5.9.3 that the pseudo-inverse of an over-determined matrix is:

$$\mathbf{A}_o^+ = \left(\mathbf{A}^T \mathbf{A}\right)^{-1} \mathbf{A}^T.$$

We use this result to find the pseudo-inverse of an under-determined matrix. We have an over-determined matrix in $\mathbf{A}^T$. The transpose of the pseudo-inverse of $\mathbf{A}^T$ is the under-determined pseudo-inverse of $\mathbf{A}$.

$$\mathbf{A}_u^+ = \left(\left(\mathbf{A}^T\right)_o^+\right)^T = \left(\left(\mathbf{A}\,\mathbf{A}^T\right)^{-1}\mathbf{A}\right)^T$$

$$= \mathbf{A}^T\left(\left(\mathbf{A}\,\mathbf{A}^T\right)^{-1}\right)^T = \mathbf{A}^T\left(\left(\mathbf{A}\,\mathbf{A}^T\right)^T\right)^{-1}$$

$$= \mathbf{A}^T\left(\mathbf{A}\,\mathbf{A}^T\right)^{-1}$$

As with the over-determined case, the direct matrix equation can be slow and prone to inaccuracy for large and poorly conditioned matrices, so orthogonal techniques such as the SVD or QR factorization algorithms are used instead. MATLAB's `pinv` function uses the SVD as described in section 6.8.5.1, while the left-divide operator uses QR factoring as described in section 5.11.3.

In addition to minimizing the length of the x vector, we also have a preference for solutions that are sparse (lots of zeros). Sparse solutions are especially desired in some machine learning applications. The smallest x in terms of the l_1–norm is known to be a sparse solution. Numerical methods for finding the sparse l_1–norm based solution is discussed in section 7.1.3.

As discussed in section 5.11.3, MATLAB's left-divide operator finds the least squares particular solution using the QR factorization algorithm. Before using QR, dependent columns of $\mathbf{A}$ are zeroed out, which yields more zeros in the solution. The left-divide operator zeros out $n - r$ columns of $\mathbf{A}$ where r is the rank of $\mathbf{A}$. The columns to zero out are found sequentially. The first column zeroed out is the column that when replaced with zeros results in the smallest l_2–norm of the solution. The evaluation is repeated until $n - r$ columns are zeroed out [46].

We conclude the discussion of under-determined systems by returning to our first under-determined example and finding solutions with MATLAB using the left-divide operator, RREF, and the pseudo-inverse with the last column zeroed out, which gives the same solution as the left-divide operator. The algorithm that the left-divide operator uses to find the solution with the QR factorization is listed in section 5.11.3.

```
>> A\b
ans =
    3.3909
    2.3864
    5.9136
         0

% Add extra rows of zeros to get the desired number of
% rows for the particular solution.
>> rref([[A b]; 0 0 0 0 0])
ans =              % particular solution in last column
    1.0000        0         0   -0.6091    3.3909
         0   1.0000         0    0.3864    2.3864
         0        0    1.0000    0.9136    5.9136
         0        0         0         0         0

>> A(:,4) = [0 0 0]'
A =
    6    0    4    0
    1    1    9    0
    5    9    5    0

>> pinv(A)*b
ans =
    3.3909
    2.3864
    5.9136
         0
```

5.9   Over-determined Systems and Vector Projections

The *over-determined* matrix equation of the form $\mathbf{A}x = b$ has more equations than unknown variables $(m > n)$. A common situation where an over-determined system occurs is in the results of an experiment. The experiment may be repeated many times as a control variable is adjusted. Thus a researcher may have many more equations relating the inputs to the outputs than unknown variables.

We saw in section 5.4.5 that such systems have a unique solution when all row equations are consistent, which we formally describe as when b is in the column space (span) of A. But otherwise, we can only approximate the solution. We find the approximation solution by *projection*, which is our first topic.

5.9.1   Projections Onto a Line

A nice aspect of linear algebra is its geometric interpretation. That is, if we consider linear algebra problems in $\mathbb{R}^2$ or $\mathbb{R}^3$, then we can plot the vectors to visually depict the geometric relationships. This is the case with project, which gives us a geometric view of over-determined systems. We will begin with the simplest of examples and later extend the concepts to higher dimensions.

Consider the following over-determined system where the unknown variable is a scalar.

$$\begin{cases} 4x = 2 \\ 2x = 1 \end{cases}$$

$$a\,x = b \tag{5.12}$$

$$\begin{bmatrix} 4 \\ 2 \end{bmatrix} [x] = \begin{bmatrix} 2 \\ 1 \end{bmatrix}$$

Figure 5.23 is a plot of the vectors a and b.

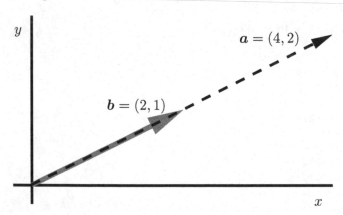

FIGURE 5.23: Vectors a and b are consistent.

Here x is a scalar, not a vector, and we can quickly see that $x = 1/2$ satisfies both rows of equation (5.12). But since a and b are both vectors, we want to use a strategy that will extend to matrices. We can multiply both sides of equation (5.12) by a^T so that the vectors turn to scalar dot products.

$$a^T a x = a^T b \qquad (5.13)$$

$$\begin{bmatrix} 4 & 2 \end{bmatrix} \begin{bmatrix} 4 \\ 2 \end{bmatrix} x = \begin{bmatrix} 4 & 2 \end{bmatrix} \begin{bmatrix} 2 \\ 1 \end{bmatrix}$$

$$x = \tfrac{10}{20} = \tfrac{1}{2}$$

Now, let's extend the problem to a more general case as shown in figure 5.24 where vector **b** is not necessarily in-line with **a**. We will find a geometric reason to multiply both sides of equation (5.12) by a^T. We wish to project a vector **b**

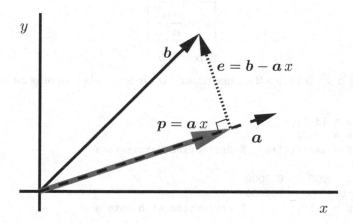

FIGURE 5.24: Vector **p** is a projection of vector **b** onto vector **a**.

onto vector **a**, such that the error between **b** and the projection is minimized. The projection, **p**, is a scalar multiple of **a**. That is, $p = a\,x$, where x is the scalar value that we want to find. The error is the vector $e = b - a\,x$.

The geometry of the problem provides a simple solution for minimizing the error. The length of the error vector is minimized when it is perpendicular to **a**. Recall that two vectors are perpendicular (orthogonal) when their dot product is equal to zero.

$$a^T e = a^T (b - a\,x) = 0$$

$$a^T a\,x = a^T b$$

$$\boxed{x = \frac{a^T b}{a^T a}}$$

Since x is a fraction of two dot products, we can think of the projection in terms of the angle, θ, between **a** and **b**.

$$x = \frac{\|a\|\,\|b\|\cos\theta}{\|a\|^2}$$

$$= \frac{\|b\|}{\|a\|}\cos\theta$$

The projection is then:

$$p = ax$$

$$= \frac{a}{\|a\|} \|b\| \cos\theta$$

$$= a \frac{a^T b}{a^T a}$$

We can also make a projection matrix, $\mathbf{P}$, so that any vector may be projected onto a by multiplying it by $\mathbf{P}$.

$$\boxed{\mathbf{P} = \frac{a\, a^T}{a^T a}}$$

$$p = \mathbf{P}\, b$$

Note that $a\, a^T$ is here a 2×2 matrix and $a^T a$ is a scalar. Here is an example.

```
>> a = [3;1];
>> b = [2;2];
>> P = a*a'/(a'*a)   % Projection Matrix to a
P =
    0.9000    0.3000
    0.3000    0.1000
>> p = P*b            % Projection of b onto a
p =
    2.4000
    0.8000
>> x = a'*b/(a'*a)    % length of projection
x =
    0.8000
>> e = b - p          % error vector
e =
   -0.4000
    1.2000
>> p'*e               % near zero dot product to
ans =                 % confirm e is perpendicular to p.
    2.2204e-16
```

5.9.2   Exact Solution or Approximation

We know that the solution to an over-determined system is exact when b is in the column space of $\mathbf{A}$, but is an approximation when it is not. The quickest way to determine if the solution is exact or an approximation is to test if b is independent or a linear combination of the columns of $\mathbf{A}$.

Vector b is in the column space of $\mathbf{A}$ when the ranks of $\mathbf{A}$ and the augmented matrix $[\mathbf{A}\ b]$ are the same, which means that b is a linear combination of the columns of $\mathbf{A}$.

MATLAB can help us see this. First, let us consider the case where b is in the column space of $\mathbf{A}$ and an exact solution exists. Notice that rank(A) is equal to rank([A b]).

```
>> f = @(t) 5 - 2.*t;      >> b = f(t)';
>> t = 1:5;                >> rank(A)
>> A = ones(5,2);          ans =
>> A(:,2) = t';               2
>> A                       >> rank([A b])
A =                        ans =
     1     1                   2
     1     2               >> x = A\b
     1     3               x =
     1     4                   5.0000
     1     5                  -2.0000
```

Now, let's add some random noise to the b vector so that it is not in the column space of $\mathbf{A}$ and the solution is an approximation. Notice that rank(A) is less than rank([A b]).

```
>> b = b + randn(1,5)';    >> rank([A b])
>> rank(A)                 ans =
ans =                         3
     2                     >> x = A\b
                           x =
                               5.6816
                              -2.1410
```

5.9.3   Projections Onto a Hyperplane

We can extend projections to $\mathbb{R}^3$ and still visualize the projection as projecting a vector onto a plane. Here, the column space of matrix $\mathbf{A}$ is two 3-dimension vectors, a_1 and a_2.

$$\mathbf{A} = \begin{bmatrix} | & | \\ a_1 & a_2 \\ | & | \end{bmatrix}$$

The span of two vectors in $\mathbb{R}^3$ forms a plane. As before, we have an approximation solution to $\mathbf{A}x = b$, which can be written as $\mathbf{A}\hat{x} = p$. The error is $e = b - p = b - \mathbf{A}\hat{x}$. We want to find $\hat{x}$ such that $b - \mathbf{A}\hat{x}$ is orthogonal to the plane, so again we set the dot products equal to zero.

$$a_1^T(b - \mathbf{A}\hat{x}) = 0$$

$$a_2^T(b - \mathbf{A}\hat{x}) = 0$$

$$\begin{bmatrix} - & a_1^T & - \\ - & a_2^T & - \end{bmatrix} \begin{bmatrix} | \\ b - A\hat{x} \\ | \end{bmatrix} = \begin{bmatrix} 0 \\ 0 \end{bmatrix}$$

The left matrix is just $\mathbf{A}^T$, which is size 2×3. The size of $\hat{x}$ is 2×1, so the size of $\mathbf{A}\hat{x}$, like b, is 3×1.

$$\mathbf{A}^T(b - \mathbf{A}\hat{x}) = 0$$

$$\mathbf{A}^T\mathbf{A}\hat{x} = \mathbf{A}^T b$$

$$\boxed{\hat{x} = \left(\mathbf{A}^T\mathbf{A}\right)^{-1}\mathbf{A}^T b}$$

In MATLAB, we can find $\hat{x}$ as:

```
x_hat = (A'*A) \ (A'*b);
```

Note: (A \ b) finds the same answer. The left-divide operator handles under-determined and over-determined systems as well as square matrix systems (section 5.11).

The projection vector is then:

$$p = \mathbf{A}\hat{x} = \mathbf{A}\left(\mathbf{A}^T\mathbf{A}\right)^{-1}\mathbf{A}^T b.$$

The projection matrix is:

$$\mathbf{P} = \mathbf{A}\left(\mathbf{A}^T\mathbf{A}\right)^{-1}\mathbf{A}^T.$$

Note: $\mathbf{A}$ is not a square matrix, and is not invertible. If it were, $\mathbf{P} = \mathbf{I}$ and there would be no need to do the projection. Try to verify why this is the case. Recall that $(\mathbf{BA})^{-1} = \mathbf{A}^{-1}\mathbf{B}^{-1}$.

Over-determined Pseudo-inverse

The matrix $\left(\mathbf{A}^T\mathbf{A}\right)^{-1}\mathbf{A}^T$ has a special name and symbol. It is called the Moore–Penrose pseudo-inverse of an over-determined matrix. It is used when the simpler $\mathbf{A}^{-1}$ can not be used. A superscript plus sign $(\mathbf{A}^+)$ is used as a short-hand math symbol for the pseudo-inverse. MATLAB has a function called pinv(A) that returns the pseudo-inverse of $\mathbf{A}$.

As with the pseudo-inverse of under-determined systems described in section 5.8.2, MATLAB uses the SVD to calculate the pseudo-inverse of over-determined systems, which is more accurate and usually faster than the direct matrix calculation (section 6.8.5.1).

5.9.3.1 Projection Example

Code listings 5.5 and 5.6 and the resulting plot in figure 5.25 demonstrate the calculation and display of the projection of a vector onto the column space of a matrix. The dot product is also calculated to verify orthogonality.

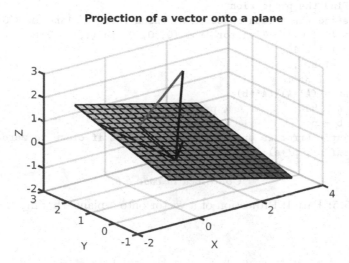

Projection of a vector onto a plane

FIGURE 5.25: The projection of a vector onto the column space of $\mathbf{A}$, which spans a plane in $\mathbb{R}^3$.

Two vectors, $\boldsymbol{a}_1$ and $\boldsymbol{a}_2$ define a plane. To find the equation for a plane, we first determine a vector, $\boldsymbol{n}$, which is orthogonal to both vectors. The cross product provides this. The vector $\boldsymbol{n}$ is defined as $\boldsymbol{n} = [a\,b\,c]^T$. Then given one point on the plane (x_0, y_0, z_0), we can calculate the equation for the plane.

$$a(x - x_0) + b(y - y_0) + c(z - z_0) + d = 0$$

In this case, we know that the point $(0, 0, 0)$ is on the plane, so the plane equation is simplified.

The points of the plane are determined by first defining a region for x and y and then using the plane equation to calculate the corresponding points for z. A simple helper function called `vector3` was used to plot the vectors. A plot of the projection is shown in figure 5.25.

5.9.3.2 Alternate Projection Equation

Not all linear algebra textbooks present the same equations for projection of a vector onto a vector space. The approach that we have used might be described as projecting a vector onto the column space of a matrix. An alternate approach projects a vector onto an equivalent vector space that is spanned by an orthonormal set of basis vectors. So the first step in the

```
% File: projectR3.m
% Projection of the vector B onto the column space of A.
% Note that with three sample points, we can visualize
% the projection in R^3.

%% Find the projection
% Define the column space of A, which forms a plane in R^3.
a1 = [1 1 0]';  % vector from (0, 0, 0) to (1, 1, 0).
a2 = [-1 2 1]';
A = [a1 a2];
b = [1 1 3]';
x_hat = (A'*A)\(A'*b);
p = A*x_hat;            % projection vector
e = b - p;             % error vector
fprintf('Error is orthogonal to the plane if close to zero:
    %f\n', p'*e);
```

Continued in code 5.6.

CODE 5.5: Part 1: Projection of a vector onto a plane—finding the projection.

alternate approach is to find the orthonormal basis vectors of the matrix columns. (See appendix A.2 for help with some of the terminology.)

Let $\mathbf{U} = \{u_1, u_2, \ldots, u_n\}$ be an orthonormal basis for the column space of matrix $\mathbf{A}$. Then the projection of vector b onto $\mathrm{Col}(\mathbf{A})$ is defined by

$$\mathrm{proj}_{\mathbf{A}} b = (b^T u_1)\, u_1 + (b^T u_2)\, u_2 + \cdots + (b^T u_n)\, u_n = \mathbf{U}\,\mathbf{U}^T b.$$

There is a trade-off between the two algorithms. The alternative approach finds the orthonormal basis vectors to the columns of $\mathbf{A}$ using either the Gram–Schmidt process, QR factorization, or the economy SVD method, which is used by the `orth` function. These algorithms are reviewed in appendix A.4 and section 6.8.5.2. Whereas, projecting onto the column space requires solving an $\mathbf{A}x = b$ problem.

The alternate projection equation comes from economy SVD described in section 6.8.5.2, where we see that we get the alternate equation when we replace $\mathbf{A}$ in the projection equation with its SVD factors. The geometry of the alternate algorithm provides a simple way to see that it also projects a vector onto a vector space.

- Any vector in the vector space must be a linear combination of the orthonormal basis vectors.

- The $b^T u_i$ coefficients in the sum are scalars from the dot products of b and each basis vector. Since the basis vectors are unit length, each term in the sum is $(\|b\|\,\cos(\theta_i))\, u_i$, where θ_i is the angle between b and the

Continued from code 5.5.

```
%% Plot the projection
% To make the equation for a plane, we need n, which is the
% orthogonal normal vector to the column space, found from
% the cross product.
n = cross(a1,a2);

% Plane is: a(x-x0) +b(y-y0) + c(z-z0) + d = 0
% Variables a, b, and c come from the normal vector.
% Keep it simple and use point (0, 0, 0) as the point on
% the plane (x0, y0, z0).
a=n(1); bp=n(2); c=n(3); d=0;
[x, y] = meshgrid(-1:0.25:3);  % Generate x and  y data
z = -1/c*(a*x + bp*y + d);     % Generate z data

figure;
surf(x,y,z);                   % Plot the plane
daspect([1 1 1]);              % consistent aspect ratio
hold on;
z0 = [0 0 0]';                 % vector starting point
vector3(z0, a1, 'k');          % column 1 of A
vector3(z0, a2, 'k');          % column 2 of A
vector3(z0, b, 'g');           % target vector
vector3(z0, p, 'b');           % projection
vector3(p, b, 'r');            % error from p to b
hold off;
xlabel('X');
ylabel('Y');
zlabel('Z');
title('Projection of a vector onto a plane');

% helper function
function vector3( v1, v2, clr )
% VECTOR3 Plot a vector from v1 to v2 in R^3.
     plot3([v1(1) v2(1)], [v1(2) v2(2)], ...
        [v1(3) v2(3)], 'Color', clr,'LineWidth',2);
end
```

CODE 5.6: Part 2: Projection of a vector onto a plane—plotting the projection.

basis vector; thus, each term in the sum is the projection of b onto a basis vector.

The example in code 5.7 finds the projection using the two methods. The `mod_gram_schmidt` function from appendix A.4.3 is used to find the orthogonal basis vectors.

```
% File: alt_project.m
%% Comparison of two projection equations
% Define the column space of A, which forms a plane in R^3.
a1 = [1 1 0]';  % vector from (0, 0, 0) to (1, 1, 0).
a2 = [-1 2 1]';
A = [a1 a2];
b = [1 1 3]';
%% Projection onto the column space of matrix A.
x_hat = (A'*A)\(A'*b);
p = A*x_hat;            % projection vector
disp('Projection onto column space: ')
disp(p)


%% Alternate projection
% The projection is the vector sum of projections onto
% the orthonormal basis vectors of the column space of A.
[U, ~] = mod_gram_schmidt(A); % Could use the orth function
u1 = U(:,1);
u2 = U(:,2);
p_alt = b'*u1*u1 + b'*u2*u2;
disp('Projection onto basis vectors: ')
disp(p_alt)
```

CODE 5.7: Example of two methods of vector projection.

The output from the example of two methods of vector projection is:

```
>> alt_project
Projection onto column space:
    0.1818
    1.8182
    0.5455
Projection onto basis vectors:
    0.1818
    1.8182
    0.5455
```

5.9.3.3   Higher Dimension Projection

If the matrix **A** is larger than 3×2, then we can not visually plot a projection as we did above. However, the equations still hold. One of the nice things about linear algebra is that we can visually see what is happening when the vectors are in $\mathbb{R}^2$ or $\mathbb{R}^3$, but higher dimensions are fine too. The projection equations are used in least squares regression where we want as many rows of data as possible to accurately fit an equation to the data.

5.9.4   A Statics Problem Solved With Projections

Another application of projections comes from geometry. Projections relate two vectors as sides of a right triangle. Thus projections can be used to solve problems that could also be solved using trigonometry. With projections, it is not necessary to compute angles between the vectors or to use the trigonometry functions.

Consider an example of a wagon sitting on an incline as shown in figure 5.26. The handle of the wagon is held at an angle. We want to calculate the force on the handle required to keep the wagon from rolling down the incline. We won't worry about inefficiency of the wheels. The force from the handle need only counter the force from gravity. The script to find the required force on the wagon handle is listed in code 5.8.

FIGURE 5.26: A wagon on an incline.

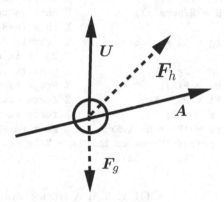

FIGURE 5.27: Free-body diagram of forces on the wagon.

NOTE: All vectors and forces are calculated as upward and to the right. The force from gravity, f_g, will be negative. Then, the important statics equation is that the sum of forces equals zero. In the calculations, projection matrices are denoted as **P**. For the sake of clarity in the code, vector variables begin with capital letters, while scalar variables are lower case.

The force toward the left (down the incline) is the projection of the gravity force onto the incline, A.

$$F_l = \mathbf{P}_a \, F_g$$

The sum of the forces to the left and right is zero.

$$F_r + F_l = 0 \longmapsto F_r = -F_l$$

The force to the right is a projection of the handle force vector onto the incline. The handle force vector is the product of the scalar force vector and the unit vector in the direction of the handle, $F_h = f_h \, H$. We need to calculate the inverse of the projection, so we separate the scalar from the unit vector to include the vector direction in the calculation.

$$F_r = \mathbf{P}_a\,F_h = \mathbf{P}_a\,\mathbf{H}\,f_h = -\mathbf{P}_a\,F_g$$

$$f_h = (\mathbf{P}_a\,\mathbf{H})^{-1}\,(-\mathbf{P}_a\,F_g)$$

```
%% File: wagon.m
%  Statics problem of a wagon on an incline solved with vector
%  projections

P = @(a) a*a'/(a'*a);    % Projection matrix equation
V = [0 1]';              % Vertical vector
A = [8 3]';              % vector of the incline slope
H = [1 1]';              % vector of handle direction
H = H/norm(H);           % unit vector of handle direction
                         % Fh = fh*H
fg = 23 * -9.81;         % Gravity force down (F = m*a)
                         % 23 kg is about 50 pounds
Fg = fg * V;             % Gravity force vector
Pa = P(A);               % Projection matrix to incline A
Fl = Pa * Fg;            % force down hill to left
Fr = -Fl;                % force to right, Fr + Fl = 0
fh = (Pa * H) \ Fr;      % Fr = Pa*Fh = Pa*H*fh
fprintf('Force on handle = %.3f newtons or %.3f kg\n', ...
    fh, fh/9.81);
```

CODE 5.8: A statics problem solved using projections.

Challenge Question

By changing the vectors of the incline and the handle, one can verify the correctness of the solution. What are the boundary conditions? Try to determine the relationship of the slope of the incline to the mechanical advantage for raising a weight vertically using a rolling wagon or cart on an incline.

5.10   Least Squares Regression

In science and engineering, we often conduct experiments where we measure a parameter of a system with respect to a control variable. The measurement and control variables might be time, voltage, weight, length, volume, temperature, pressure, or any number of parameters relative to the system under investigation. We might expect that the measurement and control variable are related by a polynomial equation, but we likely do not know the coefficients of the polynomial. We might not even know if the relationship of the measured data is linear, quadratic, or a higher order polynomial. We are conducting the experiment to learn about the system. Unfortunately, in addition to measuring the relationship between the variables, we will also measure some noise caused by volatility of the system and measurement errors.

If the x is the control variable, the measured data can be modeled as:

$$b(x) = f(x) + n(x)$$

- $b(x)$ is the measured data.

- $f(x)$ is the underlying relationship that we wish to find.

- $n(x)$ is the measurement noise, which may be correlated to x, or may be independent of x.

The accuracy of our findings will increase with more sample points. After conducting the experiment, we want to apply projection to fit the data to a polynomial that minimizes the square of the error $(E = \|e\|^2)$.

Note: Least squares regression results can be very good if the noise is moderate and the data does not have *outliers* that will skew the results. After initial regression fitting, a further analysis of the noise might lead to removal of outlier data so that a better fit can be achieved.

5.10.1 Linear Regression

In linear regression we model the data as following a line, so we represent the relationship between the measured data points, $b(t)$, and the control variable, t, with line equations. The coefficients C and D are the unknowns that we need to find.

$$b(t) = C + D\,t$$

$$C + D\,t_1 = b(t_1)$$
$$C + D\,t_2 = b(t_2)$$
$$C + D\,t_3 = b(t_3)$$
$$C + D\,t_4 = b(t_4)$$
$$\vdots$$
$$C + D\,t_n = b(t_n)$$

As a matrix equation, $\mathbf{A}$ defines the control variables. This $\mathbf{A}$ matrix is often called a *design matrix* in regression terminology.

$$\mathbf{A} = \begin{bmatrix} 1 & t_1 \\ 1 & t_2 \\ \vdots & \vdots \\ 1 & t_n \end{bmatrix}$$

Vector $\boldsymbol{u}$ represents the unknown polynomial coefficients.

$$\boldsymbol{u} = \begin{bmatrix} C \\ D \end{bmatrix}$$

$$\mathbf{A}\boldsymbol{u} = \boldsymbol{b}$$

The equations for least squares regression calculations are given in three mathematical disciplines—linear algebra, calculus, and statistics. The computations for the three are actually the same, but the equations look different depending on one's vantage point.

5.10.1.1   Linear Algebra Based Linear Regression

Because the system of equations is over-determined and the measured data, $\boldsymbol{b}$, is not likely to lie in the column space of $\mathbf{A}$, the projection equations from section 5.9.3 are used to approximate $\boldsymbol{u}$.

$$\hat{\boldsymbol{u}} = \left(\mathbf{A}^T \mathbf{A}\right)^{-1} \mathbf{A}^T \boldsymbol{b}$$

The projection $\boldsymbol{p} = \mathbf{A}\hat{\boldsymbol{u}}$ follows directly.

5.10.1.2   Calculus Based Linear Regression

The error between the projection, $\mathbf{A}\boldsymbol{u}$, and the target vector, $\boldsymbol{b}$, is $\boldsymbol{e} = \boldsymbol{b} - \mathbf{A}\boldsymbol{u}$. To account for both positive and negative errors, we wish to minimize the squares of the error vector, $E = \|\boldsymbol{e}\|^2$.
Note that the squared length of a vector is its dot product with itself.

$$E = \|\boldsymbol{b} - \mathbf{A}\boldsymbol{u}\|^2 = (\boldsymbol{b} - \mathbf{A}\boldsymbol{u})^T (\boldsymbol{b} - \mathbf{A}\boldsymbol{u}) = (\boldsymbol{b}^T - \boldsymbol{u}^T\mathbf{A}^T)(\boldsymbol{b} - \mathbf{A}\boldsymbol{u})$$

$$= \boldsymbol{b}^T\boldsymbol{b} - \boldsymbol{b}^T\mathbf{A}\boldsymbol{u} - \boldsymbol{u}^T\mathbf{A}^T\boldsymbol{b} + \boldsymbol{u}^T\mathbf{A}^T\mathbf{A}\boldsymbol{u}$$

The minimum occurs where the derivative is zero. The derivative is with respect to the vector $\boldsymbol{u}$, which is the gradient of E. Appendix C of [78] and the vector calculus chapter of [17] lists the derivative results that we need.

$$\nabla E = \frac{\partial E}{\partial \boldsymbol{u}} = -\mathbf{A}^T\boldsymbol{b} - \mathbf{A}^T\boldsymbol{b} + 2\mathbf{A}^T\mathbf{A}\boldsymbol{u} = 0$$

$$\mathbf{A}^T\mathbf{A}\boldsymbol{u} = \mathbf{A}^T\boldsymbol{b}$$

Calculus gives us the same equation that we get from linear algebra.

$$\hat{\boldsymbol{u}} = \left(\mathbf{A}^T\mathbf{A}\right)^{-1}\mathbf{A}^T\boldsymbol{b}$$

5.10.1.3 Statistical Linear Regression

We start by calculating some statistical parameters from both the results, b, and the control variable, t. The equations for calculating the mean and standard deviation are found in section 3.2.

1. Mean of b is $\bar{b}$.

2. Mean of t is $\bar{t}$.

3. Standard deviation of b is σ_b.

4. Standard deviation of t is σ_t.

5. The correlation between t and b is r.

$$r = \frac{n \sum tb - \left(\sum b\right)\left(\sum t\right)}{\sigma_t\,\sigma_b}$$

We turn to statistics textbooks for the linear regression coefficients [37, 58].

$$D = \frac{r\,\sigma_b}{\sigma_t}$$

$$C = \bar{b} - D\,\bar{t}$$

We write the matrix and vector terms as sums of the variables t and b to show that the equations for linear regression from the study of statistics and linear algebra are the same.

$$\mathbf{A}^T\mathbf{A} = \begin{bmatrix} n & \sum t \\ \sum t & \sum t^2 \end{bmatrix} \qquad \mathbf{A}^T b = \begin{bmatrix} \sum b \\ \sum tb \end{bmatrix}$$

Using elimination to solve, we get:

$$\left[\begin{array}{cc|c} n & \sum t & \sum b \\ \sum t & \sum t^2 & \sum tb \end{array}\right]$$

$$\left[\begin{array}{cc|c} n & \sum t & \sum b \\ 0 & \sum t^2 - \frac{1}{n}\left(\sum t\right)^2 & \sum tb - \frac{1}{n}\left(\sum b\right)\left(\sum t\right) \end{array}\right]$$

We get a match for the D coefficient by identifying statistical equations in the equation from elimination.

$$D = \frac{\sum tb - \frac{1}{n}\left(\sum b\right)\left(\sum t\right)}{\sum t^2 - \frac{1}{n}\left(\sum t\right)^2} = \frac{r\,\sigma_t\,\sigma_b}{\sigma_t{}^2} = \frac{r\,\sigma_b}{\sigma_t}$$

The C coefficient comes from back substitution of the elimination matrix.

$$C = \frac{\sum b - D\sum t}{n} = \bar{b} - D\,\bar{t}$$

5.10.1.4 Linear Regression Example

Code 5.9 shows a simple linear regression example. The exact output is different each time the script is run because of the random noise. A plot showing the output is shown in figure 5.28.

```
% File: linRegress.m
% Least Squares linear regression example showing the
% projection of the test data onto the design matrix A.
%
% Using control variable t, an experiment is conducted.
% Model the data as: b(t) = C + Dt.
f = @(t) 2 + 4.*t;

% A*u_hat = b, where u_hat = [ C D ]'
% Define the design matrix A.
m = 20; n = 2;
t = linspace(-2, 3, m);
A = ones(m, n);
A(:,2) = t';
b = f(t)' + 3*randn(m, 1);
u_hat = (A'*A)\(A'*b);   % least squares line
p = A*u_hat;             % projection vector
e = b - p;               % error vector

C = u_hat(1);
D = u_hat(2);
fprintf('Equation estimate = %.2f + %.2f*t\n', C, D);

%% Plot the linear regression
figure, plot(t, b, 'o');
hold on
plot(t, p, 'Color', 'k');
hold off
grid on, xlabel('t'), ylabel('b')
title('Linear Regression')
```

CODE 5.9: Least squares linear regression example.

5.10.2 Quadratic and Higher Order Regression

Application of least squares regression to quadratic polynomials only requires adding an additional column to the design matrix for the square of the control variable [68].

$$C + D\,t_1 + E\,t_1^2 = b(t_1)$$
$$C + D\,t_2 + E\,t_2^2 = b(t_2)$$
$$C + D\,t_3 + E\,t_3^2 = b(t_3)$$
$$C + D\,t_4 + E\,t_4^2 = b(t_4)$$
$$\vdots$$
$$C + D\,t_n + E\,t_n^2 = b(t_n)$$

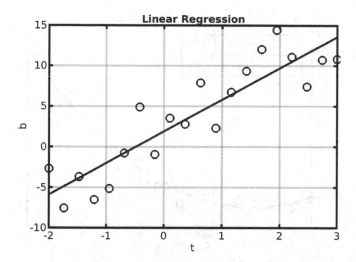

FIGURE 5.28: Simple linear least squares regression. Output from the script is Equation estimate = 1.90 + 3.89*t, which is close to the data model of $b = 2 + 4\,t$.

$$\mathbf{A} = \begin{bmatrix} 1 & t_1 & t_1^2 \\ 1 & t_2 & t_2^2 \\ \vdots & \vdots & \vdots \\ 1 & t_n & t_n^2 \end{bmatrix}$$

Higher order polynomials would likewise require additional columns in the **A** matrix.

Code 5.13 on page 233 lists a quadratic regression example. The plot of the data is shown in figure 5.29.

5.10.3 polyfit function

An erudite engineer should understand the principles of how to do least squares regression for fitting a polynomial equation to a set of data. However, as might be expected with MATLAB, it has a function that does the real work for us. The polyfit function does the regression calculation to find the coefficients. Then the polyval function can be used to apply the coefficients to create the y axis data points.

```
%% MATLAB's polyfit function
coef = polyfit(t, y, 2);
disp(['Polyfit Coefficients: ', num2str(coef)])
pvals = polyval(coef, t);
```

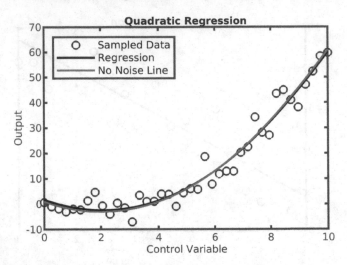

FIGURE 5.29: Least squares regression fit of quadratic data. The output of the script is `Equation estimate = 1.79 + -4.06*t + 0.99*t^2`, which is a fairly close match to the quadratic equation without the noise.

5.10.4 Goodness of a Fit

Statistics gives us a metric called the *coefficient of determination*, R^2, to measure how closely a regression fit models the data. Let y_i be the data values, p_i is the regression fit values, and μ is the mean of the data values.

$$R^2 = 1 - \frac{\sum_i (y_i - p_i)^2}{\sum_i (y_i - \mu)^2}$$

The range of R^2 is from 0 to 1. If the regression fit does not align well with the data, the R^2 value will be close to 0. If the model is an exact fit to the data, then R^2 will be 1. In reality, it will be somewhere in between, and hopefully closer to 1 than 0.

5.10.5 Generalized Least Squares Regression

The key to least squares regression success is to correctly model the data with an appropriate set of basis functions. When fitting the data to a polynomial, we use progressive powers of t as the basis functions.

$$f_0(t) = 1, \text{ a constant}$$
$$f_1(t) = t$$
$$f_2(t) = t^2$$
$$\vdots$$
$$f_n(t) = t^n$$

The regression algorithm then uses a simple linear algebra vector projection algorithm to compute the coefficients $(\alpha_1, \alpha_2, ... \alpha_n)$ to fit the data to the polynomial.

$$\begin{cases} \hat{y}(t_0) = \alpha_0\, f_0(t_0) + \alpha_1\; f_1(t_0) + ... + \alpha_n\; f_n(t_0) \\ \hat{y}(t_1) = \alpha_0\, f_0(t_1) + \alpha_1\; f_1(t_1) + ... + \alpha_n\; f_n(t_1) \\ \quad\vdots \\ \hat{y}(t_m) = \alpha_0\, f_0(t_m) + \alpha_1\, f_1(t_m) + ... + \alpha_n\, f_n(t_m) \end{cases}$$

The basis functions need not be polynomials, however. They can be any function that offers a reasonable model for the data. Consider a set of basis functions consisting of a constant offset and oscillating trigonometry functions.

$$f_0(t) = 1, \text{ a constant}$$
$$f_1(t) = sin(t)$$
$$f_2(t) = cos(t)$$

The design matrix for the regression is then,

$$\begin{bmatrix} 1 & sin(t_0) & cos(t_0) \\ 1 & sin(t_1) & cos(t_1) \\ 1 & sin(t_2) & cos(t_2) \\ & \vdots & \\ 1 & sin(t_m) & cos(t_m) \end{bmatrix}.$$

Code 5.14 on page 234 lists a function that implements generalized least squares regression. The basis functions are given as an input in the form of a cell array containing function handles.

Code 5.10 is a script to test the genregression function. A plot of the output is shown in figure 5.30.

A fun application of generalized least squares is to find the Fourier series coefficients of a periodic waveform. It turns out that the regression calculation can also find Fourier series coefficients when the basis functions are sine waves of the appropriate frequency and phase. The fourierRegress script listed in code 5.11 demonstrates this for a square wave. The plot of the data is shown in figure 5.31.

5.10.6 Fitting Exponential Data

If the data is exponential, use the logarithm function to convert the data to linear before doing regression data fitting.

- The data takes the form $y(x) = c\,e^{a\,x}$.

- Taking the logarithm of the data converts it to linear data, $Y = log(y(x)) = log(c) + a\,x$.

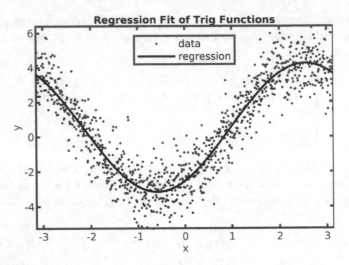

FIGURE 5.30: Least squares regression of trigonometric functions.

```
% File: trigRegress.m
f = {@(x) 1, @sin, @cos };
x = linspace(-pi, pi, 1000)';
y = 0.5 + 2*sin(x) - 3*cos(x) + randn(size(x));
[alpha, curve] = genregression(f, x, y);
disp(alpha)
figure, plot(x, y, 'b.')
hold on
plot(x, curve,'r', 'LineWidth', 2);
xlabel('x')
ylabel('y')
title('Regression Fit of Trig Functions')
legend('data', 'regression', 'Location', 'north')
hold off
axis tight
```

CODE 5.10: Least squares regression of trigonometric functions.

- Use linear regression to find $\hat{Y}$.

- Then $\hat{y}(x) = e^{\hat{Y}}$.

Code 5.12 shows an example of using a logarithm to facilitate least squares regression of exponential data. The plot of the output is shown in figure 5.32.

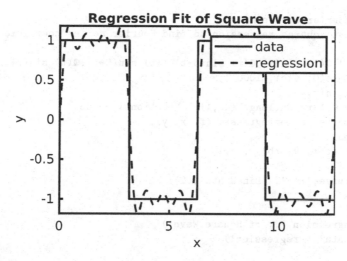

FIGURE 5.31: A few terms of the Fourier series of a square wave found by least squares regression.

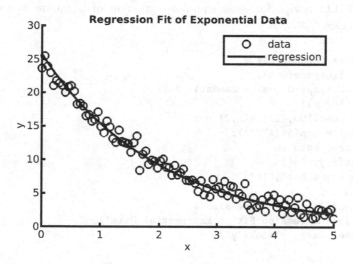

FIGURE 5.32: Least squares regression of exponential data.

5.11 Left-Divide Operator

In this chapter, we have focused primarily on solutions to systems of linear equations, which we express in terms of the $\mathbf{A}x = b$ problem. A constant reference point in the discussion has been the left-divide or backslash operator, \, which efficiently finds x for all but singular equations. The results and their

```
% File: fourierRegress.m
% Use least squares regression to find Fourier series coefficients.

f = {@(t)1,@(t) sin(t),@(t) sin(3*t),@(t) sin(5*t),@(t) sin(7*t)};
x = linspace(0, 4*pi, 1000)';
y = sign(sin(x));
% y = y + 0.1*randn(length(x),1);  % optional noise
[alpha, curve] = genregression(f, x, y);
disp(alpha)
figure, plot(x, y, 'b')
hold on
plot(x, curve,'r--', 'LineWidth', 2);
xlabel('x')
ylabel('y')
title('Regression Fit of Square Wave')
legend('data', 'regression')
hold off
axis tight
```

CODE 5.11: Script for least squares regression of a square wave by the Fourier series.

```
% File: expRegress.m
x = linspace(0,5);
y = 25*exp(-0.5*x) + randn(1, 100);
Y = log(y);
A = ones(100, 2); A(:,2) = x';
y_hat = exp(A*(A\Y'));
figure, hold on
plot(x,y, 'o')
plot(x, y_hat', 'r')
hold off
legend('data', 'regression')
title('Regression Fit of Exponential Data')
xlabel('x'), ylabel('y')
```

CODE 5.12: Least squares regression of exponential data.

application are more important to us than the details of MATLAB's algorithms, but it is illuminating to briefly discuss the algorithms in light of what we have learned about systems of equations.

The function invoked by the left-divide operator is called `mldivide`. The documentation does not tell us every secret behind the implementation of the algorithms, but it does tell us the basic strategies. Please take a look at the MATLAB documentation of `mldivide`. Near the bottom of the page is a flow chart showing how the **A** matrix is evaluated in terms of its shape and special properties to determine the most efficient algorithm. Note that `mldivide` often acts as a recursive function. It will use factorization to break a problem down to simpler problems and then call itself to solve the simpler problems. In particular, lower and upper triangular matrices are sought so that in turn the efficient triangular solver can be used to find the solution by forward and back substitution.

Right Division

As noted in section 5.2.1.4, MATLAB also has a right-divide operator (`/`). It provides the solution to $x\mathbf{A} = b$ as `x = b/A;`. We have seldom mentioned it for two reasons. First, the need for the right-divide operator is much less than that of the left-divide operator, especially in regard to systems of equations. Secondly, right division is implemented using the left-divide operator with the relationship `b/A = (A'\b')'`. So the algorithms reviewed here for the left-divide operator also apply to the right-divide operator.

5.11.1 Left-Divide of Critically-determined Systems

As we previously discussed in section 5.6, when the matrix **A** is square, the primary algorithm is LU decomposition with row exchanges (partial pivoting). Other algorithms listed for square matrices take advantage of special structure features of the matrix to use more efficient matrix factorizations. The algorithms of the Hessenberg, Cholesky, and LDL^T factorization solvers are not discussed in this text, but we briefly describe their application. Complete descriptions of the algorithms are given in [28].

A square matrix is first checked to see if it is triangular or can be multiplied by a permutation matrix to become triangular. Triangular systems are solved with forward and back substitution.

Next it is checked to see if it is *Hermitian*, which for a real matrix just means that it is symmetric. A complex square matrix is Hermitian if it is equal to its own complex conjugate transpose. In MATLAB, the ' operator is a complex conjugate transpose, so the expression `isequal(A, A')` tests if **A** is Hermitian. If **A** is Hermitian and positive definite then the symmetry allows quick factoring into the *Cholesky* factorization $\mathbf{A} = \mathbf{R}^T \mathbf{R}$, where **R** is upper triangular. If the Cholesky factorization can not be used, then the LDL^T factorization is used. The factoring is $\mathbf{A} = \mathbf{L}\,\mathbf{D}\,\mathbf{T}^T$, where the **L** matrix is lower triangular, **D** is diagonal and $\mathbf{L}^T$ is upper triangular.

Note: MATLAB checks if the diagonal values of a Hermitian matrix are all positive or all negative, which is a necessary but not sufficient requirement for being positive definite. If the first test passes, then a second test is required to see if the Cholesky factorization may be used. A Hermitian matrix is positive definite if all of its eigenvalues are positive.

Here is an example of the triangular Cholesky and LDL^T factorization of a positive definite Hermitian matrix.

```
>> A                        >> [L, D] = ldl(A)
A =                         L =
    2    -2                     1    0
   -2     5                    -1    1
>> R = chol(A)              D =
R =                             2    0
    1.4142   -1.4142            0    3
         0    1.7321        >> L*D*L'
>> R'*R                     ans =
ans =                          2   -2
    2.0000   -2.0000           -2    5
   -2.0000    5.0000
```

A square matrix that is not triangular or Hermitian is solved with either LU decomposition as presented in section 5.6 or with a variation of the LU decomposition algorithm that takes advantage of the structure of upper Hessenberg matrices. An upper Hessenberg matrix is one diagonal off from being upper triangular. It contains zeros below the first subdiagonal (just below the main diagonal). It may seem obscure to have a special solver for Hessenberg matrices, but they occur in many problems. Upper Hessenberg matrices also include the class of tridiagonal matrices that only contain nonzero elements on the main diagonal and the diagonals immediately above and below the main diagonal. Heat conduction and fluid flow problems often result in tridiagonal matrices [9].

$$\begin{bmatrix} a & b & c & d \\ e & f & g & h \\ 0 & i & j & k \\ 0 & 0 & l & m \end{bmatrix} \qquad \begin{bmatrix} a & b & 0 & 0 \\ c & d & e & 0 \\ 0 & f & g & h \\ 0 & 0 & i & j \end{bmatrix}$$

Upper Hessenberg   Tridiagonal

5.11.2   Left-Divide of Over-determined Systems

As discussed in section 5.9, the solution to over-determined systems is based on vector projections. It is an exact solution in the rare event that vector b is in the columns space of $\mathbf{A}$, but it is usually an approximation.

We have shown how the least squares solution is found from the vector projection equation $x = \left(A^T A\right)^{-1} A^T b$ or from the pseudo-inverse ($x = A^+ b$), which MATLAB finds from the economy SVD. However, the mldivide function uses QR factorization to find the least squares solution. The QR algorithm is described in appendix A.5. It is a matrix factoring returning an orthogonal matrix, Q, and a upper triangular matrix, R, such that $A = QR$. It is an efficient and accurate algorithm. Using the result to find the solution to our system of equations is also quick because Q is orthogonal and R is upper triangular. So finding the solution only requires a matrix transpose, a matrix–vector multiplication, and back substitution with the triangular solver.

$$A = QR$$
$$Ax = QRx = b$$
$$x = R^{-1} Q^T b$$

Note that for an over-determined matrix, the R matrix will end with rows of all zeros, so to remove those rows we use the *economy QR* by adding an extra zero argument to qr. Here is a quick example:

```
>> A = randi(20, 5, 2) - 10;
>> b = randi(20, 5, 1) - 10;
>> x = A\b            % The left-divide solution
x =
    0.6383
    0.1901

>> [Q, R] = qr(A, 0);     % Find the solution by QR
>> x = R\(Q'*b)
x =
    0.6383
    0.1901
```

5.11.3 Left-Divide of Under-determined Systems

As discussed in section 5.8, the solution to under-determined systems is an infinite set rather than a unique solution. So any solution satisfying the equation $Ax = b$ is correct, but some solutions are better than others. The minimum length correct solution is desired. The documentation for mldivide states that the returned solution has at most r (the rank of A) nonzero values. To produce at least $n-r$ zeros, it replaces $n-r$ columns of A with zeros before using the QR factorization. The columns to zero out are found sequentially. The first column zeroed out is the column that when replaced with zeros results in the smallest the l_2–norm of the solution. The evaluation is repeated until $n-r$ columns are zeroed out [46].

By using A^T rather than A, we can find a least squares solution that minimizes the l_2–norm of x using the economy QR factorization as done with over-determined systems. We will call the factors of $\mathbf{A}^T$ $\mathbf{Q}_t$ and $\mathbf{R}_t$. The orthogonality of $\mathbf{Q}_t$ and triangular structure of $\mathbf{R}_t$ again yield an efficient solution algorithm.

$$\mathbf{A}^T = \mathbf{Q}_t\,\mathbf{R}_t$$

$$\mathbf{A} = \left(\mathbf{A}^T\right)^T = \left(\mathbf{Q}_t\,\mathbf{R}_t\right)^T = \mathbf{R}_t^T\,\mathbf{Q}_t^T$$

$$\mathbf{A}\,x = \mathbf{R}_t^T\,\mathbf{Q}_t^T\,x$$

$$x = \mathbf{Q}_t\,\left(\mathbf{R}_t^T\right)^{-1}\,b$$

In the following example, the solution using the left-divide operator is found first to see which columns should be replaced with zeros.

```
>> A
A =
     2      4     20      3     16
     5     17      2     20     17
    19     11      9      1     18
>> b = [2; 20; 13];

>> y = A\b
y =
    0.5508
    0
    0
    0.7794
    0.0975
```

To get the same result as the left-divide operator for this example, we replace the second and third columns with zeros.

```
>> A(:,2) = zeros(3,1);
>> A(:,3) = zeros(3,1)
A =
     2      0      0      3     16
     5      0      0     20     17
    19      0      0      1     18

>> [Qt, Rt] = qr(A', 0);
>> x = Qt*(Rt'\b)
x =
    0.5508
    0
    0
    0.7794
    0.0975
```

```
% File: quadRegress.m
%% Quadratic Least Squares Regression Example
%
%  Generate data per a quadratic equation, add noise to it, use
%  least squares regression to find the equation.

%% Generate data
f = @(t) 1 - 4*t + t.^2;    % Quadratic function as a handle

Max = 10;
N = 40;         % number of samples
std_dev = 5;    % standard deviation of the noise

t = linspace(0, Max, N);
y = f(t) + std_dev*randn(1,N); % randn -> mean = 0, std = 1
%% Least Squares Fit to Quadratic Equation
%  model data as: b = C + Dt + Et^2
%  Projection is A*u_hat, where u_hat = [C D E]'

%  Determine A matrix from t.
A = ones(N,3);     % pre-allocate A - Nx3 matrix
% First column of A is already all ones for offset term.
A(:,2) = t';       % linear term, second column
A(:,3) = t'.^2;    % Quadratic term, third column

% b vector comes from sampled data
b = y';

u_hat = (A'*A)\(A'*b);  % least squares equation
p = A*u_hat;            % projection vector
% e = B - p;            % error vector
C = u_hat(1); D = u_hat(2); E = u_hat(3);
fprintf('Equation estimate = %.2f + %.2f*t + %.2f*t^2\n',C,D,E);

%% Plot the results
figure, plot(t, y, 'ob');
hold on
plot(t, p, 'k')
plot(t, f(t), 'm')
labels = {'Sampled Data', 'Regression', 'No Noise Line'};
legend(labels, 'Location', 'northwest');
hold off
title('Quadratic Regression');
xlabel('Control Variable');
ylabel('Output');
```

CODE 5.13: Quadratic least squares regression example.

```
function [ alpha, varargout ] = genregression( f, x, y )
% GENREGRESSION Regression fit of data to specified functions
%   Inputs:
%           f - cell array of basis function handles
%           x - x values of data
%           y - y values of data
%   Output:
%           alpha - Coefficients of the regression fit
%           yHat  - Curve fit data (optional)

    % A will be m-by-n
    % alpha and b will be n-by-1
    if length(x) ~= length(y)
        error('Input vectors not the same size');
    end
    m = length(x);
    n = length(f);
    A = ones(m,n);
    if size(y, 2) == 1 % a column vector
        b = y;
        cv = true;
    else
        b = y';
        cv = false;
    end
    if size(x, 2) == 1 % column vector
        c = x;
    else
        c = x';
    end
    for k = 1:n
        A(:,k) = f{k}(c);
    end
    alpha = (A'*A) \ (A'*b);
    if nargout > 1
        yHat = A * alpha;
        if ~cv
            yHat = yHat';
        end
        varargout{1} = yHat;
    end
end
```

CODE 5.14: Function implementing generalized least squares regression.

5.12 Exercises

Exercise 5.1 Vector Calculations

Use MATLAB to answer the following questions.

1. Consider the following two vectors in $\mathbb{R}^2$ that begin at the origin of the axes, $(0, 0)$.

$$u = \begin{bmatrix} 4 \\ 1 \end{bmatrix} \quad v = \begin{bmatrix} 2 \\ 3 \end{bmatrix}$$

   (a) What is the length of each vector?

   (b) What is the dot product between the vectors?

   (c) What is the angle between the vectors in radians and degrees?

   (d) Find a point, p, on vector v such that a line from point $q = (8, 2)$ to p is perpendicular to v.

   (e) Verify that $(p - q)$ is perpendicular to v.

2. Consider the following two vectors in $\mathbb{R}^3$ that begin at the origin of the axes, $(0, 0, 0)$.

$$u = \begin{bmatrix} 3 \\ 1 \\ 1 \end{bmatrix} \quad v = \begin{bmatrix} 2 \\ 1 \\ 3 \end{bmatrix}$$

   (a) What is the length of each vector?

   (b) What is the dot product between the vectors?

   (c) What is the outer product of the vectors?

   (d) What are the coordinates of the vector $w = 2\,u + 3\,v$?

   (e) What are the coordinates of $\hat{w}$, which is w as a unit vector?

Exercise 5.2 Matrix Calculations

Use MATLAB to answer the following questions.

1. Perform matrix multiplication.

   (a) $\begin{bmatrix} 9 & 10 & 3 \\ 10 & 7 & 6 \\ 2 & 1 & 10 \end{bmatrix} \begin{bmatrix} 0 & 5 & 1 \\ 1 & -2 & 3 \\ 5 & 5 & 0 \end{bmatrix}$

   (b) $\begin{bmatrix} 3 & 8 \\ 4 & 2 \end{bmatrix} \begin{bmatrix} -2 & 1 \\ 7 & 5 \end{bmatrix}$

$$(c) \begin{bmatrix} 9 & 10 & 3 \\ 10 & 7 & 6 \\ 2 & 1 & 10 \end{bmatrix} \begin{bmatrix} 5 \\ 2 \\ 1 \end{bmatrix}$$

2. Consider the matrix:

$$\mathbf{B} = \begin{bmatrix} -1 & -4 & -1 \\ 0 & 0 & -1 \\ -1 & 0 & -1 \end{bmatrix}$$

   (a) What is its transpose?

   (b) What is the determinant of $\mathbf{B}$?

   (c) What is the rank of $\mathbf{B}$?

   (d) Is the matrix symmetric?

   (e) Is the matrix singular or full rank? If full rank, what is its inverse?

3. Consider the matrix:

$$\mathbf{A} = \begin{bmatrix} 6 & -2 & 8 & 14 \\ -11 & -11 & 0 & -11 \\ -3 & -25 & 22 & 19 \end{bmatrix}$$

   (a) What is the rank of $\mathbf{A}$?

   (b) Use the CR factorization function to find which columns are independent and the relationship of the dependent columns to the independent columns.

4. Consider the matrix:

$$\mathbf{A} = \begin{bmatrix} 0.1231 & -0.4765 & 0.8705 \\ 0.9847 & -0.0502 & -0.1667 \\ 0.1231 & 0.8777 & 0.4630 \end{bmatrix}$$

   (a) Use the `vecnorm` function to find the length of each column.

   (b) Is $\mathbf{A}$ an orthogonal matrix?

   (c) What is its inverse?

Exercise 5.3   A Change of Coordinate Frames

Robotic programming environments often provide multiple coordinate frames as a convenience to the programmer. They provide a world coordinate frame where the (X, Y, Z) positions are relative to the base of the robot. A work-piece coordinate frame can also be configured, which allows a coordinate frame to be established with points relative to the piece being worked on. In this problem, we will consider a sheet of metal at a $30°$ angle along the x axis (rotated about the y axis). Another rectangular piece of metal is to be welded perpendicular to the work-piece at a fixed location and then another

piece welded to its top. Our robot needs to make two weld lines. To help
the robot weld straight lines, we need to find a set of points along the weld
lines that are 1/8 inch apart. The location of the weld lines in the work-piece
coordinate frame are shown in the code in the `change_coords.m` file.

We need to transform the points from work-piece coordinates to world
coordinates. To do this, we add the base of the work-piece coordinate frame to
a transformation of the points. A set of points parallel to the work-piece along
the x axis (fixed Z coordinate) are at a 30° angle in the world coordinate frame.
We can find the transformed points by setting the work-piece coordinates in a
3×1 column vector and multiplying by a 3×3 transform matrix. The diagram
below shows a side view of the work-piece and a vertical line perpendicular
to the work-piece. Gray lines make right triangles that will help us determine
the transformation matrix based on the trigonometry of the right triangles.

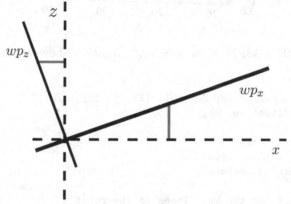

If the transformation matrix is correct, MATLAB will draw the work-piece
surface as a plane and lines where the welds are to be made. An example of
the plot is shown in figure 5.33.

Most all of the code is provided for you. Just fill in the values of the
transformation matrix.

```
% File: change_coords.m
%      Demonstration of changing the coordinate frame between a robotic
% work-piece coordinate frame and the world coordinate frame. A
% work-piece surface is defined as starting at [10; -20; 30] in the
% world coordinate frame and tilting up from there at 30 degrees along
% the x-axis.
%      The robot needs to make two 30-inch welds from points [5; 30] to
% [23; 6] on the work-piece surface and 10 inches above and parallel
% to the work-piece.  To make straight line welds, the robot needs
% a set of points 1/8 apart in the world coordinate frame.

% Parallel 30 inch lines 1/8 inch apart in work-piece coordinates,
% one on the work-piece surface and one 10 inches above.
line1_wp = linepoints([5; 30; 0], [23; 6; 0], 1/8);
```

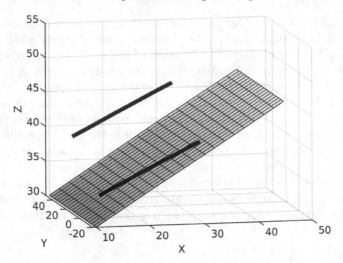

FIGURE 5.33: Weld lines on and 10 inches above a work-piece.

```
line2_wp = linepoints([5; 30; 10], [23; 6; 10], 1/8);
points = size(line1_wp, 2);

% pre-allocate storage
line1_rw = zeros(3, points);
line2_rw = zeros(3, points);

% Map points of the two weld lines to the robot's
% world coordinate frame.
Base_wp = [10; -20; 30];  % world coordinates of wp base
c30 = cosd(30);           % constants
s30 = sind(30);

%% YOU DO THIS: Fill in the 3x3 transformation matrix
% Consider how the world coordinates change with these three
% work-piece coordinates: [1 0 0]', [0 1 0]', [0 0 1]'

T = [ ]; % Orthogonal transform

%%
for p = 1:points
    line1_rw(:,p) = Base_wp + T*line1_wp(:,p);
    line2_rw(:,p) = Base_wp + T*line2_wp(:,p);
end

% points for plotting the work-piece surface
[X, Y] = meshgrid(linspace(0, 40*c30, 20), linspace(-20, 50, 20));
Z = 30*ones(20) + X*s30;
X = 10*ones(20) + X;
```

```
% points for plotting the two weld lines
x1 = line1_rw(1,:);
y1 = line1_rw(2,:);
z1 = line1_rw(3,:);
x2 = line2_rw(1,:);
y2 = line2_rw(2,:);
z2 = line2_rw(3,:);

% Make a surface plot of the work-piece and plot the two weld lines.
figure
hold on
surf(X, Y, Z, 'FaceAlpha',0.5)
plot3(x1, y1, z1, x2, y2, z2, 'LineWidth', 3)
hold off
grid on
xlabel('X');
ylabel('Y');
zlabel('Z');

function wp = linepoints(start, finish, space)
% LINEPOINTS -- Generate a set of points of a line between
% start and finish points at a specified spacing.
    v = finish - start;
    len = norm(v);
    p = 0:space:len;
    step = v/len;
    wp = start + p.*step;
end
```

Exercise 5.4   Square Matrix Systems of Equations

1. Consider the system of linear equations:

$$\begin{cases} 8\,x_1 + 9\,x_2 \qquad\qquad = \quad 5 \\ 9\,x_1 + 5\,x_2 + \;\; 4\,x_3 = -3 \\ -3\,x_1 - 3\,x_2 + 10\,x_3 = \quad 5 \end{cases}$$

   (a) Make matrices for **A** and b in MATLAB such that the system of equations can be expressed as a matrix equation in the form $\mathbf{A}x = b$.

   (b) Is matrix **A** full rank? Based on the rank, does this matrix equation have a solution for $x = (x_1, x_2, x_3)$?

   (c) If a solution exists, use MATLAB's left-divide operator to find it.

2. Consider the system of linear equations:

$$\begin{cases} -8\,x_1 + \;\;7\,x_2 - 7\,x_3 = \;\;10 \\ 9\,x_1 - 10\,x_2 + 5\,x_3 = \;-9 \\ 5\,x_1 + 10\,x_2 - \;\;\;\, x_3 = \quad 6 \end{cases}$$

Use Gaussian elimination with an augmented matrix to find the solution for $x = (x_1, x_2, x_3)$. Verify your answer with MATLAB using the left-divide operator, `rref`, and LU decomposition.

3. If you have an A matrix of size 10×10 and 100 b vectors each of size 10×1, which of the following two strategies is faster for finding the 100 solutions to $Ax = b$?

    1. Find the inverse of A once and then use matrix multiplication to find each solution.

    2. Find the L, U, and P decomposition of A once with `lu` and then use the left-divide operator to find each solution.

Write a MATLAB script to find your answer. Use the `tic` and `toc` functions to measure the execution times. Use random number generators to create the A matrix and b vectors.

Exercise 5.5   Under-determined Systems

1. Use the `rref` function on an augmented matrix for the following under-determined matrix equation and express the set of solutions for $x = (x_1, x_2, x_3, x_4)$ in parametric form.

$$\begin{bmatrix} 6 & -6 & -1 & -1 \\ -2 & 1 & -2 & 8 \\ -5 & 13 & 3 & -1 \end{bmatrix} \begin{bmatrix} x_1 \\ x_2 \\ x_3 \\ x_4 \end{bmatrix} = \begin{bmatrix} 6 \\ 3 \\ -2 \end{bmatrix}$$

2. Use the Moore-Penrose pseudo-inverse, the left-divide operator, and the `lsqminmorm` function to find solutions to the following under-determined matrix equation.

$$\begin{bmatrix} 4 & 16 & 8 & 4 \\ 8 & 14 & -2 & -4 \\ 4 & 5 & -2 & -10 \end{bmatrix} \begin{bmatrix} x_1 \\ x_2 \\ x_3 \\ x_4 \end{bmatrix} = \begin{bmatrix} 1 \\ -7 \\ 4 \end{bmatrix}$$

Exercise 5.6   Over-determined Systems

1. Consider the following two vectors in $\mathbb{R}^3$ that begin at the origin of the axes, $(0, 0, 0)$.

$$u = \begin{bmatrix} 1 \\ 1 \\ 0.5 \end{bmatrix} \qquad v = \begin{bmatrix} 2 \\ 1 \\ 2 \end{bmatrix}$$

    (a) Use the `orth` function to find a matrix, Q, whose columns form orthogonal basis vectors for the vector space (plane) spanned by the two vectors.

(b) Verify that **Q** is a unitary matrix.

(c) Use the cross function to find a perpendicular vector to the vector space.

(d) Make a surface plot of the plane spanned by the two vectors.

(e) Which of the following points are in the vector space: $(4,4,3)$, $(6,4,5)$, $(6,4,4)$, $(2.5,1.5,2.25)$, $(2,1,3)$. *Hint:* Use the rank function.

2. Consider the matrix equation of the form $\mathbf{A}x = b$, where:

$$\mathbf{A} = \begin{bmatrix} 11 & 3 \\ 14 & 3 \\ 18 & 6 \\ 20 & 17 \\ 11 & 6 \end{bmatrix}, \qquad b = \begin{bmatrix} 6 \\ 13 \\ 10 \\ 8 \\ 17 \end{bmatrix}.$$

(a) Use the vector projection equation to find x.

(b) Using the rank function determine if your solution to x is an exact solution or an approximation.

(c) Is vector b in the column space of $\mathbf{A}$?

Exercise 5.7   Projection of Static Force Vectors

Consider figure 5.34. Using projection vectors, calculate the tension forces in newtons on cables **A**, **B**, and **C** such that cable **A** is completely horizontal. Cable **B** is at a 45° angle.

Exercise 5.8   Old Faithful Wait Prediction

Old Faithful is a cone geyser located in Yellowstone National Park in Wyoming, United States. It was named in 1870 during the Washburn-Langford-Doane Expedition and was the first geyser in the park to receive a name. It is a highly predictable geothermal feature, which makes it a favorite to visitors. In 1939, the average wait time between eruptions was 66.5 minutes, but wait times have slowly increased to an average of 90 minutes today. Since 2000, the wait times between eruptions varied from 44 to 120 minutes. The duration of the last eruption is known to be a general indicator of the wait until the next eruption. A longer eruption predicts a longer wait until the next eruption starts [63].

In 1985, measurements were taken of the eruption durations and wait times until the next eruption for 272 eruptions [5]. This public domain data is on several Internet sites and is included as file faithful.txt in the downloadable resources from the book's website.

Use linear regression to fit a prediction line to the data. Then calculate the *coefficient of determination* to assess the accuracy of the prediction line (section 5.10.4).

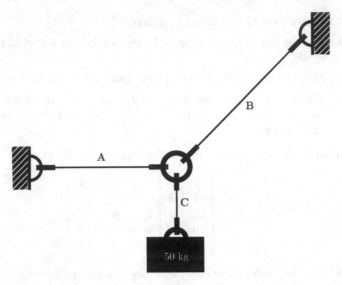

FIGURE 5.34: Static force projection.

Exercise 5.9   Left Divide Solver

(a) Write a MATLAB function to solve systems of equations. For square matrix systems, use LU decomposition. For rectangular systems, use the QR factorization equations from sections 5.11.2 and 5.11.3. Use MATLAB's left divide operator to solve triangular systems.

(b) Write MATLAB functions to solve lower and upper triangular systems of equations with back-substitution and forward-subsitution. Change your function your the previous problem to use your triangular solvers.

Chapter 6

Application of Eigenvalues and Eigenvectors

Eigenvalue / Eigenvector problems are one of the more important linear algebra topics. Eigenvalues and eigenvectors are used to solve systems of differential equations, but more generally they are used for data analysis, where the matrix $\mathbf{A}$ represents data rather than coefficients of a system of equations. They introduce a simple, yet very powerful relationship between a matrix and a set of special vectors and scalar values. This simple relationship provides elegant solutions to some otherwise difficult problems.

Eigenvalues and eigenvectors are used to find the singular value decomposition (SVD) matrix factoring, which is used by MATLAB in many linear algebra functions that we used in chapter 5 to solve systems of equations.

6.1 Introduction to Eigenvalues and Eigenvectors

To a vector, matrix multiplication is a lot of stretching and rotation. When we multiply a matrix and a vector together $(\mathbf{A}\boldsymbol{x})$ the result is usually a stretching and a rotation of the original vector $\boldsymbol{x}$. However, that is not always the case. There are special matrices that rotate a vector, but do not change its length. (See section 5.3.1 for more on rotation matrices.) There are also vectors corresponding to square matrices that are stretched by the multiplication, but not rotated. This later case is what we consider here.

Consider the following matrix and three vectors.

$$\mathbf{A} = \begin{bmatrix} 6 & 1 \\ 8 & 4 \end{bmatrix}, \quad \boldsymbol{x}_1 = \begin{bmatrix} 1 \\ 2 \end{bmatrix}, \quad \boldsymbol{x}_2 = \begin{bmatrix} 1 \\ -4 \end{bmatrix}, \quad \boldsymbol{x}_3 = \begin{bmatrix} 1 \\ -1 \end{bmatrix}$$

$$\mathbf{A}\boldsymbol{x}_1 = \begin{bmatrix} 8 \\ 16 \end{bmatrix} = 8\,\boldsymbol{x}_1, \quad \mathbf{A}\boldsymbol{x}_2 = \begin{bmatrix} 2 \\ -8 \end{bmatrix} = 2\,\boldsymbol{x}_2, \quad \mathbf{A}\boldsymbol{x}_3 = \begin{bmatrix} 5 \\ 4 \end{bmatrix}$$

Figure 6.1 shows the three vectors and the product of each vector with matrix $\mathbf{A}$. The product $\mathbf{A}\boldsymbol{x}_3$ shows the typical result of matrix to vector multiplication. The multiplication both stretches and rotates the vector. But in the cases of $\mathbf{A}\boldsymbol{x}_1$ and $\mathbf{A}\boldsymbol{x}_2$, the vectors are stretched, but not rotated. This is

DOI: 10.1201/9781003271437-6

The best kept secret of engineering

Whether you know it or not, you use eigenvalues and eigenvectors everyday. Furthermore, knowing about them can greatly expand your ability to solve interesting and challenging problems. Engineers encounter eigenvalues and eigenvectors when studying mechanics, vibrations, or when working with big data. Google's Internet search engine uses eigenvectors to rank web pages, and Netflix uses eigenvectors to predict your preference for a movie that you have not yet watched.

We use eigenvalues and eigenvectors to better understand and simplify expressions involving a matrix. Of course we do not say that a matrix is equivalent to a single scalar value, but when multiplying a matrix by one of its eigenvectors, then the matrix may be replaced by a scalar (its eigenvalue). **What an incredibly fantastic simplification!** Many problems allow us to represent a matrix in terms of its eigenvalues and eigenvectors. Making this replacement can reduce the computational complexity of a problem and reveal valuable insights about the matrix and the problems we are attempting to solve.

Admittedly, you may need to think about this for a while and see some application examples before appreciating the full value of eigenvalues and eigenvectors. The important thing to remember is that eigenvalues and eigenvectors reveal and take advantage of important properties of matrices. Using a computer to find the eigenvalues and eigenvectors makes it easy to use and apply them to various problems.

---

because vectors x_1 and x_2 are eigenvectors of matrix $\mathbf{A}$. The eigenvalues tell us the stretching (scaling) factor for each eigenvector.

When a vector x_i is one of the eigenvectors of a matrix $\mathbf{A}$, the following special relationship holds.

$$\boxed{\mathbf{A}x_i = \lambda_i\,x_i}$$

For the $n \times n$ square matrix, there are n *eigenvectors*, x_i, and n corresponding scalar *eigenvalues*, λ_i. Each pair of a coupled eigenvalue and eigenvector can be called an *eigenpair*.

The power of this relationship is that when, and only when, we multiply a matrix by one of its eigenvectors, then in a simplified equation we can *replace the matrix with just a simple scalar value*.

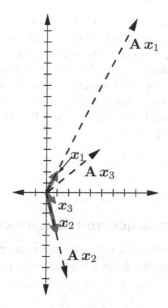

FIGURE 6.1: Eigenvectors are only scaled by matrix multiplication. Other vectors are both scaled and rotated by multiplication.

What is in the name?

In the German language, the word *eigen* means *own*, as in "my own opinion" or "my own family". It is also used for the English word *typical*, as in "that is typical of him". Perhaps the German word *eigenschaft* meaning feature, property, or characteristic is more clear as to the intent. For a matrix, the eigenvectors and eigenvalues represent the vector directions and magnitudes distinctly embodied by the matrix.

The first known writing about eigenvectors and eigenvalues is by Leonhard Euler in the 18th century. Various mathematicians studied and gave them different names in the next two centuries. German mathematician David Hilbert (1862 - 1943) is credited with naming them eigenvalues and eigenvectors.

6.2 Eigenvector Animation

Eigenvectors and eigenvalues can be difficult to understand at first, so the file `eigdemo.m` in code listings 6.12 and 6.13 on pages 302 and 303 shows an animation to help readers visualize what makes a vector an eigenvector. The

file is part of the downloadable code from the book's website. The code defines
a function that plots an animation when a 2×2 matrix is passed to it. The
matrix must have real eigenvectors and eigenvalues, which is always achieved
when the matrix is symmetric (equal to its own transpose). The animation
shows the matrix product of the matrix eigenvalues and eigenvectors $\lambda_i \, x_i$, a
rotating unit vector, and the rotating product of the matrix and vector. When
the unit vector is inline with an eigenvector, then the product of the matrix
and vector overlays the eigenvector.

6.3   Finding Eigenvalues and Eigenvectors

We will begin with the equation for eigenvectors and eigenvalues and insert
an identity matrix so that we have a matrix multiplied by a vector on both
sides of the equality.

$$\mathbf{A}x = \lambda \, x$$
$$\mathbf{A}x = \lambda \mathbf{I} x$$

Then we rearrange the equation to find what is called the characteristic eigenvalue equation.

$$\mathbf{A}x - \lambda \mathbf{I} x = 0$$

$$\boxed{(\mathbf{A} - \lambda \mathbf{I}) \, x = 0}$$

The case where $x = 0$ is a trivial solution that is not of interest to us. Eigenvectors are defined to be nonzero vectors. The solution only exists when the
columns of matrix $\mathbf{A} - \lambda \, \mathbf{I}$ form a linear combination with x yielding the zero
vector. This linear dependence of the columns of the characteristic equation
means that it is singular—having a zero determinant.

6.3.1   Finding Eigenvalues

The n scalar eigenvalues, $\{\lambda_1, \lambda_2, \ldots, \lambda_n\}$, can be viewed as the shift of
the matrix's main diagonal that will make the characteristic matrix singular.
Eigenvalues can be found by subtracting λ along the main diagonal and finding
the set of λ for which the determinant is zero.

$$det(\mathbf{A} - \lambda \, \mathbf{I}) = 0$$

$$\begin{vmatrix} a_{11} - \lambda & a_{12} & \cdots & a_{1n} \\ a_{21} & a_{22} - \lambda & \cdots & a_{2n} \\ \vdots & \vdots & & \vdots \\ a_{n1} & a_{n2} & \cdots & a_{nn} - \lambda \end{vmatrix} = 0$$

Why singular requirement?

You may be wondering why it is required that $(\mathbf{A} - \lambda\,\mathbf{I})$ be a singular matrix. Let us call this matrix $\mathbf{C}$, and the columns of $\mathbf{C}$ will be labeled as c_i. For $\mathbf{C}x = \mathbf{0}$, it must be that either:

1. $c_1 = c_2 = \ldots = c_n = \mathbf{0}$, and hence $\mathbf{C} = \mathbf{0}$, which is not the case we are looking for.

2. $x = \mathbf{0}$, which is also not what we are looking for.

3. The columns of $\mathbf{C}$ are linearly dependent and a null solution, x, exists such that $\mathbf{C}x = \mathbf{0}$. It is the linear dependent property of the columns of $\mathbf{C}$ that makes $\mathbf{C}$ a singular matrix. Recall from the discussion about the column view of matrix multiplication in section 5.4 that a matrix must be singular to have a null solution.

Note: We will use the determinant here on small matrices because it keeps things simple. But as noted in section 5.2.4, calculating determinants is computationally slow. So it is not used for large matrices. MATLAB uses an iterative algorithm based on the QR factorization. The QR algorithm for finding eigenvalues is described in appendix A.7.

The determinant yields a degree-n polynomial, which can be factored to find the eigenvalue roots.

$$\mathbf{A} = \begin{bmatrix} 2 & 2 \\ 2 & -1 \end{bmatrix}$$

$$\begin{vmatrix} (2 - \lambda) & 2 \\ 2 & (-1 - \lambda) \end{vmatrix} = 0$$

$$(2 - \lambda)(-1 - \lambda) - 4 = 0$$
$$\lambda^2 - \lambda - 6 = 0$$
$$(\lambda - 3)(\lambda + 2) = 0$$
$$\lambda_1 = -2, \qquad \lambda_2 = 3$$

For the generalized 2×2 matrix, the coefficient of the λ term in the quadratic equation is the negative of the sum of the matrix diagonal (the trace), while the constant term is the determinant of the matrix.

$$\begin{vmatrix} (a - \lambda) & b \\ c & (d - \lambda) \end{vmatrix} = 0$$

$$(a - \lambda)(d - \lambda) - bc = 0$$
$$\lambda^2 - (a + d)\lambda + (ad - bc) = 0$$

This result for 2×2 matrices is further simplified by the quadratic equation. We will define variables m as the mean of the diagonal elements and p as the determinant of the matrix. Then we have a simple equation for the two eigenvalues.

$$m = \frac{a+d}{2}$$
$$p = ad - bc$$

$$\boxed{\lambda_1, \lambda_2 = m \pm \sqrt{m^2 - p}}$$

Here is a quick example, which is verified with MATLAB's `eig` function.

```
>> A
A =
      -6    -1
      -8    -4
>> eig(A)
ans =
      -8
      -2
>> m = -5;
>> p = 24 - 8;        % p = 16
>> l1 = m - sqrt(m^2 - p) % l1 = -5 - 3
l1 =
      -8
>> l2 = m + sqrt(25 - 16)
l2 =
      -2
```

Here are four properties of eigenvalues.

- The *trace* of **A** is equal to the sum of the eigenvalues. The trace of a matrix is the sum of the values along the forward diagonal. For a 2×2 matrix: $a_{1,1} + a_{2,2} = \lambda_1 + \lambda_2$.

- The determinant of **A** is equal to the product of all of the eigenvalues.

- Eigenvalues, and hence eigenvectors, often have complex numbers. In some cases, algorithms will force real eigenvalues by using symmetric matrices, which have only real eigenvalues. In some applications, when taking products and sums of eigenvalues and eigenvectors the imaginary parts will cancel leaving only real numbers. In other cases, the presence of complex eigenvalues implies oscillation in a system (appendix B.1.2).

- Eigenvalues are not always unique—the same number may be repeated in the set of eigenvalues. Such cases are called *degenerate* because the matrix does not have linearly independent eigenvectors, and thus can not be factored using diagonalization, which is described in section 6.5.1.

6.3.2   Roots of a Polynomial by Eigenvalues

As we saw above, finding the eigenvalues of a matrix is equivalent to finding the roots of the determinant of the characteristic equation. But as noted, the algorithm that MATLAB uses to find eigenvalues neither calculates a determinate nor finds the roots of a polynomial. Instead it uses a faster, iterative algorithm based on QR factorization that is described in appendix A.7.

Rather than finding polynomial roots to calculate eigenvalues, finding the roots of a polynomial is instead an application of the eigenvalue algorithm. This is how the MATLAB `roots` function finds the roots of a polynomial. To take advantage of the eigenvalue algorithm, a matrix is cleverly found that has eigenvalues equivalent to the roots of the polynomial. Figure 6.2 shows the algorithmic relationship between finding the roots of a polynomial and finding the eigenvalues of a matrix.

An example should illustrate how this works. Consider the following polynomial equation.

$$f(x) = x^3 - 4x^2 + x + 6$$

The argument passed to the `roots` function is a row vector containing the coefficients of the polynomial.

```
>> r = roots([1 -4 1 6])
r =
    3.0000
    2.0000
   -1.0000
```

The `poly` function is the inverse of `roots`:

```
>> poly(r)
ans =
    1.0000   -4.0000    1.0000    6.0000
```

The algorithm used by the `roots` function is short and quite clever.

```
>> n = 3;             % degree of the polynomial
>> p = [1 -4 1 6];    % coefficients
```

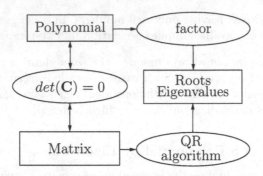

FIGURE 6.2: We first found the eigenvalues of a 2×2 matrix using the determinant of the characteristic matrix to find a polynomial which was factored to find the polynomial roots, which are the matrix eigenvalues. Such a strategy is slow for larger matrices. The iterative QR algorithm is faster for finding eigenvalues. In fact, a better strategy for finding the roots of a polynomial is to first change the polynomial to a matrix that has the same eigenvalues as the polynomial roots. Then the QR algorithm may be used to find the polynomial roots.

```
>> A = diag(ones(n-1,1),-1)
A =
     0    0    0
     1    0    0
     0    1    0
>> A(1,:) = -p(2:n+1)./p(1)
A =
     4   -1   -6
     1    0    0
     0    1    0
>> r = eig(A)
r =
     3.0000
     2.0000
    -1.0000
```

The determinant of the characteristic equation of **A** has the same coefficients and thus the same roots as $f(x)$.

$$\begin{vmatrix} (4-\lambda) & -1 & -6 \\ 1 & (-\lambda) & 0 \\ 0 & 1 & (-\lambda) \end{vmatrix} = 0$$

$$\lambda^3 - 4\lambda^2 + \lambda + 6 = 0$$

- If analytic roots are needed, then the *Symbolic Math Toolbox* can help (section 4.2.5).

- Numeric methods may be used to find the numeric roots of a non-polynomial function (section 7.1.1).

6.3.3 Finding Eigenvectors

The eigenvectors, x_i (one per eigenvalue) lie on the same line as Ax_i.

$$(A - \lambda_i I)x_i = 0$$

The solution to the above equation is called the *null* solution because we are looking for a vector, x_i, that sets the equation to zero. The eigenvectors can be found with elimination or with the `null` function (appendix A.3.2).

We now continue the previous example with elimination to find the eigenvectors.

$$\lambda_1 = -2: \quad \begin{bmatrix} 4 & 2 & | & 0 \\ 2 & 1 & | & 0 \end{bmatrix}$$

Add $-1/2$ of *row 1* to *row 2* and then divide *row 1* by 4:

$$\begin{bmatrix} 1 & 1/2 & | & 0 \\ 0 & 0 & | & 0 \end{bmatrix}$$

$$x_{1a} + 1/2\, x_{1b} = 0$$

The second row of zeros occurs because it is a singular matrix. This means that we have a *free variable*, so we can set one variable to any desired value (usually 1).

$$x_1 = \begin{bmatrix} 1 \\ -2 \end{bmatrix}$$

$$\lambda_2 = 3: \quad \begin{bmatrix} -1 & 2 & | & 0 \\ 2 & -4 & | & 0 \end{bmatrix}$$

Add 2 of *row 1* to *row 2* and then divide *row 1* by -1:

$$\begin{bmatrix} 1 & -2 & | & 0 \\ 0 & 0 & | & 0 \end{bmatrix}$$

$$x_{2a} - 2\, x_{2b} = 0$$

$$x_2 = \begin{bmatrix} 2 \\ 1 \end{bmatrix}$$

Note two things about the eigenvectors returned from `null`. First, MATLAB always normalizes the vector so that it has length of one. Secondly, eigenvectors may always be multiplied by a scalar. It is the direction of the eigenvector that matters, not the magnitude. The eigenpair equation has the same eigenvector on both sides of the equality, making them invariant to scale.

$$\begin{aligned} Ax_i &= \lambda_i\, x_i \\ c\, Ax_i &= c\, \lambda_i\, x_i \end{aligned}$$

```
>> A = [2 2; 2 -1];              >> C2 = A - 12*eye(2)
% the eigenvalues                C2 =
>> l1 = -2; l2 = 3;                 -1     2
>> C1 = A - l1*eye(2)                2    -4
C1 =
    4     2                      % normalized eigenvector
    2     1                      >> x2 = null(C2)
% normalized eigenvector         x2 =
>> x1 = null(C1)                    -0.8944
x1 =                               -0.4472
   -0.4472
    0.8944                       % scaled eigenvector
% scaled eigenvector             >> x2 = x2/x2(2)
>> x1 = x1/x1(1)                  x2 =
x1 =                                2
    1                               1
   -2
                                 >> A*x2
>> A*x1                          ans =
ans =                               6
   -2                               3
    4                            >> 12*x2
>> l1*x1                         ans =
ans =                               6
   -2                               3
    4
```

MATLAB has a function called `eig` that calculates both the eigenvalues and eigenvectors of a matrix. The results are returned as matrices. The eigenvectors are the columns of X. As with the `null` function, the `eig` function always normalizes the eigenvectors. The eigenvalues are on the diagonal of L, the MATLAB function `diag(L)` will return the eigenvalues as a row vector.

```
>> A = [2 2;2 -1];               L =
>> [X, L] = eig(A)                  -2     0
X =                                  0     3
    0.4472   -0.8944
   -0.8944   -0.4472
```

Passing the matrix as a symbolic math variable to `eig` will show eigenvectors that are not normalized.

```
>> [X, L] = eig(sym(A))        L =
X =                            [ -2, 0]
[ -1/2, 2]                     [  0, 3]
[    1, 1]
```

6.4   Properties of Eigenvalues and Eigenvectors

Here is a summary of basic properties of eigenvalues and eigenvectors.

- Eigenvalues and eigenvectors are properties of the matrix alone.

- Only square matrices have eigenvalues and eigenvectors.

- An $n{\times}n$ matrix has n eigenvalues and may have up to n independent eigenvectors if the eigenvalues are unique.

  They are in matched pairs. If there are repeating eigenvalues, then there will also be repeating eigenvectors. When the eigenvalues of a matrix are distinct, the eigenvectors form a set of linearly independent vectors, which is proven in appendix A.8.

- Eigenvectors are invariant to scale.

  The product of an eigenvector and a scalar constant, c, is also an eigenvector. So technically, a matrix has an infinite number of eigenvectors, but has at most n independent eigenvectors.

  $$\mathbf{A}\left(c\,\boldsymbol{x}\right) = \lambda\left(c\,\boldsymbol{x}\right)$$

- They are often complex.

  Eigenvalues are often found to be complex numbers, which also means that the eigenvectors will have complex numbers.

- Symmetric matrices have real eigenvalues and eigenvectors.

  A proof is given in appendix A.9.1.

- If a matrix is invertible and symmetric, then each eigenvalue of the matrix inverse is the reciprocal of an eigenvalue of the matrix. If we use the notation $\lambda_i(\mathbf{A})$ to mean the ith eigenvalue of $\mathbf{A}$, then

  $$\lambda_i(\mathbf{A}^{-1}) = \frac{1}{\lambda_i(\mathbf{A})}. \tag{6.1}$$

- Symmetric matrices have orthogonal eigenvectors.

  A proof is given in appendix A.9.2.

- Eigenvalues are not unique to a matrix.

  Matrices with the same eigenvalues are said to be *similar*. The proof of existence and requirements for similar matrices are given in appendix A.6. Similar matrices are an important part of how MATLAB finds eigenvalues. Through a sequence of matrix conversions based on the QR factorization, a similar upper triangular matrix is found with the eigenvalues on the diagonal.

- The *trace* of a matrix (sum of the diagonal elements) is the sum of the eigenvalues.

- The determinant of a matrix is the product of the eigenvalues.

- The `eig` function:

  - MATLAB returns the eigenvectors as unit vectors. This provides a convenience in some applications.
  - When no returned variables are saved from `eig`, it returns the eigenvalues.
  - To save both the eigenvectors (X) and eigenvalues (L), use:

    ```
    >> [X, L] = eig(A)
    ```
  - The columns of X are the eigenvectors.
  - The eigenvalues are on the diagonal of L. To get a vector of the eigenvalues, use `l = diag(L)`.

6.5   Diagonalization and Powers of A

Here, we will establish a factoring of a matrix based on its eigenvalues and eigenvectors. This factoring is called *diagonalization*. Then, we will use the factoring to find a computationally efficient way to compute powers of the matrix ($\mathbf{A}^k$). Then in section 6.6, we will extend this to $\mathbf{A}^k \, v$, where v is a vector. This final result has direct application to difference equations, which are used to solve several types of problems.

6.5.1 Diagonalization

The eigenpairs of the $n \times n$ matrix $\mathbf{A}$ give us n equations.

$$\mathbf{A}\,\boldsymbol{x}_1 = \lambda_1\,\boldsymbol{x}_1$$
$$\mathbf{A}\,\boldsymbol{x}_2 = \lambda_2\,\boldsymbol{x}_2$$
$$\vdots$$
$$\mathbf{A}\,\boldsymbol{x}_n = \lambda_n\,\boldsymbol{x}_n$$

We wish to combine the n equations into one matrix equation.

Consider the product of $\mathbf{A}$ and a matrix called $\mathbf{X}$ containing the eigenvectors of $\mathbf{A}$ as the columns.

$$\mathbf{A}\,\mathbf{X} = \mathbf{A}\begin{bmatrix} | & | & & | \\ \boldsymbol{x}_1 & \boldsymbol{x}_2 & \cdots & \boldsymbol{x}_n \\ | & | & & | \end{bmatrix}$$

$$= \begin{bmatrix} | & | & & | \\ \lambda_1\boldsymbol{x}_1 & \lambda_2\boldsymbol{x}_2 & \cdots & \lambda_n\boldsymbol{x}_n \\ | & | & & | \end{bmatrix}$$

$$= \begin{bmatrix} | & | & & | \\ \boldsymbol{x}_1 & \boldsymbol{x}_2 & \cdots & \boldsymbol{x}_n \\ | & | & & | \end{bmatrix}\begin{bmatrix} \lambda_1 & 0 & \cdots & 0 \\ 0 & \lambda_2 & \cdots & 0 \\ \vdots & \vdots & \ddots & \vdots \\ 0 & 0 & \cdots & \lambda_n \end{bmatrix}$$

$$= \mathbf{X}\,\Lambda$$

The matrix Λ is a diagonal eigenvalue matrix. If the matrix has linearly independent eigenvectors, which will be the case when all eigenvalues are different (no repeating λs), then $\mathbf{X}$ is invertible.

$$\mathbf{A}\,\mathbf{X} = \mathbf{X}\,\Lambda$$
$$\mathbf{X}^{-1}\mathbf{A}\,\mathbf{X} = \Lambda$$

$$\boxed{\mathbf{A} = \mathbf{X}\,\Lambda\,\mathbf{X}^{-1}}$$

This factoring of $\mathbf{A}$ is called the **diagonalization** factoring.

6.5.1.1 When does Diagonalization not work?

Notice that for the diagonalization factoring to work we need to find the inverse of the eigenvector matrix. So each eigenvector *must be independent* of the other eigenvectors.

One example of a matrix with repeating eigenvectors is a 2×2 matrix with the same values on the forward diagonal and a zero on the backward diagonal. The determinant of the characteristic matrix has repeating eigenvalues, so it will also have repeating eigenvectors. Thus the eigenvector matrix is singular and may not be inverted.

$$\begin{vmatrix} a - \lambda & n \\ 0 & a - \lambda \end{vmatrix} = (a - \lambda)(a - \lambda)$$

```
>> A = [5 7; 0 5]              L =
A =
     5     7                        5     0
     0     5                        0     5
>> [X, L] = eig(A)             >> rank(X)
X =                            ans =
     1.0000   -1.0000               1
          0    0.0000
```

6.5.1.2 Diagonalization of a Symmetric Matrix

Symmetric matrices have a simplified diagonalization because the matrix of eigenvectors of a symmetric matrix, **S**, is orthogonal (appendix A.9). Recall also from section 5.2.3.2 that from the spectral theorem, orthogonal matrices have the property $\mathbf{Q}^T = \mathbf{Q}^{-1}$. Thus the diagonalization of a symmetric matrix is

$$\boxed{\mathbf{S} = \mathbf{Q}\,\mathbf{\Lambda}\,\mathbf{Q}^T}.$$

Note: Recall that the columns of orthonormal matrices must be unit vectors (length of 1), which the `eig` function returns. Orthonormal columns are required to use a transposed matrix instead of the inverse matrix.

6.5.2 Powers of A

How does one compute matrix **A** raised to the power k $(\mathbf{A}^k)$?

The brute force approach is to simply multiply **A** by itself $k - 1$ times. This could be slow if k and the size of the matrix (n) are large.

If **A** is a $n \times n$ square matrix, the product $\mathbf{A}\,\mathbf{A}$ requires n^3 multiply and addition operations $(\mathcal{O}(n^3))$. Thus $\mathbf{A}^k$ has complexity of $\mathcal{O}\big((k-1)n^3\big)$; or if we are clever about the order of the multiplications, it could be reduced to $\mathcal{O}\big(\log_2(k)n^3\big)$. That is:

$$\mathbf{A}^k = \underbrace{\underbrace{\underbrace{\mathbf{A}\,\mathbf{A}}_{\mathbf{A}^2}\,\mathbf{A}^2\,\mathbf{A}^4}_{\mathbf{A}^4}\ldots\mathbf{A}^{k/2}}_{\mathbf{A}^k}.$$

Fortunately, diagonalization gives us a faster way to compute $\mathbf{A}^k$ as a stand-alone matrix. The complexity of this method is $\mathcal{O}(n^3)$.

$$\mathbf{A}^2 = \mathbf{X}\mathbf{\Lambda}\mathbf{X}^{-1}\mathbf{X}\mathbf{\Lambda}\mathbf{X}^{-1} = \mathbf{X}\mathbf{\Lambda}^2\mathbf{X}^{-1}$$
$$\mathbf{A}^3 = \mathbf{X}\mathbf{\Lambda}^2\mathbf{X}^{-1}\mathbf{X}\mathbf{\Lambda}\mathbf{X}^{-1} = \mathbf{X}\mathbf{\Lambda}^3\mathbf{X}^{-1}$$

$$\boxed{\mathbf{A}^k = \mathbf{X}\mathbf{\Lambda}^k\mathbf{X}^{-1}}$$

Because the $\mathbf{\Lambda}$ matrix is diagonal, only the individual λ values need be raised to the k power (element-wise exponent).

$$\mathbf{\Lambda}^k = \begin{bmatrix} \lambda_1^k & 0 & \cdots & 0 \\ 0 & \lambda_2^k & \cdots & 0 \\ \vdots & \vdots & \ddots & \vdots \\ 0 & 0 & \cdots & \lambda_n^k \end{bmatrix}$$

Example Power of A

In this example, the matrix is symmetric, so the inverse simplifies to a transpose.

```
>> S = gallery('moler', 4)
S =
      1    -1    -1    -1
     -1     2     0     0
     -1     0     3     1
     -1     0     1     4
>> [Q, L] = eig(S)
Q =
    -0.8550    0.1910    0.3406   -0.3412
    -0.4348   -0.6421   -0.6219    0.1093
    -0.2356    0.6838   -0.4490    0.5248
    -0.1562   -0.2894    0.5437    0.7722
L =
     0.0334         0         0         0
          0    2.2974         0         0
          0         0    2.5477         0
          0         0         0    5.1215

>> Q * L.^5 * Q'
ans =
        425      -162      -639      -912
       -162       110       204       273
       -639       204      1022      1389
       -912       273      1389      2138
>> S^5
ans =
```

$$
\begin{array}{rrrr}
425 & -162 & -639 & -912 \\
-162 & 110 & 204 & 273 \\
-639 & 204 & 1022 & 1389 \\
-912 & 273 & 1389 & 2138
\end{array}
$$

A few observations on computing powers of A

- Considerable computation is saved by using the element-wise exponent operator on Λ (L in the MATLAB code). This gives the same result as a matrix exponent because Λ is a diagonal matrix.

- The eigenvalues and eigenvectors of a matrix are often complex. In the example, a symmetric matrix is used because such will always have only real eigenvalues and eigenvectors. When a matrix has complex eigenvalues, the method still works. After multiplying the eigenvalues by the eigenvectors, the imaginary values may not be exactly zero due to round-off errors, but they will be very small. Use the `real` function to get only the real component of the complex numbers.

6.6   Change of Basis and Difference Equations

When we wish to multiply $\mathbf{A}^k$ by a vector ($\mathbf{A}^k\mathbf{v}$), we could diagonalize the matrix to find $\mathbf{A}^k$ and then multiply by the vector. But there is another approach that also makes apparent the *steady state* of the system (when k goes to infinity). The change of basis method of computing $\mathbf{A}^k\mathbf{v}$ has particular application to difference equations, which are commonly used as a model for discrete systems.

First we need to change the vector into a linear combination of the eigenvectors of $\mathbf{A}$.

$$
\mathbf{v} = c_1\,\mathbf{x}_1 + c_2\,\mathbf{x}_2 + \cdots + c_n\,\mathbf{x}_n,
$$

This is called a *change of basis*, because as shown in figure 6.3 we are using the eigenvectors as a new coordinate frame to describe the vector instead of the normal Cartesian basis vectors. In matrix form, the vector $\mathbf{v}$ is described as

$$
\mathbf{v} = \mathbf{X}\mathbf{c} = \begin{bmatrix} | & | & & | \\ \mathbf{x}_1 & \mathbf{x}_2 & \cdots & \mathbf{x}_n \\ | & | & & | \end{bmatrix} \begin{bmatrix} c_1 \\ c_2 \\ \vdots \\ c_n \end{bmatrix}.
$$

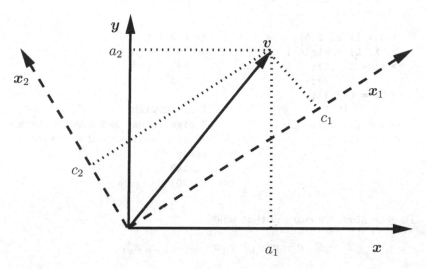

FIGURE 6.3: Vector v is changed from coordinates (a_1, a_2) using the standard basis vectors to coordinates (c_1, c_2) using the eigenvectors of $\mathbf{A}$ as the basis vectors.

Finding the linear coefficients is just a matter of solving a system of linear equations: $c = \mathbf{X}^{-1} v$, or in MATLAB: c = X\v.

Now, we will take advantage of $\mathbf{A X} = \mathbf{X} \Lambda$ along with $v = \mathbf{X} c$ to multiply v by powers of $\mathbf{A}$,

$$
\begin{aligned}
\mathbf{A}v &= \mathbf{A X} c \\
&= \mathbf{X} \Lambda c \\
&= \lambda_1 c_1 x_1 + \lambda_2 c_2 x_2 + \cdots + \lambda_n c_n x_n \\
\mathbf{A}^2 v &= \mathbf{A X} \Lambda c \\
&= \mathbf{X} \Lambda^2 c \\
&= \lambda_1^2 c_1 x_1 + \lambda_2^2 c_2 x_2 + \cdots + \lambda_n^2 c_n x_n \\
\mathbf{A}^k v &= \mathbf{A X} \Lambda^{k-1} c \\
&= \mathbf{X} \Lambda^k c \\
&= \lambda_1^k c_1 x_1 + \lambda_2^k c_2 x_2 + \cdots + \lambda_n^k c_n x_n
\end{aligned}
$$

The Λ matrix is the same diagonal eigenvalue matrix from the diagonalization factoring.

Let's illustrate this with an example. Consider the following matrix and its eigenvectors and eigenvalues. We will find $\mathbf{A}^2 v$ using the change of basis strategy, where $v = (3, 11)$ is changed to a linear combination of the eigenvectors.

```
>> A = [1 2; 5 4];              >> A^2 * v
>> [X, L] = eig(A);             ans =
>> x1 = X(:,1);                    143
>> x2 = X(:,2);                    361
>> l1 = L(1,1);
>> l2 = L(2,2);                 % check with
>> v = [3; 11];                 % eigenvalues and eigenvectors
>> c = X\v;                     >> l1^2 * x1 + l2^2 * 2 * x2
                                ans =
                                   143
                                   361
```

To generalize, we can say that when

$$v = c_1 \, x_1 + c_2 \, x_2 + \cdots + c_n \, x_n,$$

then

$$\mathbf{A}^k v = \lambda_1^k c_1 \, x_1 + \lambda_2^k c_2 \, x_2 + \cdots + \lambda_n^k c_n \, x_n. \tag{6.2}$$

Note: Eigenvalues and eigenvectors are often complex, but when the terms are added together, the result will be real (zero imaginary part).

6.6.1   Difference Equations

One of the main applications where you might want to compute the power of a matrix multiplied by a vector is for a difference equation, which is the discrete form of a differential equation. Difference equations are often used in the study of control systems, which is important to nearly every engineering discipline.

We have a difference equation when the state of a system at observation k is its state in the previous observation multiplied by a matrix. Note that the *state* of a system can be a set of any desired conditions of the system put into a vector, e.g., position, velocity, acceleration, or anything else that describes a property of the system at each observation.

$$u_k = \mathbf{A} \, u_{k-1}.$$

The state of the system at the first observation as given by its initial condition is

$$u_1 = \mathbf{A} \, u_0.$$

After the second observation, it is

$$u_2 = \mathbf{A} \, u_1 = \mathbf{A} \, \mathbf{A} \, u_0 = \mathbf{A}^2 \, u_0.$$

After k observations, it is

$$u_k = \mathbf{A}^k \, u_0.$$

Change of basis for steady state analysis

The number of computations to find $\mathbf{A}^k \boldsymbol{v}$ by diagonalization and change of basis is essentially the same. The reason to employ the change of basis method is to more easily see what the equation will do when $k \to \infty$, which is *steady-state analysis*.

- If any $0 \leq \lambda_i < 1.0$, then that term goes to zero when k goes to infinity.

- If any $\lambda_i > 1.0$, the equation goes to infinity.

- If any $\lambda_i = 1.0$, that term will have a stable, nonzero steady-state value.

- If all $0 \leq \lambda_i \leq 1.0$, then the equation has a stable steady-state value.

- If any $-1.0 < \lambda_i < 0$, then that term oscillates between positive and negative, but goes to zero when k goes to infinity.

- If any $\lambda_i = -1.0$, that term will oscillate between a nonzero positive and negative value.

- If any $\lambda_i < -1.0$, the equation oscillates between positive and negative and goes to infinity when k goes to infinity.

Applying equation (6.2) gives us an equation for the state of a system defined by a difference equation after k observations.

$$\boldsymbol{u}_k = \mathbf{A}^k \boldsymbol{u}_0 = \lambda_1^k c_1 \, \boldsymbol{x}_1 + \lambda_2^k c_2 \, \boldsymbol{x}_2 + \cdots + \lambda_n^k c_n \, \boldsymbol{x}_n, \tag{6.3}$$

The eigenvalues and eigenvectors of $\mathbf{A}$ are λ_i and $\boldsymbol{x}_i$. The c coefficients are the elements of the vector $\boldsymbol{c} = \mathbf{X}^{-1} \boldsymbol{u}_0$.

6.6.2 Application: Fibonacci Sequence

A simple application of a difference equation is the Fibonacci sequence of numbers. This is the sequence of numbers starting with 0 and 1 and continuing with each number being the sum of the previous two numbers: $\{0, 1, 1, 2, 3, 5, 8, 13, 21, \ldots \}$, Fibonacci$(k)$ = Fibonacci$(k-1)$ + Fibonacci$(k-2)$. This sequence of numbers occurs naturally in nature and is often fodder for programming problems using recursion and dynamic programming, but it can also be described as a difference equation, from which we can use the change of basis method to find a closed form solution known as Binet's formula.

The state of the system here is a vector with Fibonacci(k) and Fibonacci($k - 1$).

$$\mathbf{F}(k) = \begin{bmatrix} \text{Fibonacci}(k) \\ \text{Fibonacci}(k-1) \end{bmatrix} = \begin{bmatrix} 1 & 1 \\ 1 & 0 \end{bmatrix} \begin{bmatrix} \text{Fibonacci}(k-1) \\ \text{Fibonacci}(k-2) \end{bmatrix}$$

$$= \begin{bmatrix} 1 & 1 \\ 1 & 0 \end{bmatrix}^k \begin{bmatrix} 1 \\ 0 \end{bmatrix}$$

Let us use the change of basis strategy to compute $\mathbf{F}(k)$ for any value of k.

Using the quadratic equation, one can quickly find the two eigenvalues from the determinant of the characteristic matrix.

$$m = \frac{1}{2}, \ p = -1, \ \lambda_{1,2} = \frac{1}{2} \pm \sqrt{\frac{1}{4} + 1} = \frac{1 \pm \sqrt{5}}{2}$$

$$\lambda_1 = \frac{1 - \sqrt{5}}{2}, \text{ and } \lambda_2 = \frac{1 + \sqrt{5}}{2}.$$

However, the algebra to find the symbolic equation for the eigenvectors and the change of basis coefficients gets a little messy, but Chamberlain shows the derivation in a nice blog post [10].

We can also let MATLAB compute the eigenvalues and eigenvectors using the *Symbolic Math Toolbox* to get the exact equations so that they can be used in the solution.

```
[X, L] = eig(sym([1 1; 1 0]))
```

The eigenvectors are:

$$\boldsymbol{x}_1 = \begin{bmatrix} \left(\frac{1-\sqrt{5}}{2}\right) \\ 1 \end{bmatrix} \text{ and } \boldsymbol{x}_2 = \begin{bmatrix} \left(\frac{1+\sqrt{5}}{2}\right) \\ 1 \end{bmatrix}$$

With elimination, we find the change of basis coefficients.

$$\boldsymbol{c} = \begin{bmatrix} -\frac{1}{\sqrt{5}} \\ \frac{1}{\sqrt{5}} \end{bmatrix}$$

Then each Fibonacci number is given by the first value of the difference equation vector.

$$\text{Fibonacci(k)} = c_1 \lambda_1^k x_{1,1} + c_2 \lambda_2^k x_{1,2}$$

With this solution, Fibonacci(0) = 1, but we observe that the eigenvector terms used are the same as the eigenvalues, so we are actually raising the eigenvalues to the $(k+1)$ power. Thus we can shift the terms and get a result with Fibonacci(0) = 0.

$$\text{Fibonacci(k)} = c_1 \lambda_1^k + c_2 \lambda_2^k$$

Code 6.1 shows a closed form (constant run time) Fibonacci sequence function. The function calculates integer values. The round function just returns values of integer data type instead of floating point numbers. The commented lines show code that yields the equations used in the code.

```
function fib = fibonacci(k)
% FIBONACCI(k) - kth value of Fibonacci sequence
% (Binet's formula)

%     A = [1 1; 1 0];
%     [X, L] = eig(sym(A));
%     X = [(1 - sqrt(5))/2 (1 + sqrt(5))/2; 1 1];
    L = [(1 - sqrt(5))/2 0; 0 (1 + sqrt(5))/2];

%     u0 = [1; 0];
%     c = X\u0;
    c = [-1/sqrt(5); 1/sqrt(5)];

    % only need the first vector value
    fib = round(c(1)*L(1,1)^k + c(2)*L(2,2)^k);
end
```

CODE 6.1: Closed form Fibonacci function.

6.6.3   Application: Markov Matrices

A Markov matrix is a stochastic (statistical) matrix showing probabilities of something transitioning from one state to another in one observation period. There are many applications where things can be thought of having one of a finite number of states. We often visualize state transition probabilities with a finite state machine (FSM) diagram. Then from a FSM, we can build a Markov matrix and make predictions about the future. Markov models are named after the Russian mathematician Andrei Andreevich Markov (1856 - 1922).

Figure 6.4 shows simple FSM to give odds for the weather on the next day given the current weather conditions. We interpret the FSM diagram to say that if it is sunny today, there is a 70% chance that it will also be sunny tomorrow, a 20% percent chance that it will be cloudy tomorrow, and a 10% chance of rain tomorrow. Similarly, we can assign probabilities for tomorrow's weather for when it is cloudy or raining today.

We put the transition probabilities into a Markov matrix. Since the values in the matrix are probabilities, each value is between zero and one. The columns represents the transition probabilities for starting in a given state. The sum of each column must be one. The rows represent the probabilities for being in a state on the next observation.

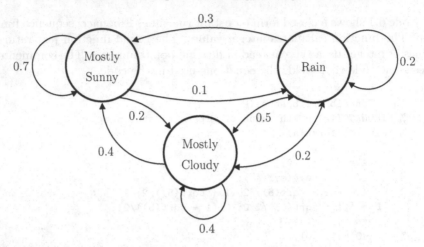

FIGURE 6.4: Weather prediction finite state machine.

The columns and rows are here ordered as sunny, cloudy, and rain.

$$\mathbf{M} = \begin{bmatrix} 0.7 & 0.4 & 0.3 \\ 0.2 & 0.4 & 0.5 \\ 0.1 & 0.2 & 0.2 \end{bmatrix}$$

Matrix multiplication tells us the weather forecast for tomorrow if it is cloudy today.

```
>> M = [0.7, 0.4, 0.3; 0.2, 0.4, 0.5; 0.1, 0.2, 0.2]
M =
    0.7000    0.4000    0.3000
    0.2000    0.4000    0.5000
    0.1000    0.2000    0.2000
>> M * [0; 1; 0]
ans =
    0.4000        % probability sunny
    0.4000        % probability cloudy
    0.2000        % probability rain
```

We need to raise $\mathbf{M}$ to the k power to predict the weather in k days.

$$p_k = \mathbf{M}^k \, p_0.$$

So if it is cloudy today, we can expect the following on the day after tomorrow.

```
>> M^2 * [0; 1; 0]
ans =
    0.5000        % probability sunny
    0.3400        % probability cloudy
    0.1600        % probability rain
```

We will always see something interesting when we look at the eigenvalues of a Markov matrix. One eigenvalue will always be one and the others will be less than one. As we saw before, this means that the forecast for the distant future converges and becomes independent to the current state of the system.

```
>> [X, L] = eig(M)
X =
    0.8529     0.7961     0.1813
    0.4713    -0.5551    -0.7801
    0.2245    -0.2410     0.5988
L =
    1.0000          0          0
         0     0.3303          0
         0          0    -0.0303
```

The steady state of the system only depends on the eigenvector corresponding to the eigenvalue of value one. The other terms will go to zero in future observations.

```
>> c_sun = X\[1; 0; 0];
>> c_clouds = X\[0; 1; 0];
>> c_rain = X\[0; 0; 1];
>> steady_state = X(:,1)*c_sun(1)
steady_state =
    0.5507
    0.3043
    0.1449
>> steady_state = X(:,1)*c_clouds(1)
steady_state =
    0.5507
    0.3043
    0.1449
>> steady_state = X(:,1)*c_rain(1)
steady_state =
    0.5507
    0.3043
    0.1449
```

We can plot the forecast to see the convergence. Code 6.2 shows the code for the change of basis calculation and plot of the weather forecast. The markers in figure 6.5 show the starting and ending of the forecast period. Each point in the 3-D plot is a probability of sun, clouds, or rain. We see from the plot that the distant forecast converges to the same probability regardless of the weather condition today.

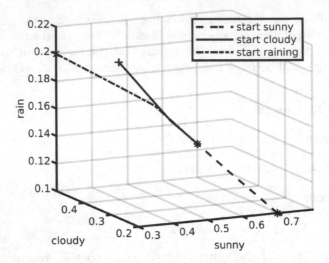

FIGURE 6.5: A 3-D plot showing the Markov matrix forecast probabilities given that today is each of sunny, cloudy, or raining.

Markov Eigenvalues

Markov matrices must have an eigenvalue of one because of the steady state, $\lim_{k\to\infty} \mathbf{M}^k \, \boldsymbol{v}$. By definition, the steady state is never zero or infinity, so there must be at least one eigenvalue of one and no eigenvalue greater than one.

Markov matrices have at least one eigenvalue of one because the columns add to one. Recall that the eigenvalues satisfy the requirement that the characteristic equation, $\mathbf{C} = \mathbf{M} - \mathbf{I}\lambda$, is a singular matrix, which will occur when $\lambda = 1$. Subtracting one from the diagonal elements makes each column sum to zero. So the rows are dependent because each row is -1 times the sum of the other rows.

```
>> M = [0.7, 0.4, 0.3; 0.2, 0.4, 0.5; 0.1, 0.2, 0.2];
>> rank(M)
ans =
     3
>> d = M - eye(3)
d =
   -0.3000    0.4000    0.3000
    0.2000   -0.6000    0.5000
    0.1000    0.2000   -0.8000
>> rank(d)
ans =
     2
```

From section 6.6, we know that if any eigenvalue is > 1 then $\mathbf{M}^k \boldsymbol{p}_0 \to \infty$ when $k \to \infty$, so it is not possible for a stochastic matrix to have an eigenvalue

```
% File: MarkovWeather.m
% Script showing change of basis calculation for
% a Markov matrix.

M = [0.7, 0.4, 0.3; 0.2, 0.4, 0.5; 0.1, 0.2, 0.2];
[X, L] = eig(M);
l = diag(L);
% coefficients from initial conditions
c_sun = X\[1; 0; 0];
c_clouds = X\[0; 1; 0];
c_rain = X\[0; 0; 1];

N = 20;
F_sun = zeros(3, N);
F_clouds = zeros(3, N);
F_rain = zeros(3, N);
% build forecast for next N days
for k = 1:N
    F_sun(:,k) = c_sun(1)*X(:,1) + c_sun(2)*l(2)^k*X(:,2) ...
                + c_sun(3)*l(3)^k*X(:,3);
    F_clouds(:,k) = c_clouds(1)*X(:,1) + c_clouds(2)*l(2)^k*X(:,2) ...
                + c_clouds(3)*l(3)^k*X(:,3);
    F_rain(:,k) = c_rain(1)*X(:,1) + c_rain(2)*l(2)^k*X(:,2) ...
                + c_rain(3)*l(3)^k*X(:,3);
end

% plot probabilities for next 20 days
figure, hold on
plot3(F_sun(1,:), F_sun(2,:), F_sun(3,:))
plot3(F_clouds(1,:), F_clouds(2,:), F_clouds(3,:))
plot3(F_rain(1,:), F_rain(2,:), F_rain(3,:))
% mark starting points
plot3(F_sun(1,1), F_sun(2,1), F_sun(3,1), 'r', 'Marker', '*')
plot3(F_clouds(1,1), F_clouds(2,1), F_clouds(3,1), 'b', 'Marker', '+')
plot3(F_rain(1,1), F_rain(2,1), F_rain(3,1), 'm', 'Marker', 'x')
% mark final point
plot3(F_sun(1,N), F_sun(2,N), F_sun(3,N), 'k', 'Marker', '*')
grid on
xlabel('sunny')
ylabel('cloudy')
zlabel('rain')
legend({'start sunny', 'start cloudy', 'start raining'})
hold off
```

CODE 6.2: Script showing the change of basis calculation and plot of the weather forecast based on a Markov model.

greater than one. A more rigorous proof is easy to show for a 2×2 matrix, and is intuitive but difficult to formally verify for larger matrices. Recall two eigenvalue properties from section 6.3:

1. The trace of a matrix (sum of the diagonal values) is the sum of the eigenvalues.

2. The determinant of a matrix is the product of the eigenvalues.

Since all entries of a Markov matrix are probabilities, every entry must be between zero and one. The identity matrix has the highest trace of any valid Markov matrix. For a $n \times n$ identity matrix, the trace is n, thus all of the eigenvalues of an identity matrix are one. All other valid Markov matrices have a trace less than n; therefore, the sum of the eigenvalues is $\leq n$. It is clear to see from a 2×2 matrix that all eigenvalues are ≤ 1 since at least one eigenvalue is equal to one.

$$trace\left(\begin{bmatrix} a & 1-b \\ 1-a & b \end{bmatrix}\right) = a + b \leq 2$$

$$1 + \lambda_2 \leq 2$$

$$\lambda_2 \leq 1$$

The maximum determinant of a Markov matrix is one, which is the case for the identity matrix. All other Markov matrices have a determinant less than one. Therefore, the product of the eigenvalues is ≤ 1.

$$\begin{vmatrix} a & 1\text{-}b \\ 1\text{-}a & b \end{vmatrix} = a + b - 1 \leq 1$$

Since the product of the eigenvalues is equal to the determinant, $\lambda_1 \lambda_2 \leq 1$, and since $\lambda_1 = 1$, then $\lambda_2 \leq 1$.

6.7   Systems of Linear ODEs

Differential equations that come up in engineering design and analysis are usually systems of equations, rather than a single equation. Fortunately, they are often first order linear equations. As discussed in section 4.4, the *Symbolic Math Toolbox* can solve many differential equations expressed by a single equation. Higher order polynomial and nonlinear systems of ODEs need numerical methods, such as discussed in section 7.5. However, systems of first order linear ODEs may be solved analytically with eigenvalues and eigenvectors.

We use vectors and matrices to describe systems of ODEs. We have have n variables, $y_1, y_2, \ldots, y_n$, and an equation for the derivative of each as a linear combination of the y variables.

$$
\begin{cases}
y_1' = a_{11}\, y_1 + a_{12}\, y_2 + \cdots + a_{1n}\, y_n \\
y_2' = a_{21}\, y_1 + a_{22}\, y_2 + \cdots + a_{2n}\, y_n \\
\vdots \\
y_n' = a_{n1}\, y_1 + a_{n2}\, y_2 + \cdots + a_{nn}\, y_n
\end{cases}
\tag{6.4}
$$

In matrix notation, this is

$$
y' = \mathbf{A}\, y.
\tag{6.5}
$$

We know that ODEs of the form

$$
\frac{dy(t)}{dt} = a\, y(t)
$$

have the solution

$$
y(t) = c\, e^{a t}.
$$

It follows that equation (6.5) has a solution of the form

$$
y = c\, e^{\mathbf{A} t}.
\tag{6.6}
$$

Equation (6.6) is an unusual equation. What does it mean to have a matrix in the exponent? We show how to put equation (6.6) into an equation that is easier to manage in appendix B.2. A power series expansion gets the matrix out of the exponent and the diagonalization factoring makes equation (6.6) a matrix equation of the product of two matrices and a vector.

As found in equation (B.6) on page 402, the general solution takes the form

$$
y(t) = c_1 e^{\lambda_1 t} x_1 + c_2 e^{\lambda_2 t} x_2 + \cdots + c_n e^{\lambda_n t} x_n
\tag{6.7}
$$

The set of scalar values $\{\lambda_1, \lambda_2, \cdots, \lambda_n\}$ are the eigenvalues of matrix $\mathbf{A}$. The vectors $\{x_1, x_2, \cdots, x_n\}$ are the eigenvectors of $\mathbf{A}$. The coefficients $c_1, c_2, \ldots, c_n$ are found from the initial conditions, $y(0)$.

Steady state

The steady state of a systems refers to the state of the system when the control variable, which is often t for time, goes to infinity. Because the eigenvalues are in an exponent, they determine the steady state. For the stead state to be stable, meaning that it does not go to infinity and does not oscillate, all eigenvalues must be real and less than or equal to zero. Some ODE systems have complex eigenvalues. When this occurs, the solution will have oscillating terms because of Euler's complex exponential equation, $e^{i x} = \cos(x) + i\, \sin(x)$ (appendices B.1 and B.1.2).

6.7.1   ODE Example

Consider the following set of ODEs and initial conditions.

$$\begin{cases} y_1(t)' = -2\,y_1(t) + \ y_2(t), & y_1(0) = 6 \\ y_2(t)' = \ y_1(t) - 2\,y_2(t), & y_2(0) = 2 \end{cases}$$

In matrix notation,

$$\mathbf{y}' = \begin{bmatrix} -2 & 1 \\ 1 & -2 \end{bmatrix} \mathbf{y}, \qquad \mathbf{y}(0) = \begin{bmatrix} 6 \\ 2 \end{bmatrix}.$$

We first use MATLAB to find the eigenvalues and eigenvectors. MATLAB always returns normalized eigenvectors, which can be multiplied by a constant to get simpler numbers if desired.

```
>> A = [-2 1; 1 -2];
>> [X, L] = eig(A)
X =
      0.7071      0.7071
     -0.7071      0.7071
L =
     -3      0
      0     -1
>> X = X*2/sqrt(2)     % Just scale the eigenvectors
X =
      1.0000      1.0000
     -1.0000      1.0000
```

The columns of the X matrix are the eigenvectors. The eigenvalues are on the diagonal of L for Lambda (Λ). Our solution has the form

$$\mathbf{y}(t) = c_1\,e^{-3t} \begin{bmatrix} 1 \\ -1 \end{bmatrix} + c_2\,e^{-t} \begin{bmatrix} 1 \\ 1 \end{bmatrix}$$

At the initial condition, the exponent terms become 1.

$$\mathbf{y}(0) = c_1 \begin{bmatrix} 1 \\ -1 \end{bmatrix} + c_2 \begin{bmatrix} 1 \\ 1 \end{bmatrix} = \begin{bmatrix} 1 & 1 \\ -1 & 1 \end{bmatrix} \begin{bmatrix} c_1 \\ c_2 \end{bmatrix} = \mathbf{X}\,\mathbf{c}, \qquad \mathbf{c} = \mathbf{X}^{-1}\mathbf{y}(0)$$

```
>> y0 = [6;2];
>> c = X\y0
c =
     2.0000
     4.0000
```

$$\begin{cases} y_1(t) = \ 2\,e^{-3t} + 4\,e^{-t} \\ y_2(t) = -2\,e^{-3t} + 4\,e^{-t} \end{cases}$$

6.7.2   Application: Closed Loop Control Systems

Here we consider an important engineering topic. Control systems use linear ODEs as discussed in this section. The coverage in this sub-section is a short and simple introduction to a vast field of study. It is hoped that this introduction will encourage further study. More complete introductions to control systems may be found in [7, 60, 68, 70]. Detailed coverage of feedback systems can be found in [62]. Corke's text on control of robotic systems [15] has good coverage of PID controllers.

Automatically controlled systems are part of our everyday experience. Electric thermostats control temperatures in indoor spaces, ovens, and freezers. A cruise control system helps maintain a constant speed when we drive on the highway. Industrial robotic and automation systems control both simple and complex systems in manufacturing plants. The list of control system applications is undoubtedly quite long.

Some systems can use gravity or other natural barriers to maintain stability. We call these systems *open-loop* because no feedback control system is required. The state-space equation of open-loop controllers take the form $\dot{x} = \mathbf{A}\,x$,[1] which is the same general equation as the ODE's that we previously considered. As before, the stability of the system requires that the real portion of the eigenvalues of $\mathbf{A}$ be negative. Most systems, however, require sensors and a feedback control system to maintain accuracy and stability. We call such systems *closed–loop* controllers. There is some variability in the configuration of closed-loop controllers, but figure 6.6 shows the principal components.

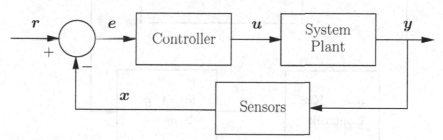

FIGURE 6.6: The input to the controller may be either the estimate of the system's state, x, or the difference, e, between the state and a reference set-point, r. The system plant includes the models of the system constraints, physical environment, and the mechanisms whereby the control signal, u can guide the state of the system. The observed output of the system is y. The sensors attempt to measure the state of the system.

Closed-loop controllers typically use one of two strategies. Systems with a single control variable, such as thermostats or automobile cruise controls often use a proportional–integral–derivative (PID) controller. These efficient

---

[1]Different branches of engineering sometimes to use different notation. In the control systems field, the derivative of x is usually written as $\dot{x}$.

controllers look at the error between the desired set-point and the measured system output. The control variable, u, is the sum of three values. A term that is proportional to the error adjusts the control variable to minimize the error. An integral term guards against long-term differences between the output and set-point. The derivative term suppresses rapid fluctuations of the control variable that the proportional control alone might produce.

The *state space* alternative to the PID controller models changes to the state of the system. The state of the system is stored in a vector, x, having terms for any variables that are significant and can be measured. The system and its control mechanisms use differential equations that track changes (derivatives) to the state. For a linear time-invariant (LTI) system, the state space equations are first-order differential equations expressed with vectors and matrices. Figure 6.7 shows how the state space equations (6.8) to (6.10) relate in the control system.

$$\dot{x}(t) = \mathbf{A}\,x(t) + \mathbf{B}\,u(t) \tag{6.8}$$

$$x(t) = \mathbf{C}\,y(t) \tag{6.9}$$

$$u(t) = -\mathbf{K}\,x(t) \tag{6.10}$$

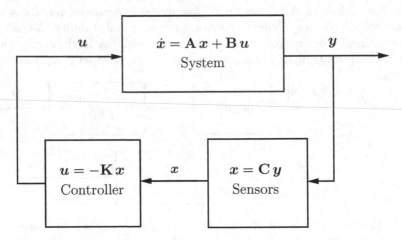

FIGURE 6.7: Closed-loop linear time-invariant control system.

In the system plant, the $\mathbf{A}$ matrix models the physics of the system. $\mathbf{B}$, which may be a matrix or a column vector, defines the ability of the control variable, u, to guide the system toward the desired behavior. $\mathbf{A}$ and $\mathbf{B}$ represent the given model of the system. The observable output of the system may define the state of the system ($x = \mathbf{C}y$). But in some cases, it may be challenging to determine the system's state from the sensors. *Kalman filters* are sometimes used when the sensors alone cannot determine the system's state. The variable $\mathbf{K}$, which may be a scalar, a row vector, or a matrix, is the tunable variable that we use to ensure a stable steady state condition.

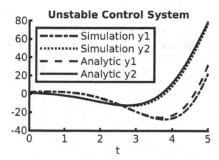

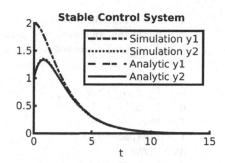

FIGURE 6.8: The state of the unstable system quickly goes to infinity.

FIGURE 6.9: The state of the stable system has a steady state of $(0,0)$.

Substituting u into the state-space equation (6.8) gives us a combined equation that we can use to test the stability of the controller.

$$\dot{x} = (\mathbf{A} - \mathbf{B}\,\mathbf{K}\,\mathbf{C})x$$

The stability is dependant on the eigenvalues of $\mathbf{A} - \mathbf{B}\,\mathbf{K}\,\mathbf{C}$.

The example in code 6.3 models a system that is unstable without some help from the controlling $\mathbf{K}$ matrix. The $\mathbf{K}$ matrix can move the eigenvalues of the combined state-space equation to values less than zero. There are, however, some subtle issues with choosing $\mathbf{K}$ such that the system is stable but not over-controlled. A complete discussion of selecting the control matrix is beyond the scope of this text. The MATLAB *Control System Toolbox* has a function called `place` that implements the algorithm described in a seminal paper by Kautsky, Nichols, and Van Dooren to find a suitable control matrix [47]. However, an acceptable control matrix for learning purposes is not difficult to calculate. Starting with negative eigenvalues and simple eigenvectors, use the diagonalization factoring to calculate a target matrix. Then, $\mathbf{K}$ may be found using a simple linear algebra calculation. Note that in this example, the $\mathbf{C}$ matrix is an identity matrix, so it is not used in the code.

The example performs a time domain simulation of a control system and also finds the equation for the state of the system based on the eigenvalues and eigenvectors. The state of the system has two values. Plots of the system response to starting state values of $(2, 1)$ are shown in figures 6.8 and 6.9 for unstable and stable systems.

To simulate the time-domain system response as a system sampled at discrete times, we recognize that the derivative of the state space is modeled as follows, which leads to a simple equation for the next x_k value.

$$\dot{x} = \frac{x_k - x_{k-1}}{(\Delta t)}$$

$$x_k = x_{k-1} + \dot{x}\,(\Delta t) \tag{6.11}$$

```
% File: closedLoopCtrl.m
% Simulation and analytic solution of a stable
% and unstable closed-loop control system.

N = 200;
A = [1 1.8; -0.7 2.5];
B = [1 0.5; 0.5 1];
T = 15;
t = linspace(0, T, N);
dt = T/N;
X = zeros(2, N);
X(:,1) = [2; 1];
K_unstable = eye(2);
Atarget = [0.5 -1; 1.5 -2];
K_stable = B\(A - Atarget);
% K = K_unstable;
K = K_stable;
% simulation
for k = 2:N
    x = X(:,k-1);
    u = -K*x;
    x_dot = A*x + B*u;
    X(:,k) = x + x_dot*dt;
end
% analytic solution
x0 = X(:,1);
[ev, L] = eig(A - B*K);
c = ev\x0;
y = real(c(1)*exp(L(1,1)*t).*ev(:,1) + ...
    c(2)*exp(L(2,2)*t).*ev(:,2));
% plot results
hold on
plot(t, X(1,:))
plot(t, X(2,:))
plot(t, y(1,:))
plot(t, y(2,:))
hold off
legend('Simulation y1', 'Simulation y2', ...
    'Analytic y1', 'Analytic y2','Location', 'north')
xlabel('t')
title('Stable Control System')
```

CODE 6.3: Simulation of a closed-loop control system. Assign variable K to K_unstable to simulate the unstable system.

Equation (6.11) is sufficient for our application here, but it assumes that $\dot{x}$ is constant between x_{k-1} and x_k. More precise algorithms for simulating the response of ODEs is presented in section 7.5.

6.8 Singular Value Decomposition (SVD)

We now consider the matrix factorization that is perhaps the most important factoring in linear algebra for three reasons.

1. The SVD can factor any matrix, even singular and rectangular matrices.

2. The SVD is used to solve many linear algebra problems and has application to many science and engineering fields of study including artificial intelligence, data analytics, and image processing. The linear algebra applications of the SVD include low-rank matrix approximations, finding the rank of a matrix, finding the pseudo-inverse of rectangular matrices, finding orthogonal basis vectors for the columns of a matrix, finding the condition number of a matrix, and finding the null space and left null space of a matrix.

3. The SVD is now fast and numerically accurate. That was not always the case. An improved algorithm for finding the SVD became available in 1970. The history of the SVD is briefly described below.

This fast algorithm and the vast application domain of the decomposition has resulted in the SVD being a capital algorithm in software such as MATLAB. For several applications, it has replaced elimination based algorithms in the internal workings of MATLAB and other software.

Note: Our discussion of the SVD algorithm focuses on its interpretation and application rather than its newer, more efficient implementation in software such as MATLAB. The complexity of that algorithm is beyond the scope of this text and is actually less important to understanding and applying it than the classic algorithm.

6.8.1 The Geometry of the SVD

As we have seen before, we can visualize the geometry of linear algebra equations in $\mathbb{R}^2$ or $\mathbb{R}^3$ and the relationships hold in higher dimensions where we can not see the relationships in a diagram. Recall from our introduction to eigenvalue problems at the beginning of chapter 6 that multiplication of a matrix and vector usually stretches and rotates the vector. As before, we

---

The history of the SVD

The official introduction of the SVD came in 1873 when Eugenio Beltrami published a paper describing it. Unaware of Beltrami's paper, Camille Jordan independently published a paper describing it a year later. Jordan's paper is more complete and accurate in the details presented. Beltrami and Jordan built on previous works related to the eigenvalues and eigenvectors of symmetric matrices, which are central to the classic SVD algorithm. Augustin-Louis Cauchy established the properties of eigenvalues and eigenvectors of symmetric matrices in 1827. Then in 1846, Carl Jacobi presented the diagonalization factorization of symmetric matrices. Stewart gives a nice summary of the early history of SVD and related topics [67].

Cleve Moler, co-founder of MathWorks, observed in a blog post that the SVD was of little practical value until 1970 [56]. Even when computers became available, the algorithm was too slow to be useful. Then in the time frame of 1965 to 1970, Gene Golub and Christian Reinsch developed an algorithm to quickly find the SVD [27]. The algorithm that they developed is not only faster than previous algorithms, but it also provides results that are numerically more accurate [79, 51].

---

want to factor out the stretching and rotating components of the matrix. For a $n \times n$ square matrix $\mathbf{A}$ and a $n \times 1$ unit vector $\mathbf{v}$, we have

$$\mathbf{A}\mathbf{v} = \sigma\, \mathbf{u}, \qquad (6.12)$$

where σ is a scalar that is the length of the resulting vector and $\mathbf{u}$ is a unit vector that shows the direction of the resulting vector. There are n vector–matrix relationships like equation (6.12) that make the SVD factorization of a $n \times n$ square matrix. We will soon see that the SVD can also be used to factor rectangular matrices.

Consider the following matrix and two orthogonal unit vectors.

$$\mathbf{A} = \begin{bmatrix} -0.1895 & 2.6390 \\ 3.1566 & -1.7424 \end{bmatrix} \qquad v_1 = \frac{\sqrt{2}}{2}\begin{bmatrix} -1 \\ 1 \end{bmatrix} \qquad v_2 = \frac{\sqrt{2}}{2}\begin{bmatrix} 1 \\ 1 \end{bmatrix}$$

In figure 6.10, we show a plot of the two vectors and the product of each with the $\mathbf{A}$ matrix. Both vectors are rotated and scaled by what are called the singular values: $\sigma_1 = 4$, and $\sigma_2 = 2$.

6.8.2 Finding the Classic SVD

As with diagonalization factoring, a square matrix has n equations like equation (6.12). Similarly, the n vector equations can be combined into a

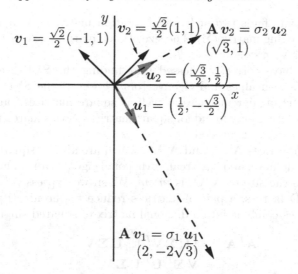

$$v_1 = \tfrac{\sqrt{2}}{2}(-1,1)$$

$$v_2 = \tfrac{\sqrt{2}}{2}(1,1) \quad \mathbf{A}\,v_2 = \sigma_2\,u_2$$
$$(\sqrt{3},1)$$

$$u_2 = \left(\tfrac{\sqrt{3}}{2}, \tfrac{1}{2}\right)$$

$$u_1 = \left(\tfrac{1}{2}, -\tfrac{\sqrt{3}}{2}\right)$$

$$\mathbf{A}\,v_1 = \sigma_1\,u_1$$
$$(2, -2\sqrt{3})$$

FIGURE 6.10: Multiplying the orthogonal v vectors by the $\mathbf{A}$ matrix rotates and scales the vectors, $\mathbf{A}v = \sigma\,u$.

matrix equation.

$$\mathbf{A}\mathbf{V} = \mathbf{A} \begin{bmatrix} | & | & & | \\ v_1 & v_2 & \cdots & v_n \\ | & | & & | \end{bmatrix} = \begin{bmatrix} | & | & & | \\ \sigma_1\,u_1 & \sigma_2\,u_2 & \cdots & \sigma_n\,u_n \\ | & | & & | \end{bmatrix}$$

$$= \begin{bmatrix} | & | & & | \\ u_1 & u_2 & \cdots & u_n \\ | & | & & | \end{bmatrix} \begin{bmatrix} \sigma_1 & 0 & \cdots & 0 \\ 0 & \sigma_2 & \cdots & 0 \\ \vdots & \vdots & \ddots & \vdots \\ 0 & 0 & \cdots & \sigma_n \end{bmatrix}$$

$$= \mathbf{U}\mathbf{\Sigma}$$

Note: This matrix equation is written as for a square matrix $(m = n)$. For rectangular matrices, the $\mathbf{\Sigma}$ matrix needs to be padded with either rows or columns of zeros to accommodate the matrix multiplication requirements. We will see this in the MATLAB examples. The number of nonzero σs is r, the rank of $\mathbf{A}$.

Then a factoring of $\mathbf{A}$ is $\mathbf{A} = \mathbf{U}\mathbf{\Sigma}\mathbf{V}^{-1}$, but a requirement of the v vectors is that they are orthogonal unit vectors such that $\mathbf{V}^{-1} = \mathbf{V}^{T}$, so the SVD factoring of $\mathbf{A}$ is

$$\boxed{\mathbf{A} = \mathbf{U}\mathbf{\Sigma}\mathbf{V}^{T}.} \tag{6.13}$$

The factoring finds two orthogonal rotation matrices ($\mathbf{U}$ and $\mathbf{V}^T$) and a diagonal stretching matrix ($\mathbf{\Sigma}$). The size of each matrix is: $\mathbf{A}$: $m \times n$, $\mathbf{U}$: $m \times m$, $\mathbf{\Sigma}$: $m \times n$, and $\mathbf{V}$: $n \times n$.

You may have correctly guessed that finding the SVD factors from $\mathbf{A}$ makes use of eigenvalues and eigenvectors. However, the SVD factorization works with rectangular matrices as well as square matrices, but eigenvalues and eigenvectors are only found for square matrices. So a square matrix derived form the values of $\mathbf{A}$ is needed.

Consider matrices $\mathbf{A}^T \mathbf{A}$ and $\mathbf{A} \mathbf{A}^T$, which are always square, symmetric, semi-positive definite, and have real, orthogonal eigenvectors. The size of $\mathbf{A}^T \mathbf{A}$ is $n \times n$, while the size of $\mathbf{A} \mathbf{A}^T$ is $m \times m$. When we express $\mathbf{A}^T \mathbf{A}$ and $\mathbf{A} \mathbf{A}^T$ with the SVD factors, a pair of matrices reduce to the identity matrix and two $\mathbf{\Sigma}$ matrices combine into a diagonal matrix of squared singular values.

$$
\begin{aligned}
\mathbf{A}^T \mathbf{A} &= \left(\mathbf{U} \mathbf{\Sigma} \mathbf{V}^T \right)^T \mathbf{U} \mathbf{\Sigma} \mathbf{V}^T \\
&= \mathbf{V} \mathbf{\Sigma}^T \mathbf{U}^T \mathbf{U} \mathbf{\Sigma} \mathbf{V}^T \\
&= \mathbf{V} \mathbf{\Sigma}^T \mathbf{\Sigma} \mathbf{V}^T \\
&= \mathbf{V} \mathbf{\Sigma}^2 \mathbf{V}^T \\
&= \mathbf{V}
\begin{bmatrix}
\sigma_1^2 & 0 & \cdots & 0 \\
0 & \sigma_2^2 & \cdots & 0 \\
\vdots & \vdots & \ddots & \vdots \\
0 & 0 & \cdots & \sigma_n^2
\end{bmatrix}
\mathbf{V}^T
\end{aligned}
$$

This factoring is the diagonalization of a symmetric matrix (section 6.5.1.2). It follows that the $\mathbf{V}$ matrix comes from the eigenvectors of $\mathbf{A}^T \mathbf{A}$. Likewise, the $\mathbf{\Sigma}$ matrix is the square root of the diagonal eigenvalue matrix of $\mathbf{A}^T \mathbf{A}$.

Similarly, the $\mathbf{U}$ matrix is the eigenvector matrix of $\mathbf{A} \mathbf{A}^T$.

$$
\begin{aligned}
\mathbf{A} \mathbf{A}^T &= \mathbf{U} \mathbf{\Sigma} \mathbf{V}^T \left(\mathbf{U} \mathbf{\Sigma} \mathbf{V}^T \right)^T \\
&= \mathbf{U} \mathbf{\Sigma}^T \mathbf{V}^T \mathbf{V} \mathbf{\Sigma} \mathbf{U}^T \\
&= \mathbf{U} \mathbf{\Sigma}^T \mathbf{\Sigma} \mathbf{U}^T \\
&= \mathbf{U} \mathbf{\Sigma}^2 \mathbf{U}^T \\
&= \mathbf{U}
\begin{bmatrix}
\sigma_1^2 & 0 & \cdots & 0 \\
0 & \sigma_2^2 & \cdots & 0 \\
\vdots & \vdots & \ddots & \vdots \\
0 & 0 & \cdots & \sigma_m^2
\end{bmatrix}
\mathbf{U}^T
\end{aligned}
$$

Although $\mathbf{U}$ comes from the eigenvectors of $\mathbf{A} \mathbf{A}^T$, calculating $\mathbf{U}$ as such is a poor choice for square matrices and may not be necessary for rectangular matrices.

6.8.2.1   Ordering the Columns of the SVD

The columns of the sub-matrices are ordered according to the values of the singular values ($\sigma_1 > \sigma_2 > \ldots > \sigma_n$). The columns of $\mathbf{U}$ and $\mathbf{V}$ are ordered to match the sorted singular values.

6.8.2.2 SVD of Square Matrices

Once $\mathbf{V}$ and $\boldsymbol{\Sigma}$ are known, $\mathbf{U}$ may be found directly from equation (6.13). Since $\mathbf{V}^T$ is orthogonal, its inverse is just $\mathbf{V}$. The diagonal structure of $\boldsymbol{\Sigma}$ makes its inverse the diagonal matrix with the reciprocals of the σs on the diagonal.

$$\mathbf{U} = \mathbf{A}\,\mathbf{V}\,\boldsymbol{\Sigma}^{-1}$$

In MATLAB, $\boldsymbol{\Sigma}^{-1}$ may be found with either the pseudo-inverse (`pinv`) function or the right-divide operator. For full rank matrices the `diag` function could quickly find the inverse of $\boldsymbol{\Sigma}$ as (`diag(1./diag(Sigma))`), but care would be needed to prevent a division by zero for singular matrices that have singular values of zero.

```
U = A*V*pinv(Sigma);
    or
U = (A*V)/Sigma;
```

6.8.2.3 SVD of Rectangular Matrices

To satisfy the size requirements for multiplying the SVD factors, the $\boldsymbol{\Sigma}$ matrix must be padded with rows of zeros for over-determined matrices and with columns of zeros for under-determined matrices. Figures 6.11 and 6.12 show the related sizes of the sub-matrices for over-determined and under-determined matrices.

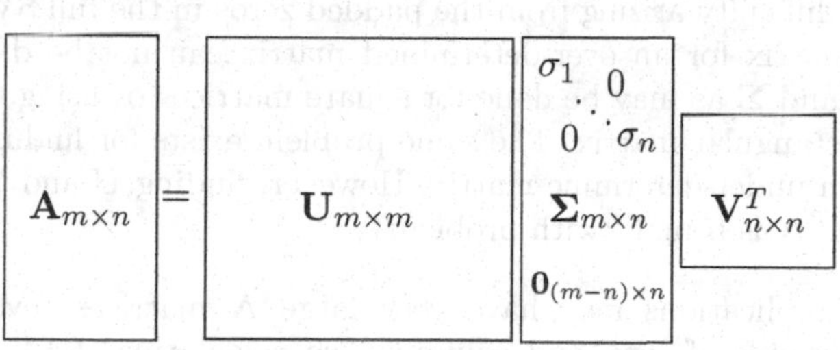

FIGURE 6.11: Shape of the SVD factors of an over-determined matrix.

Note that in the multiplication of the factors to yield the original matrix, $(m-n)$ columns of $\mathbf{U}$ for an over-determined matrix and $(n-m)$ rows of $\mathbf{V}^T$ for an under-determined matrix are multiplied by zeros from $\boldsymbol{\Sigma}$. They are not needed to recover $\mathbf{A}$ from its factors. Many applications of the SVD do not require the unused columns of $\mathbf{U}$ or the unused rows of $\mathbf{V}^T$. So the *economy* SVD is often used instead of the full SVD. The economy SVD removes the unused elements. Figure 6.13 shows the related sizes of the economy sub-matrices for over-determined matrices. The primary difference to be aware of when apply the economy SVD is a degradation of the unitary properties of $\mathbf{U}$

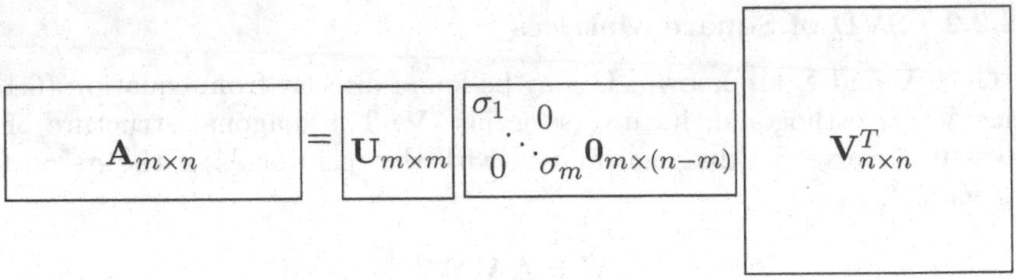

$$\mathbf{A}_{m \times n} = \mathbf{U}_{m \times m} \begin{bmatrix} \sigma_1 & 0 \\ & \ddots & \\ 0 & \sigma_m \end{bmatrix} \mathbf{0}_{m \times (n-m)} \quad \mathbf{V}^T_{n \times n}$$

FIGURE 6.12: Shape of the SVD factors of an under-determined matrix.

and V. For an over-determined matrix $\tilde{\mathbf{U}}^T \tilde{\mathbf{U}} = \mathbf{I}$, but $\tilde{\mathbf{U}}^T \tilde{\mathbf{U}} \neq \mathbf{I}$. Similarly, for an under-determined matrix $\tilde{\mathbf{V}}^T \tilde{\mathbf{V}} = \mathbf{I}$, but $\tilde{\mathbf{V}} \tilde{\mathbf{V}}^T \neq \mathbf{I}$.

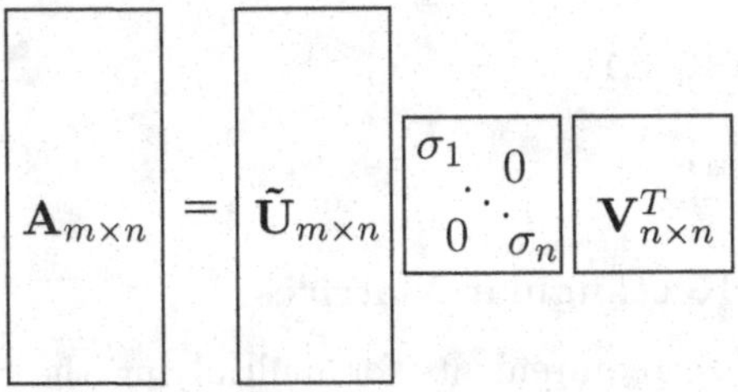

$$\mathbf{A}_{m \times n} = \tilde{\mathbf{U}}_{m \times n} \begin{bmatrix} \sigma_1 & 0 \\ & \ddots & \\ 0 & \sigma_n \end{bmatrix} \mathbf{V}^T_{n \times n}$$

FIGURE 6.13: Shape of the economy SVD factors of an over-determined matrix.

Another difficulty arising from the padded zeros in the full SVD $\boldsymbol{\Sigma}$ is that the full $\mathbf{U}$ matrix for an over-determined matrix can not be directly found from $\mathbf{A}$, $\mathbf{V}$, and $\boldsymbol{\Sigma}$ as may be done for square matrices or using the economy SVD of a rectangular matrix. The same problem exists for finding the full $\mathbf{V}$ matrix for an under-determine matrix. However, finding $\mathbf{U}$ and $\mathbf{V}$ from both $\mathbf{A}\mathbf{A}^T$ and $\mathbf{A}^T\mathbf{A}$ is fraught with problems

- Some applications may have very large $\mathbf{A}$ matrices (even hundreds of thousands of rows and columns), so calculating $\mathbf{A}\mathbf{A}^T$ for an over-determined matrix might take a very long time and risks having unacceptable round-off errors.

- If the eigenvectors of $\mathbf{A}^T\mathbf{A}$ and $\mathbf{A}\mathbf{A}^T$ are computed independent of each other, there can be a problem with certain columns of $\mathbf{U}$ or $\mathbf{V}$ needing to be multiplied by -1 for the factoring to be correct. It is best to do only one eigenvector calculation.

Of course, MATLAB's svd function and similar software do not use the classic algorithm to find the SVD, so MATLAB does not have the same challenges when computing the full SVD factors.

6.8.2.4   Classic Implementation

Code 6.4 lists a simple function that uses the classic algorithm to correctly find the SVD of a square matrix and the economy SVD of a rectangular matrix. As the comments indicate, the code shown here is not how MATLAB computes the SVD. The function is for educational purposes only.

```
function [U, S, V] = econSVD(A)
% ECONSVD - an implementation of Singular Value Decomposition (SVD)
%           using the classic algorithm.
%           It finds the full SVD when A is square and the economy
%           SVD when A is rectangular.
% [U,S,V] = econSVD(A) ==> A = U*S*V'
%
% MATLAB comes with a svd() function, which should normally be used.

% We use the smaller of A'*A or A*A' for eig calculation.
    [m, n] = size(A);
    if m < n                    % under-determined
        [U, S] = order_eigs(A*A');
        V = (A'*U)/S;
    else                        % square or over-determined
        [V, S] = order_eigs(A'*A);
        U = (A*V)/S;
    end
end

function [X, S] = order_eigs(B)
% helper function, B is either A'*A or A*A'
    [X2, S2] = eig(B);
    [S2, order] = sort(diag(S2), 'descend');
    X = X2(:, order);
    % There are no negative eigenvalues but we still need abs()
    % in the next line just because a 0 can be -0, giving an undesired
    % complex result.
    S = diag(sqrt(abs(S2)));
end
```

CODE 6.4: An implementation of Singular Value Decomposition (SVD) using the classic algorithm.

6.8.2.5   SVD Example: Square, full rank

```
                                   -3     3
>> A = [4 4; -3 3]           >> [U,S,V] = svd(A)
A =                          U =
       4      4                  -1.0000   -0.0000
```

```
       0.0000    1.0000           V =
  S =                              -0.7071   -0.7071
       5.6569         0            -0.7071    0.7071
            0    4.2426         >> U*S*V'
                                 ans =
                                   4.0000    4.0000
                                  -3.0000    3.0000
```

6.8.2.6   SVD Example: Square, singular

As expected for a singular matrix, one of the singular values is zero.

```
>> A = [2 3; 4 6]           S =
A =                            8.0623        0
   2    3                           0   0.0000
   4    6                    V =
>> rank(A)                     -0.5547   -0.8321
ans =                          -0.8321    0.5547
   1                        >> U*S*V'
>> [U,S,V] = svd(A)         ans =
U =                            2.0000    3.0000
     -0.4472   -0.8944         4.0000    6.0000
     -0.8944    0.4472
```

6.8.2.7   SVD Example: Rectangular

Note in the following over-determined $(m > n)$ example that with the full SVD the Σ matrix is padded with rows of zeros so that it has m rows. Similarly, an under-determined matrix $(m < n)$ is padded with columns of zeros so that it has n columns.

The more compact economy SVD, from the **econ** option to the **svd** function, is also shown.

```
                                      0     0.7077
>> A = [2 3; 4 10; 5 12]              0         0
A =                              V =
   2     3                         -0.3871    0.9220
   4    10                         -0.9220   -0.3871
   5    12                      >> U * Sigma * V'
>> [U, Sigma, V] = svd(A)       ans =
U =                               2.0000    3.0000
   -0.2053    0.9649   -0.1638    4.0000   10.0000
   -0.6243   -0.2580   -0.7373    5.0000   12.0000
   -0.7537   -0.0490    0.6554
Sigma =
   17.2482            0
```

```
                                          0      0.7077
                                   V =
>> [U, Sigma, V] = svd(A, 'econ')       -0.3871    0.9220
U =                                      -0.9220   -0.3871
   -0.2053     0.9649            >> U * Sigma * V'
   -0.6243    -0.2580            ans =
   -0.7537    -0.0490                     2.0000    3.0000
Sigma =                                   4.0000   10.0000
   17.2482         0                      5.0000   12.0000
```

6.8.3 How SVD Changes Vectors

With the exception of rotation matrices and eigenvectors, we normally see both rotation and stretching when a point or vector is multiplied by a matrix. When the matrix is represented by its SVD factors, we see rotation, stretching, and again rotation when we multiply sequentially by V^T, Σ, and U. Use the demonstration function in code 6.5 to experiment with several 2×2 matrices to see the rotation, stretching, and rotation. The demonstration starts with a set of points on a circle. In the end, the circle pattern will be changed to show a rotated ellipse. Figures 6.14 and 6.15 show example plots. The plots will be different for each matrix supplied.

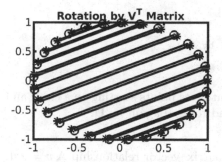

FIGURE 6.14: When each point is multiplied by V^T the points rotate around the circle. The lines show the initial and final positions of the points.

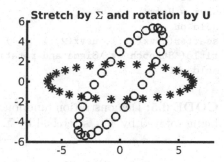

FIGURE 6.15: When each point is multiplied by Σ, the points around the circle are moved out from the center to form an ellipse. Then the ellipse is rotated when the points are multiplied by U.

6.8.4 Dimensionality Reduction

When a matrix contains important information that is obscured by noise or less significant details, the SVD factors can reveal the most important aspects of the data while removing the less significant details. We began our

```
function show_SVD(A)
% SHOW_SVD - A demonstration of how the U, S, and V matrices
%    from SVD rotate, stretch, and rotate vectors making a circle.
%    Try several 2x2 matrices to see how each behaves.

theta = linspace(0, 2*pi, 30);
x = [cos(theta); sin(theta)];

[U,S,V] = svd(A);

Vx = V'*x;
figure, plot(x(1,:), x(2,:), '*')
hold on
scatter(Vx(1,:), Vx(2,:))
for z = 1:30
    line([x(1,z), Vx(1,z)], [x(2,z), Vx(2,z)], 'Color', 'k')
end
title('Rotation by V^T Matrix')
hold off

svx = S*Vx;
figure, scatter(svx(1,:), svx(2,:), '*')
daspect([1 1 1])
usvx = U*svx;
hold on
scatter(usvx(1,:), usvx(2,:), 'o')
title('Stretch by \Sigma and rotation by U')
hold off
```

CODE 6.5: Demonstration function that plots points starting on a circle and being rotated by $\mathbf{V}^T$, stretched by $\mathbf{\Sigma}$ and rotated again by $\mathbf{U}$.

discussion of the SVD with the simple matrix–vector relationship $\mathbf{A}\,\boldsymbol{v} = \sigma\,\boldsymbol{u}$. Each set of vectors $\boldsymbol{v}$ and $\boldsymbol{u}$ along with the corresponding singular value σ are combined to build an *outer product* matrix of $\sigma\,\boldsymbol{u}\,\boldsymbol{v}^T$. The singular values are our guide to finding the most significant information. For most matrices, a few singular values will be much larger than the other singular values. Thus, in the complete matrix multiplication by outer products, the product terms with larger singular values hold the significant information and the lesser terms can be discarded with only minimal loss of information.

The *outer product* view of matrix multiplication builds the final matrix from a sum of the rank one matrices. The σ values are scalar multiples of each matrix (section 5.2.1.3).

$$\mathbf{A} = \sigma_1\,\boldsymbol{u}_1\,\boldsymbol{v}_1^T + \sigma_2\,\boldsymbol{u}_2\,\boldsymbol{v}_2^T + \cdots + \sigma_r\,\boldsymbol{u}_r\,\boldsymbol{v}_r^T$$

The first few σ values are larger than later values, which may be zero, or near zero. The *Eckart–Young* theorem developed in 1936 says that the closest rank $k < r$ matrix to the original comes from the SVD [7, 21]. The choice of k should be based on the singular values. It might be chosen such that the sum of the first k singular values constitute 95%, or another percentage, of the sum of all singular values. Gavish and Donoho [26] suggest that if the noise can be modeled independent of the data then k should be the number of singular values from the data that are larger than the largest singular value of the noise. They also suggest algorithms and equations for picking k when the noise can not be modeled. The rank-k approximation of $\mathbf{A}$ is given by the following sum of rank-1 matrices.

$$\mathbf{A}_k = \sigma_1 \, \boldsymbol{u}_1 \, \boldsymbol{v}_1^T + \sigma_2 \, \boldsymbol{u}_2 \, \boldsymbol{v}_2^T + \cdots + \sigma_k \, \boldsymbol{u}_k \, \boldsymbol{v}_k^T$$

Eckart–Young Theorem

Consider two different matrices, $\mathbf{A}$ and $\mathbf{B}$, of the same size. $\mathbf{A}$ has rank $r > k$. Define $\mathbf{A}_k$ as the sum of the first k rank-1 outer product matrices from the SVD. Both $\mathbf{A}_k$ and $\mathbf{B}$ have rank k. Then the Eckart–Young theorem tells us that $\mathbf{A}_k$ is a closer match to $\mathbf{A}$ than any other rank k matrix, $\mathbf{B}$ [21]. We express this as the norm (magnitude) of differences between matrices—$\|\mathbf{A} - \mathbf{B}\| > \|\mathbf{A} - \mathbf{A}_k\|$. This relationship holds for all matrix norms listed in appendix A.1.3.

The example in code 6.6, builds a simple pattern in the matrix and then adds noise to the matrix. We see that the best representation of the original matrix is found in the first three terms of the SVD outer product. The remaining two terms restore the noise, which we would just as well do without.

Code 6.7 shows another dimensionality reduction example. This one examines the SVD of an image. With the sum of a few rank-1 matrices from the SVD, the major parts of the image start to take shape. The higher terms add fine details. Something to think about: how is a low rank image from SVD different or the same as a low-pass filtered image? Figure 6.16 shows the singular values of the image, and the images with progressive rank are shown in figure 6.17.

6.8.5 Other Applications of the SVD

6.8.5.1 Pseudo-inverse

Since the SVD works for any matrix, it can also be used to calculate the inverse of a square matrix and the Moore-Penrose pseudo-inverse of both underdetermined and over-determined matrices (sections 5.8.2 and 5.9.3). Note that

```matlab
% File: bestKsvd.m
% Demonstrate the Eckart-Young theorem that says that the
% closest rank k matrix to the original comes from the SVD.

% Because of the random noise added, you may want to run this
% more than once. You should see that after about 3 SVD terms,
% the SVD matrix is closest (lowest mean-squared error) to the
% clean matrix before the noise was added. The later terms
% mostly add the noise back into the matrix. We can see why a
% rank k (k < r) matrix from SVD may be preferred to the
% full rank r matrix.

A1 = 5*eye(5) + 5*rot90(eye(5));
Aclean = [A1 A1];                   % A test pattern
A = [A1 A1] + 0.5*randn(5, 10);     % add some noise
[U, S, V] = svd(A);
figure, plot(diag(S), '*'), title('Singular Values')
Ak = zeros(5, 10);
figure

% Look at each rank k matrix from SVD
for k = 1:5
    disp(['SVD terms: ', num2str(k)])
    Ak = Ak + U(:,k) * S(k,k) * V(:,k)';
    disp( Ak )
    subplot(2, 3, k), imshow(Ak, []), title(['rank = ', num2str(k)])
    disp(['rank: ', num2str(rank(Ak))])
    disp(['Mean-squared Error: ', num2str(immse(A, Ak))])
    disp(['No noise Mean-squared Error: ', num2str(immse(Aclean, Ak))])
    disp(' ')
end
subplot(2,3,6), imshow(A, []), title('Original')
disp('Original matrix with noise')
disp(A)

% NOTE:
% >> rank(A1)
% ans =
%      3
% >> rank(A)
% ans =
%      5
```

CODE 6.6: Demonstration showing that a rank-k SVD can retain the essence of data with some noise removed.

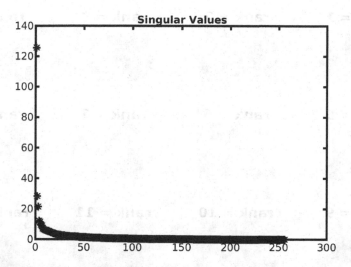

FIGURE 6.16: The singular values of the *cameraman* image. Most of the important data to the image is contained in a small number of rank 1 images.

```
% File: bestKsvdImage.m
% Use an image to demonstrate the Eckart-Young theorem that says
% that the closest rank k matrix to the original comes from the SVD.


% Use a public domain image from the Image Processing Toolbox.
A = im2double(imread('cameraman.tif'));
[m, n] = size(A);

[U, S, V] = svd(A);
figure, plot(diag(S), '*'), title('Singular Values')
%%
Ak = zeros(m, n);
figure

% Look at each rank k matrix from SVD
for k = 1:15
    disp(['SVD terms: ', num2str(k)])
    Ak = Ak + U(:,k) * S(k,k) * V(:,k)';
    subplot(4, 4, k), imshow(Ak, []), title(['rank = ', num2str(k)])
    disp(['rank: ', num2str(rank(Ak))])
    disp(['Mean-squared Error: ', num2str(immse(A, Ak))])
    disp(' ')
end
subplot(4,4,16), imshow(A, []), title('Original')
```

CODE 6.7: Script showing SVD dimensionality reduction of an image.

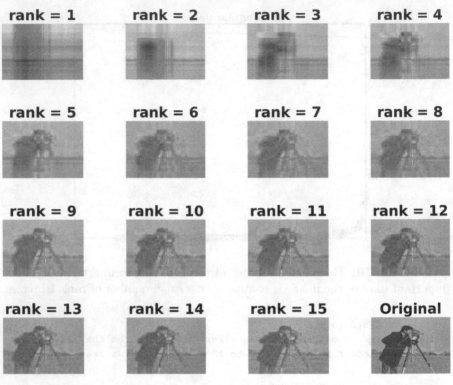

FIGURE 6.17: Images with progressive rank from the sum of SVD outer products.

unlike the matrix equations, the same equation of the SVD sub-matrices finds the pseudo-inverse for both under-determined and over-determined matrices. Recall that since $\mathbf{U}$ and $\mathbf{V}$ are orthogonal, their inverse are just their transpose.

$$\mathbf{A}^{+} = \left(\mathbf{U}\,\mathbf{\Sigma}\,\mathbf{V}^{T} \right)^{+} = \mathbf{V}\,\mathbf{\Sigma}^{+}\,\mathbf{U}^{T}$$

Since $\mathbf{\Sigma}$ is a diagonal matrix its inverse is the reciprocal of the nonzero elements on the diagonal. The zero elements remain zero.

```
>> S = diag([2 3 0])
S =
     2     0     0
     0     3     0
     0     0     0
>> pinv(S)
ans =
    0.5000         0         0
         0    0.3333         0
         0         0         0
```

Here is an example of the pseudo-inverse of an under-determined matrix, which is calculated with the SVD, the matrix equation, and the pinv function.

```
>> A = [2 4 5; 3 8 9]          % The right-divide operator
A =                            >> A'/(A*A')
     2     4     5             ans =
     3     8     9                 1.4390   -0.7561
>> [U,S,V] = svd(A);             -1.1707    0.6829
>> V*pinv(S)*U'                   0.5610   -0.2439
ans =                          >> pinv(A)
    1.4390   -0.7561           ans =
   -1.1707    0.6829               1.4390   -0.7561
    0.5610   -0.2439              -1.1707    0.6829
                                   0.5610   -0.2439
```

Here is an example of the pseudo-inverse of an over-determined matrix, which is again calculated with the SVD, the matrix equation, and the pinv function.

```
>> A = [2 3; 4 8; 5 9]         >> (A'*A)\A'
A =                            ans =
     2     3                       1.4390   -1.1707    0.5610
     4     8                       -0.7561   0.6829   -0.2439
     5     9                    >> pinv(A)
>> [U, S, V] = svd(A);         ans =
>> V*pinv(S)*U'                    1.4390   -1.1707    0.5610
ans =                             -0.7561    0.6829   -0.2439
    1.4390   -1.1707    0.5610
   -0.7561    0.6829   -0.2439
```

MATLAB uses the SVD method for computing the pseudo-inverse for both over-determined and under-determined matrices.

6.8.5.2 Projection and the Economy SVD

The SVD has application to vector projections as described in section 5.9. We begin with an observation related to the padded rows of zeros in the Σ matrix for over-determined systems. As discussed in section 6.8.2.3, some columns of the U matrix do not contribute to the final SVD product because they get multiplied by zeros in Σ.

Over-determined Full SVD

$$\mathbf{A} = \begin{bmatrix} \tilde{\mathbf{U}} & \mathbf{U}_{unused} \end{bmatrix} \begin{bmatrix} \tilde{\Sigma} \\ \mathbf{0} \end{bmatrix} \mathbf{V}^T$$

Economy SVD

The economy SVD removes the unused columns of $\mathbf{U}$ and the zero rows of $\mathbf{\Sigma}$.

$$\mathbf{A} = \tilde{\mathbf{U}}\tilde{\mathbf{\Sigma}}\mathbf{V}^T$$

The economy SVD is a valid factoring. The only noticeable application difference is that $\tilde{\mathbf{U}}$ is not unitary: $\tilde{\mathbf{U}}^T\tilde{\mathbf{U}} = \mathbf{I}$, but $\tilde{\mathbf{U}}\tilde{\mathbf{U}}^T \neq \mathbf{I}$. The economy SVD is used to solve over-determined systems of equations ($\mathbf{A}\boldsymbol{x} = \boldsymbol{b}$) and projection approximations.

$$\hat{\boldsymbol{x}} = \mathbf{A}^+\boldsymbol{b} = \mathbf{V}\,\tilde{\mathbf{\Sigma}}^+\tilde{\mathbf{U}}^T\boldsymbol{b}$$

Two pairs of matrices in the projection equation reduce to identity matrices.

$$\begin{aligned} \boldsymbol{p} \ &= \mathbf{A}\,\hat{\boldsymbol{x}} \\ &= \tilde{\mathbf{U}}\,\tilde{\mathbf{\Sigma}}\,\mathbf{V}^T\mathbf{V}\,\tilde{\mathbf{\Sigma}}^+\tilde{\mathbf{U}}^T\boldsymbol{b} \\ &= \tilde{\mathbf{U}}\,\tilde{\mathbf{U}}^T\boldsymbol{b} \end{aligned}$$

As mentioned in section 5.9.3.2, orthonormal basis vectors of the $\mathbf{A}$ matrix are needed for the projection. Either the modified Gram–Schmidt algorithm, QR factorization, or the $\tilde{\mathbf{U}}$ from the economy SVD may be used. The MATLAB function `orth` uses the economy SVD method to compute orthonormal basis vectors.

Code 6.8 lists example code showing four ways to achieve projection of an over-determined system. The plot of the projections is shown in figure 6.18.

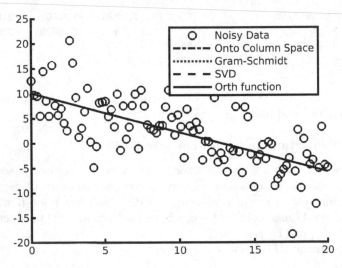

FIGURE 6.18: Four vector projection alternatives. The projection lines appear as one line because they are on top of each other.

```
% File: four_projections.m
% Comparison of 4 ways to compute vector projections of an
% over-determined system.
%
%% Over-determined System with noise
t = linspace(0,20);
y = 10 - 0.75.*t + 5*randn(1,100);
b = y';
scatter(t,b)
A = ones(100,2); % A is the design matrix
A(:,2) = t';


%% basic pseudo-inverse projection onto the column space of A
x_hat = (A'*A)\(A'*b);
p1 = A*x_hat;


%% Alternate Gram-Schmidt
G = mod_gram_schmidt(A);
% u1 = G(:,1);      % could use vectors for projection
% u2 = G(:,2);
% p2 = b'*u1*u1 + b'*u2*u2;
p2 = G*G'*b;        % or matrix multiplication accomplishes the same


%% Econ SVD projection
[U, ~, ~] = svd(A, 'econ');
p3 = U*U'*b;


%% MATLAB's Orth function
O = orth(A);  % O and U should be the same
p4 = O*O'*b;


%% Plot
figure, hold on, scatter(t, b)
plot(t, p1), plot(t, p2), plot(t, p3) plot(t, p4)
hold off
legend('Noisy Data', 'Onto Column Space', 'Gram-Schmidt', ...
    'SVD', 'Orth function')
```

CODE 6.8: Script showing four ways to calculate vector projection of an over-determined system.

6.8.5.3 Rank from the SVD

There are multiple ways to compute the rank of a matrix. An example of using Gaussian elimination to find the rank of a matrix is show in appendix A.2.3. The MATLAB **rank** function uses the SVD. The number of singular values that are greater than a small tolerance is the rank of the matrix.

According to the MATLAB documentation, the SVD algorithm requires more computation than some alternatives, but it is the most reliable.

In the following example, we start with a full rank matrix and make the third row the difference of the first two rows. The final singular value is close to zero and the rank of the singular matrix is 2.

```
>> A(3,:) = A(1,:) - A(2,:)     >> svd(A)
A =                             ans =
      6      3      6               11.1167
      4     -3      3                6.3576
      2      6      3                0.0000
>> rank(A)
ans =                           >> fprintf("%g\n", ans(3))
      2                         1.32734e-15
```

6.8.5.4   Fundamental Subspaces from the SVD

We can use rank determined regions of the SVD sub-matrices to identify the four fundamental subspaces of a matrix as described below and in appendix A.3 [71]. The identified regions in figure 6.19 show the four fundamental subspaces from $\tilde{\mathbf{U}}$, $\mathbf{U}_{null}$, $\tilde{\mathbf{V}}^T$, and $\mathbf{V}_{null}^T$.

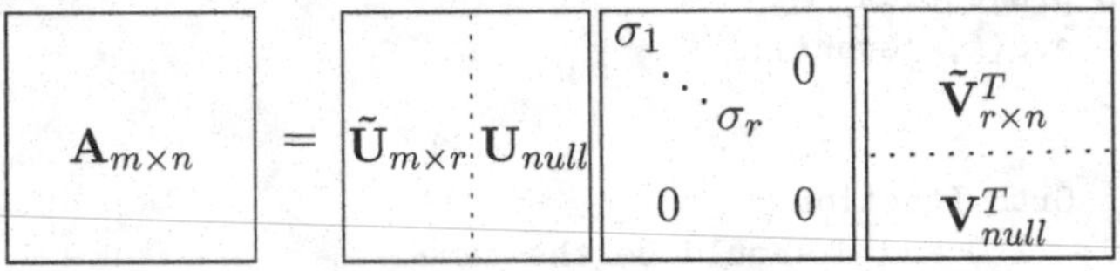

FIGURE 6.19: The SVD has r (matrix rank) nonzero singular values on the diagonal of $\boldsymbol{\Sigma}$. The columns of $\mathbf{U}$ and rows of $\mathbf{V}^T$ that are multiplied by either nonzero or zero valued singular values define the column space, row space, left null space, and the null space of the matrix.

Column Space
   The first r columns of $\mathbf{U}$ ($\tilde{\mathbf{U}}$) are the basis vectors for the vector space spanned by the columns of $\mathbf{A}$.

Row Space
   The first r rows of $\mathbf{V}^T$ ($\tilde{\mathbf{V}}^T$) are the basis vectors for the space spanned by the rows of $\mathbf{A}$.

Left Null Space
   Columns r to m of $\mathbf{U}$ ($\mathbf{U}_{null}$) are the transpose of the basis vectors for the space spanned by the row vectors $\boldsymbol{u}$ such that $\boldsymbol{u}\,\mathbf{A} = 0$. Note that square, full-rank matrices do not have a left null space.

Null Space

Rows r to n of $\mathbf{V}^T$ ($\mathbf{V}_{null}^T$) are the transpose of the basis vectors for the space spanned by the column vectors v such that $\mathbf{A}v = 0$. Note that square, full-rank matrices do not have a null space. MATLAB's `null` function returns these vectors.

6.8.5.5   Condition Number

The singular values of a singular matrix will contain one or more zeros. Likewise, matrices that are close to singular will contain singular values that are very close to zero. As described in section 5.5.4.3, the solution to a poorly conditioned matrix equation is sensitive to perturbations of the elements of $\mathbf{A}$ or b. Viewing the solution to $\mathbf{A}x = b$ from the perspective of the outer product of the SVD gives us an intuition into the sensitivity of x to perturbations in $\mathbf{A}$ or b [28].

$$x = \mathbf{A}^{-1}b = \left(\mathbf{U}\,\Sigma\,\mathbf{V}^T\right)^{-1} b$$

$$= \sum_{i=1}^{n} \frac{u_i^T b}{\sigma_i} v_i$$

The scalar fractions $\left(u_i^T b\right)/\sigma_i$ are dot products divided by singular values. Thus the magnitude of the singular values has a significant impact on the sensitivity of the $\mathbf{A}x = b$ problem. A matrix with singular values close to zero is poorly conditioned.

The $\|\cdot\|_2$ condition number of a matrix may be estimated by the ratio of the largest and smallest singular values.

$$C = \frac{\sigma_{max}}{\sigma_{min}}$$

A full rank matrix will have a fairly small condition number. Singular and near singular matrices will have condition numbers of infinity or very large (several thousand). Thus the condition number is a quick invertible test. To avoid division by zero, MATLAB uses the reciprocal of the condition number. The `rcond` function calculate an estimate for the reciprocal $\|\cdot\|_1$ condition number. If $\mathbf{A}$ is well conditioned, `rcond(A)` is near 1.0. If $\mathbf{A}$ is poorly conditioned, `rcond(A)` is near 0.

When using the left-divide operator to find the solution to a matrix equation, one may occasionally see a warning message such as follows.

Warning: Matrix is close to singular or badly scaled. Results may be inaccurate. RCOND = 3.700743e-18.

Here is some code to demonstrate zero or near zero singular values for singular and near singular matrices. Due to the use of random number generators, the output changes with each run of the example. But the first SVD calculation yields one singular value of zero, while one singular value from the second set will be close to zero.

```
>> A = randi(10, 4);   % Start with a random, usually full rank matrix
>> A(:,4) = sum(A(:,1:3), 2)/3; % make singular
>> svd(A)                        % zero singular value
ans =
   24.5089
    6.1966
    4.3492
    0.0000

>> A(:,4) = A(:,4) + 0.1*randn(4,1); % close to singular
>> svd(A)                        % near zero singular value
ans =
   24.4969
    6.2017
    4.3498
    0.1082
```

6.8.5.6 Polar Decomposition

There is another factoring of a matrix that uses the sub-matrices of the SVD. The *polar decomposition* splits the matrix up into a symmetric matrix and an orthogonal matrix. The factoring is $\mathbf{A} = \mathbf{R}\,\mathbf{Q}$, which is intended to be a generalization to matrices of the polar representation of vectors on a complex plane, $z = r\,e^{i\theta}$, where r is the scalar length of z and $e^{i\theta}$ gives the direction of the vector according to Euler's complex exponential formula. In the polar decomposition, $\mathbf{Q}$ is a unitary rotation matrix, and $\mathbf{R}$ has the same $\|\cdot\|_2$ matrix norm as $\mathbf{A}$. But with multiplication by a vector, the $\mathbf{R}$ matrix will both scale and rotate the vector. It can be found by simply inserting an identity matrix in the form of $\mathbf{U}^T\mathbf{U}$ into the SVD equation.

$$
\begin{aligned}
\mathbf{A} &= \mathbf{U}\,\boldsymbol{\Sigma}\,\left(\mathbf{U}^T\mathbf{U}\right)\,\mathbf{V}^T \\
&= \left(\mathbf{U}\,\boldsymbol{\Sigma}\,\mathbf{U}^T\right)\left(\mathbf{U}\,\mathbf{V}^T\right) \\
&= \mathbf{R}\,\mathbf{Q}
\end{aligned}
$$

$$
\mathbf{R} = \mathbf{U}\,\boldsymbol{\Sigma}\,\mathbf{U}^T
$$

$$
\mathbf{Q} = \mathbf{U}\,\mathbf{V}^T
$$

The polar decomposition has application to computer graphics and materials engineering where it is used to decompose stress tensors [8].

6.9 Principal Component Analysis (PCA)

As the name implies, Principal Component Analysis (PCA) accentuates the principal components of data such that the salient features of the data are more apparent. Like the SVD, PCA reduces the dimensionality of data emphasizing the distinguishing characteristics while removing the uncorrelated noise. The PCA algorithm maps the data into new axes called the PCA space. From the PCA space, the data may be reconstructed with reduced dimension. However, it is more common to use the PCA space data for classification and recognition applications. In the PCA space, similar items are grouped together and separated from clusters of dissimilar items.

PCA can help if you wish to automate the classification of items into groups based on measurable features. PCA uses correlations between the features to find a linear combination of the features such that similar items are clustered together in the PCA space. PCA uses the most significant eigenvectors found from a covariance matrix to find the linear combinations of the features. Use of a covariance matrix is significant to PCA as it reveals how similar and dissimilar (at variance) the data samples are to each other. Then, when we multiply sample data by the most significant eigenvectors (determined by the eigenvalues), the differences and similarities between the items become more obvious.

If you have n measured features from m sources (samples) then the data can be arranged in a matrix. Examples of potential data sets include the stock values from m companies taken over n days, opinions of m customers regarding n competing products, or the grades of m students in n courses. In the examples presented here, the matrix columns hold the measured features and each row holds the data taken from one source. Although with adjustments to the equations, the data may be transposed with the features in rows and the samples in columns.

The first step in the algorithm is to subtract the mean of each feature from the data so that each feature has a zero mean. This is needed to make a covariance matrix.

$$\mathbf{D} = \begin{bmatrix} | & | & & | \\ (\boldsymbol{x}_1 - \mu_1) & (\boldsymbol{x}_2 - \mu_2) & \cdots & (\boldsymbol{x}_n - \mu_n) \\ | & | & & | \end{bmatrix}$$

Recall from section 3.6.3 that a covariance matrix is symmetric with the variance of each feature on the diagonal and the covariance between features in the off-diagonal positions.

$$\mathbf{Cov_x} = \frac{\mathbf{D}^T \mathbf{D}}{m - 1}$$

$$\mathbf{Cov_x} = \begin{bmatrix} Var(x_1, x_1) & Cov(x_1, x_2) & \cdots & Cov(x_1, x_n) \\ Cov(x_2, x_1) & Var(x_2, x_2) & \cdots & Cov(x_2, x_n) \\ \vdots & \vdots & \ddots & \vdots \\ Cov(x_n, x_1) & Cov(x_n, x_2) & \cdots & Var(x_n, x_n) \end{bmatrix}$$

Next, the eigenvectors, $\mathbf{V}$, and eigenvalues of the covariance matrix are found. The eigenvalues are used to sort the eigenvectors. Because we wish to achieve dimensionality reduction, only the eigenvectors corresponding the k largest eigenvalues are used to map the data onto the PCA space.

$$\mathbf{W} = \begin{bmatrix} \boldsymbol{v}_1 & \cdots & \boldsymbol{v}_k \end{bmatrix}$$

Note: The $\mathbf{V}$ sub-matrix from the SVD of the $\mathbf{D}$ matrix could be used as the eigenvectors (`[~, ~, V] = svd(D)`).

$\mathbf{W}$ gives us the linear combinations of the data features to project the data onto the PCA space, which is found by multiplying the $\mathbf{W}$ matrix by each sample vector.

$$\mathbf{Y} = \mathbf{D}\mathbf{W}.$$

For purposes of classification and object recognition, the data in the PCA space ($\mathbf{Y}$) is used to display or compare the data.

To reconstruct the original data with reduced dimensionality, we just reverse the steps.

$$\begin{aligned} \tilde{\mathbf{X}} &= \mathbf{D} + \mu \\ &= \mathbf{W}(\mathbf{W}^T \mathbf{D}) + \mu \\ &= \mathbf{W}\mathbf{Y} + \mu \end{aligned}$$

Transposed data

If the measured features are held in the rows with the samples in the columns, then the adjusted equations are as follows.

$$\mathbf{D} = \begin{bmatrix} — & (\boldsymbol{x}_1 - \mu_1) & — \\ — & (\boldsymbol{x}_2 - \mu_2) & — \\ \vdots & \vdots & \vdots \\ — & (\boldsymbol{x}_n - \mu_n) & — \end{bmatrix}$$

$$\mathbf{Cov_x} = \frac{\mathbf{D}\mathbf{D}^T}{n-1}.$$

$$\mathbf{Y} - \mathbf{W}^T \mathbf{D}.$$

6.9.1 PCA for Data Analysis

To illustrate what PCA does with a simple plot, the code listing in code 6.9 shows a PCA example with only two data features. Noise was added to the data to show how dimensionality reduction separates the correlated data from the uncorrelated noise. In this example, the data was reconstructed with reduced dimensionality. Notice the lines in figure 6.20 showing the principal and minor axes. The orientation of the principal axis is aligned with the correlation of the data, while the minor axis is aligned with the uncorrelated noise. Thus, as we also saw in section 6.8.4 for the SVD, the lower rank data is preferred to the full rank data that contains noise. Figure 6.20 shows a plot demonstrating that the reconstructed data is aligned with the correlation between the columns of the original data.

Note:

- The MATLAB *Statistics and Machine Learning Toolbox* has a PCA function, but the direct calculation is only a few lines of code.

- The example in code 6.9 uses the `repmat` MATLAB function that has not been previously used. The `repmat` function creates a matrix with copies of a scalar, vector, or matrix. In this case, the variable `mu` is a row vector, so `repmat(mu, m, 1)` makes a matrix with m rows, each holding one copy of the `mu` vector. This allows use of element-wise subtraction of the mean from each column.

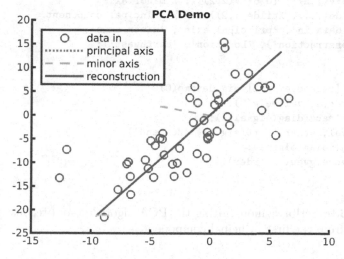

FIGURE 6.20: The principal component of the reconstructed data from PCA is aligned with the correlation between the columns of the original data.

```
% File: PCA_demo.m
% PCA - 2D Example
n = 2;
m = 50;
% two columns random but correlated data
X = zeros(m, n);
X(:,1) = 5*randn(m, 1) - 2.5;
X(:,2) = -1 + 1.5 * X(:,1) + 5*randn(m,1);
X = sortrows(X,1);


%% find PCA from eigenvectors of covariance
mu = mean(X);
D = X - repmat(mu, m, 1);
covx = D'*D/(m - 1);        % sources in rows, features in columns
[V, S] = eig_decomp(covx);
% project onto PCA space
W = V(:,1);                 % only use dominant eigenvector
Y = D * W;
disp('Eigenvalue ratios to total')
disp(S/trace(S))
% Data reconstruction
Xtilde = Y*W' + repmat(mu, m, 1);
% plots
figure, scatter(X(:,1), X(:,2), title('PCA Demo');
hold on
plot([0 5*V(1,1)], [0 5*V(2,1)]);  % principal axis
plot([0 5*V(1,2)], [0 5* V(2,2)]); % minor axis
plot(Xtilde(:,1), Xtilde(:,2));    % principal component
legend({'data in', 'principal axis', 'minor axis', ...
    'reconstruction'}, 'Location', 'northwest');
hold off

function [eigvec,eigval]=eig_decomp(C)
    [eigvec,eigval]=eig(C);
    eigval=abs(diag(eigval)');
    [eigval, order]=sort(eigval, 'descend');
    eigval=diag(eigval);
    eigvec=eigvec(:, order);
end
```

CODE 6.9: Script demonstrating the PCA algorithm and reconstruction of the data from the first principal component.

6.9.2   PCA for Classification

An important application of PCA is classification, which reduces the dimensions of the data for the purpose of making it easier to see how the

---

Is PCA the same as least squares regression?

The plot in figure 6.20 makes it appear that the reconstructed data from PCA is the same data fit as achieved from least squares regression; however, it differs in two ways.

1. In least squares regression, the minimized error is parallel to the y axis; whereas, in PCA the minimized error is orthogonal to the data projection line (principal axis).

2. In least squares regression, the data fit is an extension of solving a system of linear equations for over-determined systems. In PCA, the data fit is a by-product of dimensionality reduction.

---

attributes of items of the same type are similar and differ from items of other types. In this example, we will look at a public domain dataset and produce a scatter plot showing how groups of the data are clustered in the PCA space.

The public domain data is in a file named `nndb_flat.csv`, which is part of the downloadable files from the book's website. The file can also be downloaded from the Data World website.[2] The data comes from the USDA National Nutrient Database and was made available by Craig Kelly. It lists many different foods and identifies food groups, nutrition, and vitamin data for each food. We will use the nutrition information in the analysis. Code listings 6.10 and 6.11 show the script for the PCA algorithm and plotting of the results. The scatter plot in figure 6.21 shows the first two principal components for 8,618 foods in seven groupings. The plot shows clusters of the foods within the groups and which food groups are nutritionally similar or different. Some of the food groups that are contained in the data are not plotted.

To see a list of all food groups in the data set, type the following in the command window after the `FoodGroups` variable is created by the script.

```
>> categories(FoodGroups)
```

6.9.3   PCA for Recognition

A fun example of how PCA can be used is for face recognition. The eigenvectors found from PCA give us a matrix for multiplying images of peoples faces. The PCA space data give a more succinct representation of whom is pictured in each image. The Euclidean distances between the images in the PCA space are calculated to find the most likely match of who is in each

---

[2]https://data.world/

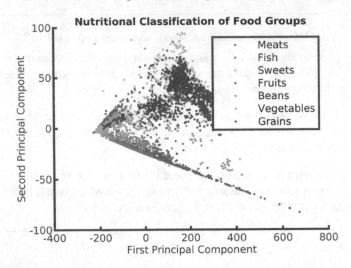

FIGURE 6.21: Scatter plot of the first two principal components from nutrition data for 8,618 foods from seven food groups. The data clusters are easier to see with color, so be sure to run the script.

picture. An example implementation with quite accurate results is found on the MathWorks File Exchange web site [2].

Recommended Additional Reading

Alaa Tharwat wrote a very good tutorial on PCA and posted it on MathWorks File Exchange [73]. The tutorial presents both the SVD and covariance PCA algorithms.

```
% File: PCA_food.m
% PCA classification and visualization of nutrition in various
% food groups. Food types are in rows.
% Nutrition attributes (features) are in columns, which are:
% Energy_kcal, Protein_g, Fat_g, Carb_g, Sugar_g, Fiber_g
food = readtable('nndb_flat.csv');
X = food{:,8:13}; % nutrition columns
k = 2;
```

Continued in code 6.11.

CODE 6.10: Part 1: Script for PCA classification of food groups by nutritional content.

Continued from code 6.10.

```
%% find PCA using covariance method
[m, n] = size(X);
mu = mean(X);
D = X - repmat(mu, m, 1);
W = V(:,1:k);                    % only use top two dominant eigenvectors
Y = D * W;

%% Define food group data
FoodGroups = categorical(food.FoodGroup);
meats = Y(any(FoodGroups == 'Beef Products', ...
    'Lamb, Veal, and Game Products', 'Pork Products', ...
    'Poultry Products', 'Sausages and Luncheon Meats',2), :);
fish = Y(FoodGroups == 'Finfish and Shellfish Products', :);
sweets = Y(FoodGroups == 'Sweets', :);
fruit = Y(FoodGroups == 'Fruits and Fruit Juices', :);
beans = Y(FoodGroups == 'Legumes and Legume Products', :);
veg = Y(FoodGroups == 'Vegetables and Vegetable Products', :);
grains = Y(any(FoodGroups == 'Baked Products', 'Breakfast Cereals', ...
    'Cereal Grains and Pasta',2), :);

%% plot food groups on the PCA space
figure
hold on
scatter(meats(:,1), meats(:,2), '.')
scatter(fish(:,1), fish(:,2), '.')
scatter(sweets(:,1), sweets(:,2), '.')
scatter(fruit(:,1), fruit(:,2), '.')
scatter(beans(:,1), beans(:,2), '.')
scatter(veg(:,1), veg(:,2), '.')
scatter(grains(:,1), grains(:,2), '.')
hold off
legend('Meats', 'Fish', 'Sweets', 'Fruits', 'Beans', ...
    'Vegetables', 'Grains', 'Location', 'northeast');
title('Nutritional Classification of Food Groups');
xlabel('First Principal Component');
ylabel('Second Principal Component');
%% Private function
function [eigvec,eigval]=eig_decomp(C)
    [eigvec,eigval]=eig(C);
    eigval=abs(diag(eigval)');
    [eigval, order]=sort(eigval, 'descend');
    eigval=diag(eigval);
    eigvec=eigvec(:, order);
end
```

CODE 6.11: Part 2: Script for PCA classification of food groups by nutritional content.

6.10   Eigenvector Animation Code

The code listed in code listings 6.12 and 6.13 is the animation function mentioned in section 6.2.

```
function eigdemo(A)
%% Eigenvector, eigenvalue Demo for input 2x2 matrix.
%
%  This function shows an animation demonstrating that the product
%  (A*x) is inline with the vector x when, and only when, x is an
%  eigenvector. The A (input) matrix must be a 2x2 matrix with real
%  eigenvalues.
%
%  The red rotating line shows possible unit vectors, x.
%  The blue moving line is v = A*x.
%  The green fixed arrows are the product (lambda*eigenvectors == A*ex),
%
%  Notice that when the red rotating vector is inline win an
%  eigenvector %  that the blue line matches the green arrows. The blue
%  line is also (-1) of the green arrow when the red rotating line is
%  (-1) of an eigenvector.
%
if size(A,1) ~= 2 || size(A,2) ~= 2
    disp('Matrix A must be 2x2')
    return;
end
theta = linspace(2*pi,0,100);
xv = cos(theta);
yv = sin(theta);
v = A*[xv; yv];
[ex, lambda] = eig(A);
if ~isreal(ex) || ~isreal(lambda)
    disp('Complex Eigenvector results, use another A matrix');
    return;
end
ep = A*ex;        % same as A*ex == lambda*ev
fprintf('eigenvector 1: %f %f\n', ex(1,1), ex(2,1));
fprintf('eigenvector 2: %f %f\n', ex(1,2), ex(2,2));
fprintf('A*eigenvector 1: %f %f\n', ep(1,1), ep(2,1));
fprintf('A*eigenvector 2: %f %f\n', ep(1,2), ep(2,2));
fprintf('lambda: %f %f\n', lambda(1,1), lambda(2,2));
M = ceil(max(vecnorm(v)));
```

continued in code 6.13.

CODE 6.12: Part 1: Animation function showing the relationship between the product of a matrix with a vector and the matrix's eigenvectors.

continuation of code 6.12.

```
figure; axis([-M M -M M]);
daspect([1 1 1]);
hold on
l1 = line([0 1], [0 0], 'LineWidth', 3, 'Color', 'r');
l2 = line([0 1], [0 0], 'LineWidth', 3, 'Color', 'b');
plot2vector(ep(:,1), 'g', 2); % lambda 1 * eigenvector 1
plot2vector(ep(:,2), 'g', 2); % lambda 2 * eigenvector 2
hold off
for k = 1:3          % make three loops
    for i = 1:100
        l1.set('XData', [0 xv(i)], 'YData', [0 yv(i)]);
        l2.set('XData', [0 v(1,i)], 'YData', [0 v(2,i)]);
        drawnow;
        pause(0.1);
    end
end
end

function plot2vector(x, color, LineWidth)
    if iscolumn(x)
        x = x';
    end
    plot([0, x(1)], [0, x(2)], 'Color', color, 'LineWidth', LineWidth)
end
```

CODE 6.13: Part 2: Animation function showing the relationship between the product of a matrix with a vector and the matrix's eigenvectors.

6.11   Exercises

Exercise 6.1   Finding Eigenvalues and Eigenvectors

1. Consider the following matrix.

$$\mathbf{A} = \begin{bmatrix} 9 & 0 \\ 7 & 5 \end{bmatrix}$$

(a) Use MATLAB to find its eigenvalues and eigenvectors.

(b) Form the two characteristic equations and use MATLAB's **null** command to verify the eigenvectors.

(c) Use MATLAB to verify that $\mathbf{A}\mathbf{x} = \lambda\mathbf{x}$ for each eigenvalue, eigenvector pair.

2. Use MATLAB's `roots` function to find the roots to the following polynomial.
$$x^4 + 4x^3 - 7x^2 - 22x + 24 = 0$$

3. Find the diagonalization factors of the following two matrices. Verify the factorization by multiplying the factors. Notice that the second matrix is symmetric. How is its diagonalization simplified?

(a)
$$\mathbf{A} = \begin{bmatrix} -16 & -3 & 49 \\ 5 & 2 & -15 \\ -5 & -1 & 16 \end{bmatrix}$$

(b)
$$\mathbf{S} = \begin{bmatrix} 25 & 4 & -4 \\ 4 & -9 & 8 \\ -4 & 8 & 18 \end{bmatrix}$$

Exercise 6.2   Powers of a Matrix

1. Making use of the diagonalization of the following matrix, find $\mathbf{A}^{20}$. You will find that the eigenvalues and eigenvectors are complex.

$$\mathbf{A} = \begin{bmatrix} -3 & 4 & 2 \\ -2 & 6 & 3 \\ 2 & -5 & 2 \end{bmatrix}$$

2. A system's state is defined by a difference equation given by the following matrix and initial state, $\mathbf{u}_k = \mathbf{A}^k \mathbf{u}_0$.

$$\mathbf{A} = \begin{bmatrix} 1.25 & 0.5 \\ -0.25 & 0.5 \end{bmatrix} \qquad \mathbf{u}_0 = \begin{bmatrix} 2 \\ 1 \end{bmatrix}$$

(a) Using the change of basis method, find a general equation for the difference equation, $\mathbf{u}_k$.

(b) What is the steady state of the system?

(c) Make a plot of the state of the system from the $\mathbf{u}_0$ to $\mathbf{u}_{20}$.

Exercise 6.3   Markov Matrix of Land Use

Stochastic Markov matrices are used by city planners to analyze trends in land use. Researchers collect data of how each parcel of land is used and track changes. A city conducted such a study and published a Markov matrix showing changes in land use between the years 2015 and 2020.

The categories are:

1. Residential.

2. Office.

3. Commercial.

4. Parking.

5. Vacant.

$$M = \begin{bmatrix} 0.85 & 0.05 & 0.05 & 0.05 & 0.10 \\ 0.05 & 0.55 & 0.20 & 0.10 & 0.10 \\ 0.05 & 0.15 & 0.65 & 0.25 & 0.20 \\ 0.05 & 0.20 & 0.05 & 0.55 & 0.20 \\ 0.05 & 0.05 & 0.05 & 0.05 & 0.40 \end{bmatrix}$$

The columns represents the transition probabilities for the initial usage of the land in 2015. The rows represent the probabilities for the land transitioning to a particular usage in 2020. For example, land that was in residential use in 2015 (column 1) had a 0.85 probability of still being in residential use in 2020 (row 1) and a 0.05 probability of being vacant in 2020 (row 5).

Let the ratios of land use in 2015 be as follows.

$$start = \begin{bmatrix} 0.25 \\ 0.3 \\ 0.2 \\ 0.1 \\ 0.15 \end{bmatrix}$$

For convenience, the file LandUse.mat may be loaded to create the M matrix and *start* vector.

We will assume that the conditions driving the changes are consistent. Use MATLAB to determine the following. See section 6.6.3 as needed.

(a) Calculate the vectors showing the expect land use in years 2025, 2030, and 2035. Just find powers of M to determine this.

(b) Find the eigenvectors and eigenvalues of the Markov matrix.

(c) Calculate the change of basis coefficients needed to express $M^k start$ as a sum of products of the eigenvectors and eigenvalues.

(d) Using the eigenvectors, eigenvalues, and coefficients calculate (matrix multiplication) the expected land usage in year 2040 (5 sample periods from 2015).

(e) Using the eigenvectors, eigenvalues, and coefficients find the steady state land usage.

Exercise 6.4   Systems of Linear ODEs

Consider the following system of differential equations and initial conditions.

$$\begin{cases} y_1'(t) = -1.4\,y_1(t) - 0.2\,y_2(t), & y_1(0) = 1 \\ y_2'(t) = -1.2\,y_1(t) - 1.6\,y_2(t), & y_2(0) = 0 \end{cases}$$

(a) Express the system of ODEs in matrix form. Use MATLAB to find the eigenvalues, eigenvectors, and the equation coefficients. What is the solution to the systems of ODEs?

(b) What is the steady state of the system?

Exercise 6.5   SVD

1. Find the singular values of the following matrix and determine the matrix's rank from the singular values.

$$\mathbf{A} = \begin{bmatrix} 1 & -2 & -4 & 8 \\ 14 & 4 & 12 & 12 \\ 7 & 8 & 1 & 5 \\ -2 & 1 & -8 & 3 \end{bmatrix}$$

2. Find the singular values of the following matrix. Find the condition number of $\mathbf{A}$ from the ratio of the largest to smallest singular values (σ_1/σ_3). Verify your answer with the cond function.

$$\mathbf{A} = \begin{bmatrix} -8 & 13 & 3 \\ -1 & 11 & 5 \\ -7 & 14 & 4 \end{bmatrix}$$

3. Write a MATLAB script to test the Eckart–Young theorem presented in section 6.8.4. You might use an image or a matrix with correlated data values. Add random noise to the data and use the immse function to test the error between successive sums of rank-1 outer product matrices from the SVD and the original data.

Exercise 6.6   PCA Data Classification

An important application of PCA is classification, which reduces the dimensions of the data for the purpose of making it easier to see how the attributes of items of the same type are similar and different from items of other types. The result of the analysis is often a plot showing the data in the PCA space.

In this assignment, we will look at a public domain dataset and produce a scatter plot showing how groups of the data are clustered in the PCA space.

Refer to the example in section 6.9.2 as a guide to see how the PCA algorithm is applied.

The data for this project is the *Iris Plant Dataset*, which is a quite old dataset. It contains four attributes of three types of Iris flower plants. The data was collected by R.A. Fisher for a paper published in 1936 [23]. Information about the dataset is found on the UCI Machine Learning Repository [20]. The dataset is in the file `iris_data.txt`, which is in the downloadable files from the book's website.

Use PCA to reduce the dimensionality of the data from four attributes to two principal components. Make a scatter plot of the samples in two dimensional PCA space.

Chapter 7

Computational Numerical Methods

This chapter covers methods for calculating numerical solutions to problems. The term *numerical* is used as the antithesis to *analytical*. Analytical solutions are sought after in mathematics courses. Analytic solutions seem assuring because they give us an equation, but they are not always available. The analytic solution may be very difficult to determine, and may not be as helpful as we had hoped. In the final analysis of most applications we need a number. Thus, computational numerical methods are often required. As we will see, numerical methods can yield accurate results.

With the algorithms considered in this chapter, there are two sources of error, the *intrinsic* or *truncation* error due to algorithmic approximations, and *round-off* error from the computation. There is a degree of control over the truncation error by selection of the algorithm and appropriate use of tolerance values. Good algorithms are usually able to find solutions with an acceptable level of truncation error for most applications. However, efforts to completely eliminate the truncation error can have diminishing returns requiring excessive computation and possibly encountering unacceptable levels of round-off errors. For example, the algorithms for numerical differentiation, integration, and solving differential equations covered in sections 7.3 to 7.5 evaluate functions at a sequence of evenly spaced points, $\{x_0, x_1, \ldots, x_n\}$. If n is excessively increased to reduce the spacing between the points and reduce the truncation error, then the amount of computation is increased, and the round-off error may also increase. The best strategy is to establish accuracy requirements appropriate to the application that are ensured with acceptable truncation error tolerance values.

7.1 Optimization

In this section we consider numerical algorithms that find specific points that are either a root of a function ($f(x_r) = 0$) or locate a minimum value of the function ($\arg\min_{x_m} f(x)$). We are primarily concerned with the simpler case where x is a scalar variable, but will use vector variables with some algorithms.

DOI: 10.1201/9781003271437-7

We also introduce an open-source framework for modeling optimization problems that allows for more flexibility and control in the specification of convex optimization problems.

7.1.1   Numerical Roots of a Function

The roots of a function is where the function is equal to zero. The numerical methods discussed here find a root of a function within a specified range, even if the function is nonlinear. If the function (equation) is a polynomial, then the `roots` function described in section 6.3.2 can be used to find the roots.

Three classic numerical algorithms find where a function evaluates to zero. The first function, Newton's method, works best when we have an equation for the function and can take the derivative of the function. The other two algorithms only require that we can find or estimate the value of the function over a range of points.

7.1.1.1   Newton-Raphson Method

As illustrated in figure 7.1, the Newton-Raphson method, or just Newton's method, uses lines that are tangent to the curve of a function to estimate where the function crosses the x axis.

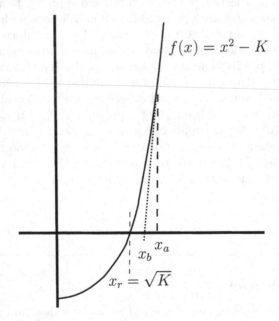

FIGURE 7.1: The Newton-Raphson method extends a tangent line to the x axis to estimate of a root and repeats the process until the root is found within an allowed tolerance.

The algorithm evaluates the function at a sequence of estimates until the root is found.

1. Evaluate the function and its derivative at the first estimate, $f(x_a)$ and $f'(x_a)$.

2. The equation for a tangent line passing through the x axis is

$$f(x_a) - f'(x_a)(x_b - x_a) = 0.$$

3. The estimate for the root is then

$$x_b = x_a - \frac{f(x_a)}{f'(x_a)}.$$

4. If x_b is not close enough to the root at x_r, then let $x_a = x_b$ and repeat the steps.

The illustration in figure 7.1 shows finding $x_r = \sqrt{K}$. Let $f(x) = x^2 - K$, then $f'(x) = 2x$.

$$\begin{aligned} x_b &= x_a - \frac{(x_a^2 - K)}{2x_a} \\ &= \frac{x_a + \frac{K}{x_a}}{2} \end{aligned}$$

Code 7.1 lists a short function that uses the Newton-Raphson method to find the square root of a number.

```
function x = newtSqrt(K)
% NEWTSQRT - Square root using the Newton-Raphson method
%    x = newtQqrt(K) returns the square root of K
%                    when K is a real number, K > 0.

    if ~isreal(K) || K < 0
        error('K must be a real, positive number.')
    end
    tol = 1e-7; % reduce tolerance for more accuracy
    x = K;      % starting point
    while abs(x^2 - K) > tol
        x = (x + K/x)/2;
    end
```

CODE 7.1: Square root function using the Newton-Raphson method.

Here we test the square root function listed in code 7.1.

```
>> newtSqrt(64)
ans =
    8.0000
>> newtSqrt(0.5)
ans =
    0.7071
```

In the case of a polynomial function such as the square root function, one can also use the `roots` function described in section 6.3.2. For example, the square root of 64 could be found as follows.

```
>> roots([1 0 -64])
ans =
     8.0000
    -8.0000
```

We can also write a general function to implement the Newton-Raphson method for finding the roots of a function. The code listed in code 7.2 needs function handles to evaluate the function and its derivative.

```
function x = newtRoot(f, df, start)
% NEWTROOT - Find a root of function f using the Newton-Raphson
method.
%    x = newtQqrt(f, df , start) returns x, where f(x) = 0.
%        f - function, df - derivative of function,
%        start - initial x value to evaluate.
%    Tip: First plot f(x) with fplot and pick a start value where
%        the slope of f(start) points towards f(x) = 0.


tol = 1e-7; % reduce tolerance for more accuracy
x = start;  % starting point
while abs(f(x)) > tol
    d = df(x);
    d(d==0) = eps; % prevent divide by 0
    x = x - f(x)/d;
end
```

CODE 7.2: A root finding function using the Newton-Raphson method.

```
>> newtRoot(((@(x) cos(x) + sin(x) - 0.5), ...
   (@(x) cos(x) - sin(x)), 1.5)
ans =
    1.9948
>> cos(ans) + sin(ans) - 0.5
ans =
   -1.4411e-13
```

7.1.1.2 Bisection Method

The bisection algorithm converges slowly to find the root of a function, but has the advantage that it will always find a root. The bisection algorithm requires two x values where the function crosses the x axis between the two points. As shown in figure 7.2, a new point is found that is half way between the points.

$$c = \frac{a+b}{2}$$

If c is not within the allowed tolerance of the root, then one on the end points is changed to c such that we have a new, smaller, range spanning the x axis. Code 7.3 lists a function implementing the bisection root finding algorithm.

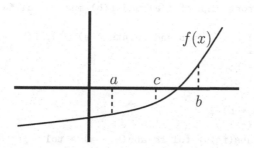

FIGURE 7.2: The bisection algorithm cuts the range between a and b in half until the root is found midway between a and b.

Here are a couple tests of the function in code 7.3.

```
>> f = @(x) x^2 - 64;
>> bisectRoot(f, 5, 10)
ans =
    8.0000
>> bisectRoot((@(x) cos(x) - 0.5), 0, pi/2)
ans =
    1.0472
```

7.1.1.3 Secant Method

The secant algorithm is similar to the bisection algorithm in that we consider a range of x values that get smaller after each iteration of the algorithm. As illustrated in figure 7.3, the secant method finds its next point of consideration by making a line between the end points and finding where the line crosses the x axis. A function implementing the secant root finding algorithm is listed in code 7.4

Here are a few tests of the secant root finding function in code 7.4.

```
>> f = @(x) x^2 - 64;
>> secantRoot(f, 5, 10)
ans =
    8.0000
>> secantRoot((@(x) cos(x) - 0.3), 0, pi/2)
ans =
    1.2661
```

```
function x = bisectRoot(f, a, b)
% BISECTROOT  find root of function f(x), where a <= x <= b.
%    x = bisectRoot(f, a, b)
%    f - a function handle
%    sign(f(a)) not equal to sign(f(b))

    tol = 1e-7; % reduce tolerance for more accuracy
    if sign(f(a)) == sign(f(b))
        error('sign of f(a) and f(b) must be different.')
    end
    if f(a) > 0 % swap end points f(a) < 0, f(b) > 0
        c = a;
        a = b;
        b = c;
    end
    c = (a + b)/2;
    fc = f(c);
    while abs(fc) > tol && abs(b - a) > tol
        if fc < 0
            a = c;
        else
            b = c;
        end
        c = (a + b)/2;
        fc = f(c);
    end
    x = c;
```

CODE 7.3: A root finding function using the bisection algorithm.

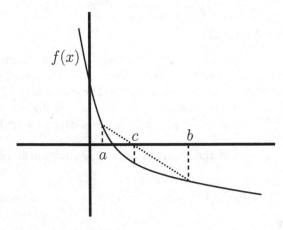

FIGURE 7.3: The secant algorithm cuts the range between a and b by finding a point c where a line between $f(a)$ and $f(b)$ crosses the x axis. The range is repeatedly reduced until $f(c)$ is within an accepted tolerance of zero.

```
function x = secantRoot(f, a, b)
% SECANTROOT  find root of function f(x), where a <= x <= b.
%    x = secantRoot(f, a, b)
%    f - a function handle
%    sign(f(a)) not equal to sign(f(b))

    tol = 1e-7; % reduce tolerance for more accuracy
    if sign(f(a)) == sign(f(b))
        error('sign of f(a) and f(b) must be different.')
    end
    if a > b % swap end points a < b
        c = a;
        a = b;
        b = c;
    end
    sign_a = sign(f(a));
    c = a + abs(f(a)*(b - a)/(f(b) - f(a)));
    fc = f(c);
    while abs(fc) > tol && abs(b - a) > tol
        if sign(fc) == sign_a
            a = c;
        else
            b = c;
        end
        c = a + abs(f(a)*(b - a)/(f(b) - f(a)));
        fc = f(c);
    end
    x = c;
```

CODE 7.4: A root finding function using the secant algorithm.

```
>> secantRoot((@(x) sin(x) - 0.3), 0, pi/2)
ans =
    0.3047
```

7.1.1.4 Fzero

MATLAB has a function called `fzero` that uses numerical methods to find a root of a function. It uses an algorithm that is a combination of the bisection and secant methods.

The arguments passed to `fzero` are a function handle and a row vector with the range of values to search. The function should be positive at one of the vector values and negative at the other value. It may help to plot the function to find good values to use. The example in figure 7.4 uses the `fplot` function to display a plot of the values of a function over a range.

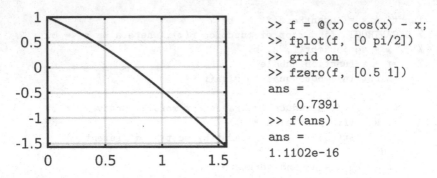

```
>> f = @(x) cos(x) - x;
>> fplot(f, [0 pi/2])
>> grid on
>> fzero(f, [0.5 1])
ans =
      0.7391
>> f(ans)
ans =
1.1102e-16
```

FIGURE 7.4: The plot from `fplot` helps one select the search range to pass to `fzero`.

7.1.2   Finding a Minimum

Finding a minimum value of a function can be a harder problem than finding a function root (where the function is zero). Ideally, we would like to use analytic techniques to find the optimal (minimal) values. We learn from calculus that a minimum or maximum of a function occurs where the first derivative is zero. Indeed, derivatives are always an important part of finding where a minimum occurs. But when the function is nonlinear and involves several variables, finding an analytic solution may not be possible, so instead we make use of numerical search algorithms.

There are several numerical algorithms for finding where a minimum occurs. Our coverage here is restricted to algorithms operating on restricted convex regions. By convex, we mean like a bowl. If we pour water over a convex surface, the minimum point is where the water stops flowing and begins to fill the bowl–like surface. Some functions are strictly convex while others have local minimum points that are not the global minimum point. In the later case, we need to restrict the domain of the search. MATLAB includes two minimum search function that are discussed below. MathWorks also sales an *Optimization Toolbox* with enhanced algorithms that are not discussed here. There are also third party and open source resources with enhanced capabilities. One such resource is discussed below in section 7.1.3.

The mathematics behind how these algorithms work is a little complicated and beyond the scope of this text. Thus the focus here is restricted to what the algorithms are able to do and how to use them. As with algorithms that find a root of a function, the minimum search algorithms call a function that is passed as an argument many times as they search for the minimum value. The first of MATLAB's two minimum search functions is `fminbnd`, which finds the minimum of functions of one variable. While the second function, `fminsearch` handles functions with multiple variables.

7.1.2.1 Fminbnd

The MATLAB function x = fminbnd(f, x1, x2) returns the minimum value of function $f(x)$ in the range $x_1 < x < x_2$. The algorithm is based on golden section search and parabolic interpolation and is described in a book co-authored by MathWork's chief mathematician and co-founder, Cleve Moler [24]. We use fminbnd to find the minimum value of a function in figure 7.5.

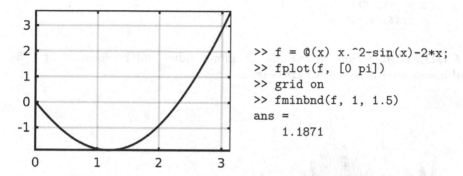

```
>> f = @(x) x.^2-sin(x)-2*x;
>> fplot(f, [0 pi])
>> grid on
>> fminbnd(f, 1, 1.5)
ans =
       1.1871
```

FIGURE 7.5: The plot from fplot shows that the minimum value of the function $f(x) = x^2 - \sin(x) - 2\,x$ is between 1 and 1.5, fminbnd then searched the region to find the location of the minimum value.

7.1.2.2 Fminsearch

The MATLAB function fminsearch uses the Nelder-Mead simplex search method [50] to find the minimum value of a function of one or more variables. Unlike the other searching functions, fminsearch takes a starting point for the search rather than a search range. The starting point may be a scalar or an array (vector). If it is an array, then the function needs to use array indexing to access the array elements. Best results are achieved when the supplied starting point is close to the minimum location.

Need a maximum instead of a minimum?

Just use the negative of the function and find the minimum.

The script in code 7.5 evaluates a function that defines a 3-dimensional surface to find the location of its minimum. The overall shape of the surface is a bowl, but with local hills and valleys from a cosine term. The surface is plotted in figure 7.6 along with output from the script.

```
% File: fminsearch1.m
% fminsearch example
f = @(x) x(1).^2 + x(2).^2 + 10*cos(x(1).*x(2));
m = fminsearch(f, [1 1]);
disp(['min at: ', num2str(m)]);
disp(['min value: ', num2str(f(m))]);
% plot it now
g = @(x, y) x.^2 + y.^2 + 10*cos(x.*y);
[X, Y] = meshgrid(linspace(-5, 5, 50), linspace(-5, 5, 50));
Z = g(X, Y);
surf(X, Y, Z)
```

CODE 7.5: Script to find the minimum value and its location of a 3-dimensional surface with `fminsearch`.

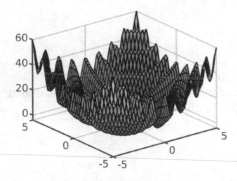

```
min at: 1.7147    1.7147
min value: -3.9175
```

FIGURE 7.6: Surface searched by `fminsearch` to find the location of the minimum value. The output of the script in code 7.5 shows the location and value of the minimum.

A common application of `fminsearch` is to define a function handle that computes an error function that we wish to minimize. For example, we can define an error function as the distance between a desired robot end effector location and a forward kinematics calculation. Then `fminsearch` can calculate the inverse kinematics problem to find the joint angles needed to position the robot end effector at the specified location. The forward kinematics calculation uses matrix multiplication to calculate the end position of a serial-link robotic arm. For robots with only a few joints, it may be possible to find an analytic inverse kinematic solution, but more complex robots often require numerical methods to find the desired joint angles. The starting point value is critical for finding the desired joint angles because multiple joint angle sets can position the robot arm at the same location. For example, we can consider either *elbow-up* or *elbow-down* configurations. So it is important to pass a starting point to fminsearch such that the nearest set of correct joint angles puts the robot in the desired configuration.

A simple example for a serial-link robot arm like shown in figure 5.17 on page 167 with two joints and two rigid arm segments (a and b) is listed in code 7.6. The minimized function returns the length of the error vector between the forward kinematic calculation and the desired location of the end effector. The forward kinematics function uses functions from Peter Corke's *Spatial Math Toolbox* [16] to produce two dimensional homogeneous coordinate rotation and translation matrices as used in section 5.3.4.

```
% File: fminsearch2.m
% Script using fminsearch for inverse kinematic
% calculation of a 2 joint robot arm.
E = [5.4 4.3]';
error = @(Q) norm(E - fwdKin(Q));
Qmin = fminsearch(error, [0 0]);
disp(['Q = ', num2str(Qmin)]);
disp(['Position = ', num2str(fwdKin(Qmin)')]);

function P = fwdKin(Q)
    a = 5;
    b = 2;
    PB = trot2(Q(1)) * transl2(a,0) * trot2(Q(2)) * transl2(b,0);
    P = PB(1:2,3);
end
```

CODE 7.6: Script to find joint angles for a two joint robot arm. The inverse kinematic calculation uses MATLAB's fminsearch function.

The output of the script is the joint angles for the robot and forward kinematic calculation of the end effector position with those joint angles.

```
>> fminsearch2
Q = 0.56764      0.3695
Position = 5.4           4.3
```

7.1.3   CVX for Disciplined Convex Programming

CVX[1] is a MATLAB-based modeling system for convex optimization. CVX allows constraints and objectives to be specified using standard MATLAB syntax. This allows for more flexibility and control in the specification of optimization problems. Commands to specify and constrain the problem are placed between cvx_begin and cvx_end statements. CVX defines a set of commands and functions that can be used to describe the optimization

---

[1] http://cvxr.com/cvx/

problem. Most applications of CVX will include a call to either the `minimize` or `maximize` function [29].

An example can best illustrate how to use CVX. Recall from section 5.8 that the preferred solution to an under-determined system of equations is one that minimizes x. MATLAB's left-divide operator finds the least squares solution using the QR algorithm as discussed in section 5.11, which minimizes the l_2-norm of x. The left-divide operator returns a solution with $(n - r)$ zeros by zeroing out columns of the matrix before finding the least squares solution. But it is also known that a minimization of the l_1-norm tends to yield a solution that is sparse (has several zeros). We turn to CVX to replace the inherent l_2-norm that we get from our projection equations with an l_1-norm.

```
>> b                            >> x
b =                             x =
      5                              -0.0000
     20                               1.1384
     13                               0.0000
>> cvx_begin                          0.0103
>>   variable x(5);                   0.0260
>>   minimize( norm(x, 1) );    >> A*x
>>   subject to                 ans =
>>       A*x == b;                    5.0000
>> cvx_end                           20.0000
                                     13.0000
```

Sparse Matrices

Very large but sparse matrices are common in some application domains. MATLAB supports a *sparse* storage class to efficiently hold large matrices that have mostly zeros. A sparse matrix contains the indices and values of only the nonzero elements. All functions in MATLAB work with sparse matrices. The `sparse` command converts a standard matrix to a sparse matrix. The `full` command restores a sparse matrix to a standard matrix [55].

```
S = sparse(A);
...
A = full(S);
```

7.2 Data Interpolation

Data interpolation algorithms are used to fill-in unknown data points between data points that are known.

7.2.1 One-Dimensional Interpolation

MATLAB has two functions that will interpolate between one-dimensional data samples to estimate unknown sample points. The difference between the two functions relates only to the purpose for calling them. The `fillmissing` function is used when values that should be in the data are missing—often replaced with the NaN symbol. The `interp1` function is used when you wish to add data points not contained in the data. We made use of `fillmissing` in section 1.10.4 when data imported into MATLAB had some missing values. So we will use `interp1` in the examples in this section. We call `interp1` with vectors for the given x and y data values along with a longer vector of x values where we want to find corresponding interpolated y values. An optional fourth argument is a string listing the interpolation algorithm.

```
y2 = interp1(x1, y1, x2, 'algorithm')
```

Several interpolation algorithms are available. Five commonly used algorithms are:

`'nearest'`	nearest-neighbor interpolation
`'linear'`	linear interpolation
`'pchip'`	piecewise cubic Hermite interpolation
`'makima'`	modified Akima cubic Hermite interpolation
`'spline'`	cubic spline interpolation

MATLAB also has functions that directly invoke the data interpolation algorithms rather than invoking them via `fillmissing` or `interp1`. These functions include `pchip`, `spline`, and `makima`.

The default interpolation algorithm is `'linear'`, which does not usually provide smooth transitions between points but is good when the given data changes slowly between sample points. The methods `'pchip'`, `'makima'`, and `'spline'` yield good results. The `'spline'` method requires the most computation because it uses a matrix computation for each point to determine polynomial coefficients that not only match the given data, but maintain constant first and second order derivatives at each data point added. However, the `'spline'` method may overshoot data points causing and an oscillation. The results from the `'pchip'` and `'makima'` algorithms are similar. They maintain consistent first order derivatives at each point but not consistent second order derivatives. They avoid overshoots and accurately connect the flat regions. Another advantage of the 'pchip' and 'makima' algorithms is that if the given data points are either monotonically increasing or decreasing, so will the interpolated data points [34]. Because of oscillations, the 'spline' interpolation may not always maintain the monotonicity of the data. Since 'pchip' and 'makima' have similar results, only the 'pchip' interpolation is used in the following comparison code and plots in figure 7.7.

```
f = @(x) 1 + x.^2 - x.^3 + 20*sin(x);
x1 = (-3:1.5:3)';  % Limited data points
y1 = f(x1);
x2 = (-3:0.5:3)';  % More data points
subplot(2,2,1)
plot(x1,y1,'o',x2,interp1(x1,y1,x2,'nearest'),'+:')
title('nearest')
subplot(2,2,2)
plot(x1,y1,'o',x2,interp1(x1,y1,x2,'linear'),'+:')
title('Linear')
subplot(2,2,3)
plot(x1,y1,'o',x2,interp1(x1,y1,x2,'pchip'),'+:')
title('pchip')
subplot(2,2,4)
plot(x1,y1,'o',x2,interp1(x1,y1,x2,'spline'),'+:')
title('spline')
```

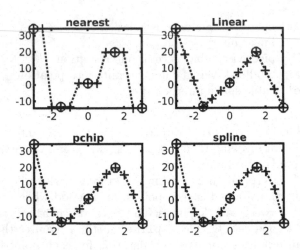

FIGURE 7.7: Comparison of four data interpolation algorithms. The circles show the given data points, while the interpolated points are marked with plus signs.

The following example illustrates the difference between 'pchip' and 'spline' interpolation results. As figure 7.8 shows, the 'pchip' result may not be as smooth, but the monotonicity of the data is preserved.

```
x = -3:3;
y = [-1 -1 -1 0 1 1 1];
t = -3:.01:3;
hold on
plot(x, y, 'o')
plot(t, interp1(x, y, t, 'pchip'), 'g')
plot(t, interp1(x, y, t, 'spline'), '-.r')
legend('data','pchip','spline', 'Location', 'NorthWest')
hold off
```

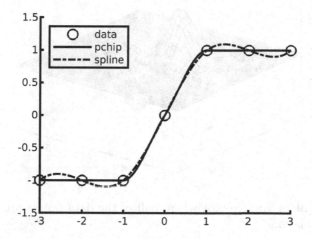

FIGURE 7.8: Comparison of 'pchip' and 'spline' data interpolation.

7.2.2 Two-Dimensional Interpolation

The two-dimensional interpolation functions `griddata` and `interp2` follow in the same pattern as the one-dimensional functions. The inputs are matrices of grid points such as returned from `meshgrid`, which we used in section 2.3 to make surface plots. The distinction between `griddata` and `interp2` is that `interp2` requires that the grid data be monotonically increasing as from `meshgrid`, while `griddata` does not have that requirement for cases when the available data is not as tidy. The available interpolation methods are 'linear', 'nearest', 'cubic', 'makima', or 'spline' with the default method again being 'linear'. In the following example, we will use the default interpolation. The appearance of the surface plot in figure 7.9

is indistinguishable from a surface plot where all of the points come from an equation rather than having some interpolated points.

```
>> [X, Y] = meshgrid(linspace(-10, 10, 20));
>> T = X.^2 + Y.^2;
>> Z = (Y-X).*exp(-0.12*T);
>> [Xi, Yi] = meshgrid(linspace(-10, 10, 40));
>> Zi = interp2(X, Y, Z, Xi, Yi);
>> surf(Xi, Yi, Zi)
```

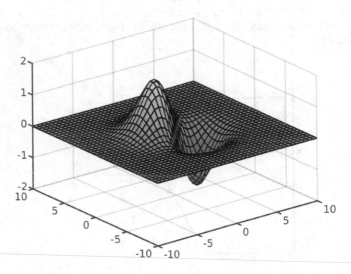

FIGURE 7.9: A surface plot where half of the data points are found from linear interpolation.

Note: MATLAB also has functions `interp3`, `griddata3`, `interpn`, and `griddatan` for interpolation in higher dimensions.

7.3   Numerical Differentiation

Computing a derivative numerically is typically not difficult and can yield accurate results. To achieve the desired accuracy, we need to pay attention to the sampling rate of the data and the limits of the computer to process very small numbers without round-off errors.

7.3.1 Euler Derivative

Calculus defines the derivative of $y(x)$ with respect to x as

$$\frac{dy(x)}{dx} = \lim_{\Delta x \to 0} \frac{\Delta\, y(x)}{\Delta\, x}.$$

For discretely sampled data, this becomes an approximation.

$$\frac{dy_i}{dx} \approx \frac{y_{i+1} - y_i}{h}$$

The variable h represents the distance between the samples with respect to x (δx) and is proportional to the error, which may be acceptable for some applications, but not acceptable for other applications. The algorithm that makes use of the simple derivative definition is called the *forward Euler finite-difference derivative*.

For sampled data values, the MATLAB function `diff` returns the difference between consecutive values, which is used to compute the derivative `dy_dx = diff(y)/h;`. Note that the size of the array returned by `diff` is one element smaller than the array passed to it.

To take the derivative of a function, first sample the data at intervals of h; and then use `diff` as with pre-sampled data. The challenge is to determine an appropriate value for h. It should be small enough to give good granularity. The evaluation of *small enough* depends entirely on the data being evaluated. However, it should not be so small as to require excessive computation or to cause excessive rounding errors from the division by h. As a rule of thumb, h should not be much smaller than 10^{-8} [66]. Some experimental testing with different values of h may be required for some applications.

Try the example in code 7.7 and see how changing the value of h, affects the accuracy of the numerical derivative. Figure 7.10 shows the plot when $h = \pi/40$.

7.3.2 Spectral Derivative

The fast Fourier transform (FFT) provides a more accurate numerical method for estimating a derivative. The Fourier series, Fourier transform, and the inverse Fourier transform are algorithms proposed in 1807 by French mathematician Jean Baptiste Joseph Fourier (1768-1830) to characterize the frequencies present in continuous functions or signals. Later, the discrete Fourier transform (DFT) was used with sampled signals; however, the DFT can be rather slow for large data sets ($\mathcal{O}(n^2)$). Then in 1965, James W. Cooley (IBM) and John W. Tukey (Princeton) developed the *fast* Fourier transform (FFT) algorithm that has run-time complexity of $\mathcal{O}(n \log n)$. The improved performance of the FFT stems from taking advantage of the symmetry found in the complex, orthogonal basis functions ($\omega_n = e^{-2\pi i/n}$). In our coverage, we will stick with our objective of applying the FFT to finding derivatives. Many

```
% File: deriv_cos.m
% Derivative example, Compute derivative of cos(x)

% sample step size, please experiment with h values
h = pi/40;
x = 0:h:2*pi;
y = cos(x);
dy_dx = diff(y)/h;
x1 = x(1:end-1);        % note one less to match dy_dx
actual_dy = -sin(x1);
error = abs(dy_dx - actual_dy);
me = max(error);
ex = x1(error == me);


p1 = plot(x1, actual_dy, 'k', x1, dy_dx, 'r:');
title('Derivatives of cos(x)');
xticks(0:pi/2:2*pi);
xticklabels({'0','\pi/2','\pi','3\pi/2','2\pi'});
line([ex ex], [-1 1]);
text(ex+0.1, -0.4, 'Max error location');
legend(p1,{'analytic derivative','numeric derivative', },...
    'Location', 'northwest');
disp(['Max error = ',num2str(me)]);
```

CODE 7.7: Script implementing the Euler derivative of the cosine function.

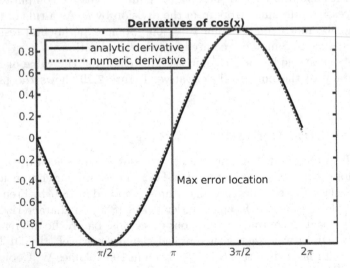

FIGURE 7.10: Even though the h value is fairly large, the Euler numerical derivative of the cosine function is still close to the analytic derivative.

resources provide detailed descriptions of the Fourier transform and the FFT. A rigorous, but still understandable coverage can be found in [7], pages 47 to 63.

The FFT is of interest to us here because of the simple relationship between the Fourier transform of a function and the Fourier transform of the function's derivative. We can easily find the derivative in the frequency domain and then use the inverse FFT to put the derivative into its original domain.

$$\mathcal{F}\left(\frac{d}{dt}f(t)\right) = i\,\omega\,\mathcal{F}(f(t))$$

Here, i is the imaginary constant $\sqrt{-1}$ and ω is the frequency in radians per second for time domain data or just radians for spatial data. We can use $\omega = 2\pi f_s k/n$ for time domain data and $\omega = 2\pi k/T$ for spatial domain data, where k is the sample number in the range $\{-n/2 \text{ to } (n/2-1)\}$, and n is the number of data samples (usually a power of 2). For time domain data, f_s is the sampling frequency (samples/second). For spacial domain data, T is the length of the data.

Code 7.8 gives an example of computing the spectral derivative using time domain data. The FFT treats frequency data as having both positive and negative values, so our ω needs to do the same. We can use our k variable to build both the time, t, values and the frequency, w, values. Both the FFT and inverse FFT algorithms reorder the data, so we need to use the fftshift function to align the frequency values with the frequency domain data. Figure 7.11 shows a plot of the time domain spectral derivative, which aligns with the analytic derivative.

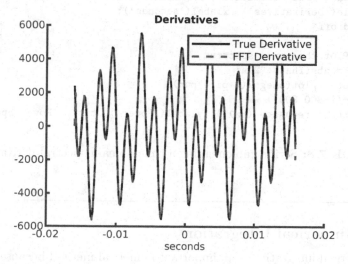

FIGURE 7.11: Plot of the analytic derivative and spectral derivative of a time domain signal.

```
%% SpectralTimeDerivative.m
%  Numerical derivative using the spectral derivative
%  (FFT based) algorithm --  time domain data
n = 256; fs = 8000; % samples/second
T = n/fs;           % time span of samples (seconds)
k = -n/2:(n/2 - 1);
t = T*k/n;          % time values
f1 = 200; f2 = 500; % Two frequency components (Hz)

% The function and its analytic derivative
f = 2*cos(2*pi*f1*t) + cos(2*pi*f2*t);
df = -4*f1*pi*sin(2*pi*f1*t) - 2*f2*pi*sin(2*pi*f2*t);

%% Derivative using FFT (spectral derivative)
fhat = fft(f);
w = 2*pi*fs*k/n;       % radians / second
omega = fftshift(w);  % Re-order fft frequencies
i = complex(0, 1);    % Define i just to be sure it is not
                      % used for something else, or could use 1i.
dfhat = i*omega.*fhat;      % frequency domain derivative
dfFFT = real(ifft(dfhat));  % time domain derivative
%% Plotting commands
figure, plot(t, f, 'b')  % derivatives too large for same plot
title('Function'), xlabel('seconds');
figure, hold on
plot(t, df,'k','LineWidth',1.5)
plot(t, dfFFT,'r--','LineWidth',1.2)
legend('True Derivative','FFT Derivative')
title('Derivatives'), xlabel('seconds');
hold off

%% Plot the power spectrum
pwr = abs(fhat).^2/n;
figure, plot(omega/(2*pi), pwr);
xlim([-800 800])
xlabel('Frequency (Hz)'), ylabel('Power'), title('Power Spectrum')
```

CODE 7.8: Script implementing a time domain spectral derivative.

7.4 Numerical Integration

Numerical integration is an important numerical method because analytic solutions are often not available. Surprisingly, the seemingly simple equation $y = e^{-x^2}$ can not be integrated analytically.

Figure 7.12 illustrated that the value of a definite integral is the area bounded by the function, the x axis, and the lower and upper bounds of the integral. Numerical techniques estimate the area by summing the areas of a

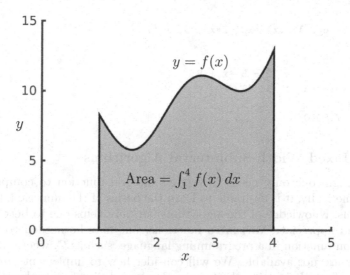

FIGURE 7.12: A definite integral is an area calculation.

sequence of thin slices (subintervals). Three such techniques are commonly used to estimate the area of each subinterval. The *Reimann Integral* uses rectangular regions to estimate the area of each small subinterval. The *Trapezoid Rule* improves the results by modeling the area of each subinterval as a trapezoid. *Simpson's Rule* provides a more accurate estimate by modeling the curve over a pair of subintervals as a quadratic polynomial.

7.4.1 MATLAB's Integral Function

MATLAB has a function for computing definite integrals called `integral`, which is a replacement for an older function called `quad`. The `integral` function uses a *global adaptive quadrature algorithm* where the integration region is divided into piecewise smooth regions before integrating each subinterval. The widths of the subintervals are adapted so that each subinterval is accurately modeled by an equation.

The `integral` function is called with arguments of a function handle and the interval of the integral (`I = integral(f, a, b)`). Be sure to use element-wise arithmetic operators when defining the function so that it can accept and return vectors. The integration interval is normally specified with a pair of real scalars, but negative and positive infinity (`-Inf` and `Inf`) may be used. See MATLAB's documentation for information about complex functions, and the use of tolerances and waypoints.

Here are a few usage examples.

```
>> integral(@(x) sin(x), 0, pi)
ans =
    2.0000
>> integral(@(x) exp(-2*x), 0, Inf)
ans =
    0.5000
>> f = @(x) x.^2 - 5.*x + 8;
>> integral(f, 1, 4)
ans =
    7.5000
```

7.4.2   Fixed Width Subinterval Algorithms

Although, one only needs MATLAB's integral function to compute integrals numerically, it is desirable to learn the basics of the numeric integration algorithms. Knowledge of the algorithms not only helps one to better understand and appreciate MATLAB's functions, but may be needed to write an integration function in a programming language, such as C, where MATLAB's functions are not available. We will consider how to implement numeric integration algorithms using the trapezoid rule and Simpson's rule. We do not discuss results from Reimann integrals because integrals from trapezoids are only slightly more complex than Reimann integrals, but are more accurate.

7.4.2.1   Trapezoid Rule Integrals

The geometry of a trapezoid is illustrated in figure 7.13. When an integration region is divided into equal width trapezoids separated by grid points $\{x_0, x_1, \ldots, x_n\}$, then the width of each trapezoid is $h = x_{i+1} - x_i$. The area the trapezoid is the product of the width with the average height of the two sides.

$$\text{Area} = h \, \frac{f(x_i) + f(x_{i+1})}{2}$$

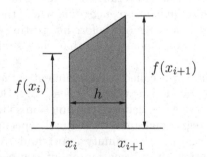

FIGURE 7.13: The shape of a trapezoid and its size measurements.

The definite integral of a function over a region from a to b is approximated by the sum of trapezoids found by splitting the region into n equal width subintervals (also called panels), where $n+1$ is the number of grid points that divide the interval into the subintervals. [2]

$$h = \frac{b-a}{n}$$

$$\int_a^b f(x)\,dx \approx \sum_{i=0}^{n-1} h\,\frac{f(x_i)+f(x_{i+1})}{2}$$

In the summation, each term except the first and last appear twice because they are part of the calculation of two trapezoids.

$$\int_a^b f(x)\,dx \approx \frac{h}{2}\left(f(x_0) + 2\sum_{i=1}^{n-1} f(x_i) + f(x_n)\right)$$

As shown in figure 7.14, the area of each trapezoid is sometimes more, less, or nearly the same as the true area of each subinterval. Using more subintervals, which reduces the size of h, will improve the accuracy. With narrower subintervals, the spanned region of $f(x)$ from x_i to x_{i+1} will more closely resemble a straight line. Of course, we need to be sensitive to the size of h to avoid excessive round-off errors that can occur when h is very small.

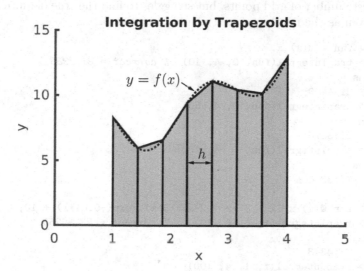

FIGURE 7.14: The trapezoid rule models each subinterval as a trapezoid. The definite integral is the sum of the trapezoid areas.

---

[2]Since MATLAB array indices start at 1, not 0, our code implementations set the grid points at $\{x_1, x_2, \ldots, x_{n+1}\}$.

The `trapIntegral` function in code 7.9 implements the trapezoid rule to calculate an estimate of a definite integral. Testing of the trapezoid rule

```
function I = trapIntegral(f, a, b, n)
% TRAPINTEGRAL - Definite integral by trapezoid area summation.
%
% f - vector aware function
% a - lower integration boundary
% b - upper integration boundary, a < b
% n - number of subinterval trapezoids

    if a >= b
        error('Variable b should be greater than a.')
    end
    h = (b - a)/n;
    x = linspace(a, b, n+1);
    y = f(x);
    I = (h/2)*(y(1) + 2*sum(y(2:end-1)) + y(end));
```

CODE 7.9: A function implementing a definite integral by the trapezoid rule.

integration function shows that it finds an answer that is somewhat close with a modest number of grid points, but struggles to find the true definite integral value even as the number of grid points is increased.

```
>> fun = @(x) x.^2;
>> trapIntegral(fun, 0, 4, 10)  % correct = 21.3333
ans =
    21.4400
>> trapIntegral(fun, 0, 4, 50)
ans =
    21.3376
>> trapIntegral(fun, 0, 4, 100)
ans =
    21.3344

>> f = @(x) x.^2 - 3.*x + 2*sin(3*x).*exp(-0.01*x) + 10;
>> trapIntegral(f, 1, 4, 10)    % correct = 27.30753
ans =
    27.4343
>> trapIntegral(f, 1, 4, 100)
ans =
    27.3088
>> trapIntegral(f, 1, 4, 500)
ans =
    27.3076
```

7.4.2.2   Simpson's Rule Integrals

Rather than using a straight line to approximate the function over an integration subinterval, Simpson's rule estimates the function as a parabola represented with a quadratic equation. Each integration region consists of two adjacent subintervals and three grid points. The three grid points are used to find the quadratic equation from which the area is computed. The development of the quadratic equation is covered in calculus textbooks and can also be found on several web pages. It is typically found from a Taylor series expansion of a quadratic equation or using *Lagrange polynomial interpolation* [66].

Lagrange polynomial interpolation finds a polynomial that goes through a set of data points. The Lagrange polynomial, $P(x)$, has the property that $P(x_i) = y_i$ for all given (x_i, y_i) data points. We first find a Lagrange basis polynomial, $P_i(x)$, for each of given x_i value. The basis polynomials are products of factors such that $P_i(x_i) = 1$ and $P_i(x_j) = 0$ when $i \neq j$. Then the Lagrange polynomial is a sum of products between the basis polynomials and the given y_i values. The properties of the basis polynomials ensure that the Lagrange polynomial passes through each point.

$$P_i(x) = \prod_{j=1, j \neq i}^{n} \frac{x - x_j}{x_i - x_j}$$

$$P(x) = \sum_{i=1}^{n} y_i \, P_i(x)$$

Figure 7.15 illustrates one region of integration. The quadratic equation $P(x)$ passes through the three points $\{(x_{i-1}, f(x_{i-1})), (x_i, f(x_i)),$ and $(x_{i+1}, f(x_{i+1}))\}$, which are the intersections between the three grid points

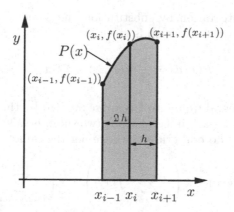

FIGURE 7.15: Simpson's rule subinterval integration.

Lagrange Polynomial Example

Let us use Lagrange polynomial interpolation to find the quadratic equation that passes through the points $\{(0,2),(1,5),\text{and}(2,6)\}$.

$$P_1(x) = \frac{(x-x_2)(x-x_3)}{(x_1-x_2)(x_1-x_3)} = \frac{(x-1)(x-2)}{(0-1)(0-2)} = \tfrac{1}{2}(x^2 - 3x + 2)$$

$$P_2(x) = \frac{(x-x_1)(x-x_3)}{(x_2-x_1)(x_2-x_3)} = \frac{(x-0)(x-2)}{(1-0)(1-2)} = -x^2 + 2x$$

$$P_3(x) = \frac{(x-x_1)(x-x_2)}{(x_3-x_1)(x_3-x_2)} = \frac{(x-0)(x-1)}{(2-0)(2-1)} = \tfrac{1}{2}(x^2 - x)$$

$$\begin{aligned} P(x) &= 2\,P_1(x) + 5\,P_2(x) + 6\,P_3(x) \\ &= (1 - 5 + 3)\,x^2 + (-3 + 10 - 3)\,x + 2 \\ &= -x^2 + 4x + 2 \end{aligned}$$

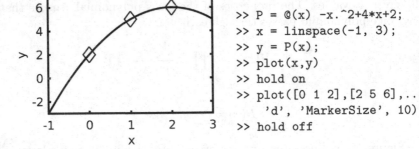

```
>> P = @(x) -x.^2+4*x+2;
>> x = linspace(-1, 3);
>> y = P(x);
>> plot(x,y)
>> hold on
>> plot([0 1 2],[2 5 6],...
        'd', 'MarkerSize', 10)
>> hold off
```

The Lagrange polynomial passes through the points $\{(0,2)$, $(1,5)$, and $(2,6)\}$.

and $f(x)$. Using integration by substitution and some algebra, the definite integral of each integration region is given by

$$I_i = \int_{x_{i-1}}^{x_{i+1}} P(x)\,dx = \frac{h}{3}\left(f(x_{i-1}) + 4f(x_i) + f(x_{i+1})\right).$$

The definite integral from a to b is approximated by the sum of the subinterval integrals. Since each integral spans two subintervals, the count of function values at even and odd grid points are not the same.

$$\int_a^b f(x)\,dx \approx \frac{h}{3}\left[f(x_0) + 4\left(\sum_{i=1,\,i\,\text{odd}}^{n-2} f(x_i)\right) + 2\left(\sum_{i=2,\,i\,\text{even}}^{n-1} f(x_i)\right) + f(x_n)\right]$$

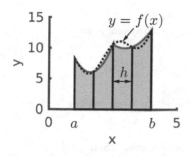 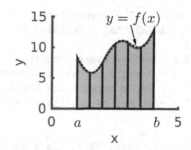

FIGURE 7.16: *Left:* Simpson's rule integration with 5 grid points, 4 subintervals, and 2 integration regions. *Right:* Simpson's rule integration with 7 grid points, 6 subintervals, and 3 integration regions.

Figure 7.16 shows that Simpson's rule can find accurate definite integral results if the subintervals are small enough to accurately model the function with quadratic equations.

Note: Because Simpson's rule computes integrals over two subintervals, the number of subintervals must be an even number. So the number of grid points must be an odd number. To satisfy the indexing requirements of the sum functions in the MATLAB code, the minimum number of grid points is 5 (4-panel integration).

The simpsonIntegral function in code 7.10 implements Simpson's rule integration. Testing results show that Simpson's rule integration is more accurate than trapezoid rule integration even with fewer subintervals.

```
>> fun = @(x) x.^2
>> simpsonIntegral(fun, 0, 4, 8)
ans =            % correct = 21.3333
    21.3333

>> f = @(x) x.^2-3.*x+2*sin(3*x).*exp(-0.01*x)+10
>> simpsonIntegral(f, 1, 4, 8)
ans =            % correct = 27.30753
    27.2950
>> simpsonIntegral(f, 1, 4, 18)
ans =
    27.3071
>> I = simpsonIntegral(f, 1, 4, 28)
I =
    27.3075
>> error = abs(I - integral(f, 1, 4))
error =          % compare to MATLAB's integral
    7.1945e-05
```

```
function I = simpsonIntegral(f, a, b, n)
% TRAPINTEGRAL - Definite integral by Simpson's rule.
%
% f - vector aware function
% a - lower integration boundary
% b - upper integration boundary, a < b
% n - number of subintervals, must be an even number and >= 4

    if a >= b
        error('Variable b should be greater than a.')
    end
    if mod(n, 2) ~= 0 || n < 4
        error('Variable n, should be an even number, n >= 4.')
    end
    h = (b - a)/n;
    x = linspace(a, b, n+1);
    y = f(x);
    I = (h/3)*(y(1) + 4*sum(y(2:2:end-1)) ...
                    + 2*sum(y(3:2:end-2)) + y(end));
    % MATLAB indexing starts at 1, so even and odd are reversed.
```

CODE 7.10: Function implementing Simpson's rule integration.

7.4.3   Recursive Adaptive Integral

The recursive function adaptIntegral listed in code 7.11 investigates the concept of an adaptive algorithm, like MATLAB's integral function, that adjusts the width of each subinterval based on the accuracy of intermediate results. The basic concept of how the algorithm works is simple. If the function to be integrated fits the integration model (trapezoid or quadratic equation) over the integral region, then we have an accurate definite integral. However, if the integration model is a poor data fit, then the interval is divided into smaller subintervals.

For each interval, the algorithm splits the interval from a to b in half and calculates the area of three intervals. The interval from a to b is divided equally into subintervals from a to c and from c to b. If the area from a to b is within a small tolerance of the sum of the areas of the subintervals, then the algorithm is satisfied with the area of the interval. Otherwise, the algorithm repeats by making recursive function calls for each subinterval. The algorithm continues to split each subinterval until the integration model accurately fits the function over the subinterval. The returned subinterval areas are added together to find the definite integral of the function from a to b.

The adaptive function can use either the trapezoid rule or Simpson's rule algorithm. The second argument should be either 't' or 's' to select the integration algorithm.

Testing shows that the adaptive interval algorithm is an accurate estimator of definite integrals. Results from using Simpson's rule are more accurate than using the trapezoid rule algorithm.

```
>> f = @(x) x.^2-3.*x+2*sin(3*x).*exp(-0.01*x)+10
>> I = integral(f, 1, 4);
>> I - adaptIntegral(f, 's', 1, 4)
    ans =            % compare to MATLAB's integral
    2.4833e-12
>> I - adaptIntegral(f, 't', 1, 4)
ans =
  -8.4254e-10

>> s = @(x) sin(x);
>> adaptIntegral(s, 's', 0, pi)
ans =
    2.0000
>> 2 - ans
ans =
  -1.2683e-12
>> 2 - adaptIntegral(s, 't', 0, pi)
ans =
    1.5821-09
```

7.5   Numerical Differential Equations

An ordinary differential equation (ODE) is an equation of the form

$$y'(t) = \frac{dy}{dt} = f(t, y), \text{ with } y(t_0) = y_0.$$

We are given the rate of change (the derivative function $f(t, y)$) of $y(t)$ from which we solve for $y(t)$. Because we are also given the initial value $y(t_0) = y_0$, this class of ODE problem is called an initial value problem (IVP). We will consider cases of the function $y(t)$ that represents a single equation and also a system of equations.

Differential equations are important to engineers and scientists because changes to the state of dynamic systems are described by differential equations. Examples of these dynamic systems include objects that move; wires carrying fluctuating electrical current; reservoirs of fluid with measurable pressure, temperature, and volume; or anything where the state of the system can be quantified numerically.

Numerical methods are especially needed to solve higher order and non-linear ODEs. As is the case with integration, our ability to properly evaluate

```
function I = adaptIntegral(f, intFun, a, b)
% ADAPTINTEGRAL - Adaptive interval definite integral.
%
% f - vector aware function
% intFun - 't' for trapIntegral or 's' for simpsonIntegral
% a - lower integration boundary
% b - upper integration boundary, a < b
%
% This code just demonstrates adaptive intervals.
% Use MATLAB's integral function, not this function.

    if a >= b
        error('Variable b should be greater than a.')
    end
    tol = 1e-12;
    c = (a + b)/2;

    % Three integrals
    if intFun == 't'
        Iab = trapIntegral(f, a, b, 2);
        Iac = trapIntegral(f, a, c, 2);
        Icb = trapIntegral(f, c, b, 2);
    elseif intFun == 's'
        Iab = simpsonIntegral(f, a, b, 4);
        Iac = simpsonIntegral(f, a, c, 4);
        Icb = simpsonIntegral(f, c, b, 4);
    else
        error("Specify 's' or 't' for intFun")
    end

    if abs(Iab - (Iac + Icb)) < tol
        I = Iac + Icb;
    else
        I = adaptIntegral(f, intFun, a, c) ...
            + adaptIntegral(f, intFun, c, b);
    end
```

CODE 7.11: Adaptive interval definite integral function

some ODEs using analytic methods is less than satisfying [70]. So numerical methods for solving ODEs is an important tool in the engineer's computational toolbox.

This is our third encounter with ODEs. In section 4.1, we used MATLAB's *Symbolic Math Toolbox* to find the algebraic equations from the differential equations associated with the effect of gravity on a ball thrown upwards. Then in section 6.7 we considered systems of first order linear ODEs that may be solved analytically with eigenvalues and eigenvectors. Now, we will consider numerical methods to find and plot the solutions to ODEs. We will cover the basic algorithmic concepts of how numerical ODE solvers work and review the efficient suite of ODE solver provided by MATLAB.

Recall the definition of a derivative:

$$\frac{dy}{dt} = \lim_{h \to 0} \frac{y(t+h) - y(t)}{h}.$$

Considered in terms of discrete samples of y, the derivative becomes

$$y'_k = \frac{y_{k+1} - y_k}{h},$$

where h is the distance between sample points, $h = t_{k+1} - t_k$. Rearranged to find y_{k+1} from the previous value, y_k, and the slope of the function at t_k, we have

$$y_{k+1} = y_k + h\, y'_k. \tag{7.1}$$

This equation is the basis for the simplest ODE solver, Euler's method. However, due to concerns about excessive computation and round-off errors, we are limited on how small h can be. So the accuracy of equation (7.1) is not acceptable for most applications.

Since differential equations are given in terms of an equation for the derivative of y, $y' = f(t, y)$, and the initial value, y_0, the general discrete solution comes from the fundamental theorem of calculus, which applies the antiderivative of a function to a definite integral.

$$y_{k+1} = y_k + \int_{t_k}^{t_{k+1}} f(t, y)dt \tag{7.2}$$

Then the numerical approximation comes from the algorithm used to calculate the definite integral.

$$I_k = \int_{t_k}^{t_{k+1}} f(t, y)dt \tag{7.3}$$

At each step, the ODE solver uses a straight line with a slope of I_k/h to advance from y_k to y_{k+1}.

7.5.1   Euler's Method

Euler's method uses a Reimann integral to approximate equation (7.3) with constant height rectangular subintervals.[3] It is a reasonable solution if $f(t, y)$ is constant or nearly constant. However, the results from Euler's methods applied to most ODEs are not accurate enough for most applications unless

---

[3]This is the *forward* Euler method. A variant to this algorithm is the implicit backward Euler method, which we will use in section 7.5.6.

the step size is small enough relative to the rate of change of the $f(t, y)$ function.

$$\int_{t_k}^{t_{k+1}} f(t, y) dt \approx h\, f(t_k, y_k)$$

$$y_{k+1} = y_k + h\, f(t_k, y_k)$$

Euler's method is a perfect match if we let $f(t, y) = a$, then

$$y_{k+1} = y_k + h\, a.$$

Then by induction,

$$\begin{cases} y_1 = & y_0 + h\, a \\ y_2 = & y_1 + h\, a = y_0 + 2\, h\, a \\ \vdots \\ y_k = & y_0 + k\, h\, a \end{cases}.$$

The `eulerODE` function listed in code 7.12 is an implementation of Euler's method. The first three arguments passed to `eulerODE` are the same as is used with the MATLAB ODE solvers. An additional argument for the number of values to find is also passed to `eulerODE`. First, define a function or function handle. The function defines the derivative of the solution. It should take two arguments, t and y. Although in many cases, the function only operates on one variable—usually the second, y, variable. The function should accept and return vector variables.

$$\frac{dy(t)}{dt} = f(t, y)$$

Next, specify a `tspan` argument, which is a row vector of the initial and final t values[4] (`[t_0 t_final]`). Then, the value of y at the initial time is specified $y_0 = y(t_0)$. The final argument is n, the number of values to find. The returned data from all of the ODE solvers are the t evaluation points and the y solution.

```
>> f = @(t, y) 5
>> [t, y] = eulerODE(f, [0 10], 5, 4)
t =
         0    3.3333    6.6667   10.0000
y =
    5.0000   21.6667   38.3333   55.0000
```

Figure 7.17 on page 343 shows a plot comparing the relative accuracy of Euler's method to other fixed step size algorithms.

---

[4]We use t for time as a variable of the ODE function, but a spatial variable, such as x, could also be used.

```
function [t, y] = eulerODE(f, tspan, y0, n)
% EULERODE - ODE solver using Euler's method.
%
%   f - vectorized derivative function, y' = f(t, y)
%   tspan - row vector [t0 tf] for initial and final time
%   y0 - initial value of y
%   n - number of y values to find
    h = (tspan(2) - tspan(1))/(n-1);
    t = linspace(tspan(1), tspan(2), n);
    y = zeros(1, n);
    y(1) = y0;
    for k = 1:n-1
        y(k+1) = y(k) + h*f(t(k), y(k));
    end
```

CODE 7.12: A function implementing Euler's method ODE solver.

7.5.2 Heun's Method

Heun's method uses trapezoid rule integration to approximate equation (7.3).

$$y_{k+1} = y_k + \frac{h}{2}\left[f(t_k, y_k) + f(t_{k+1}, y_{k+1})\right] \tag{7.4}$$

However, equation (7.4) makes use of y_{k+1}, which is not yet know. So Euler's method is first used to make an initial estimate $(\tilde{y}_{k+1})$. Because the algorithm makes an initial prediction of $\tilde{y}_{k+1}$ and then updates the estimate with a later equation, Heun's method belongs to a class of algorithms called *predictor-corrector methods*.

$$\tilde{y}_{k+1} = y_k + h f(t_k, y_k)$$

The complete Heun method approximation equation is given by equation (7.5).

$$y_{k+1} = y_k + \frac{h}{2}\left[f(t_k, y_k) + f(t_{k+1}, (y_k + h f(t_k, y_k)))\right] \tag{7.5}$$

The `heunODE` function listed in code 7.13 is called with the same arguments as the `eulerODE` function.

7.5.3 Runge-Kutta Method

The Runge-Kutta method uses Simpson's rule integration to approximate equation (7.3). There are actually several algorithms that are considered to be part of the Runge-Kutta family of algorithms. The so-called *classic* algorithm used here is an order–4 algorithm with a comparatively simple expression. Runge-Kutta algorithms are the most accurate of the fixed step size algorithms. They are only surpassed by adaptive step size algorithms, which typically make use of Runge-Kutta equations.

```
function [t, y] = heunODE(f, tspan, y0, n)
% HEUNODE - ODE solver using Heun's method.
%
%    f - vectorized derivative function, y' = f(t, y)
%    tspan - row vector [t0 tf] for initial and final time
%    y0 - initial value of y
%    n - number of y values to find
     h = (tspan(2) - tspan(1))/(n-1);
     t = linspace(tspan(1), tspan(2), n);
     y = zeros(1, n);
     y(1) = y0;
     for k = 1:n-1
         fk = f(t(k), y(k));
         yk1 = y(k) + h*fk;
         y(k+1) = y(k) + (h/2)*(fk + f(t(k+1), yk1));
     end
```

CODE 7.13: A function implementing the Heun's method ODE solver.

Runge-Kutta algorithms are weighted sums of coefficients calculated from evaluations of $f(t, y)$ (the derivative function). Later coefficients make use of previous coefficients, so the coefficients at each evaluation point are first calculated sequentially before the solution value can be determined.

$$s_1 = f(t_k, y_k)$$

$$s_2 = f(t_k + \tfrac{h}{2}, y_k + \tfrac{h}{2} s_1)$$

$$s_3 = f(t_k + \tfrac{h}{2}, y_k + \tfrac{h}{2} s_2)$$

$$s_4 = f(t_k + h, y_k + h s_3)$$

$$y_{k+1} = y_k + \tfrac{h}{6} (s_1 + 2s_2 + 2s_3 + s_4)$$

The `rk4ODE` function listed in code 7.14 is called with the same arguments as the `eulerODE` and `heunODE` functions.

7.5.4   Algorithm Comparison

We will use the well known decaying exponential ODE to compare the three methods that use a fixed step size. We will also reference this simple ODE when considering stability constraints on the step size in section 7.5.5.

The ODE

$$y' = \frac{dy(t)}{dt} = -a\,y, \text{ where } a > 0, \text{ and } y(0) = c$$

has the solution $y(t) = c\,e^{-at}$. For our comparison, $a = 1$, and $y(0) = 1$.

```
function [t, y] = rk4ODE(f, tspan, y0, n)
% RK4ODE - ODE solver using the classic Runge-Kutta method.
%
%   f - vectorized derivative function, y' = f(t, y)
%   tspan - row vector [t0 tf] for initial and final time
%   y0 - initial value of y
%   n - number of y values to find
    h = (tspan(2) - tspan(1))/(n - 1);
    t = linspace(tspan(1), tspan(2), n);
    y = zeros(1, n);
    y(1) = y0;
    for k = 1:n-1
        s1 = f(t(k), y(k));
        s2 = f(t(k)+h/2, y(k)+s1*h/2);
        s3 = f(t(k)+h/2, y(k)+s2*h/2);
        s4 = f(t(k+1), y(k)+s3*h/2);
        y(k+1) = y(k) + (h/6)*(s1 + 2*s2 + 2*s3 + s4);
    end
```

CODE 7.14: A function implementing the Runge-Kutta method ODE solver.

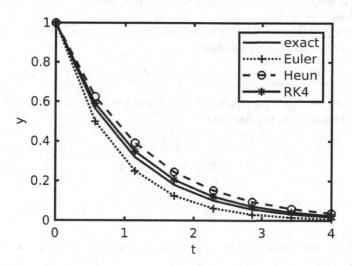

FIGURE 7.17: Comparison of fixed interval ODE solvers.

We can see in figure 7.17 that, as expected, the classic Runge-Kutta method performs the best followed by Heun's method and then Euler's method.

```
>> % Code for figure 7.17
>> f = @(t, y) -y;
>> [t, y_euler] = eulerODE(f, [0 4], 1, 8);
```

```
>> y_exact = exp(-t);
>> [t, y_heun] = heunODE(f, [0 4], 1, 8);
>> [t, y_rk4] = rk4ODE(f, [0 4], 1, 8);
```

Note: All of the algorithms shown in figure 7.17 have good performance if h is significantly reduced. Finding noticeable differences in the plot is only achieved by using a larger step size.

7.5.5   Stability Constraints

An unstable algorithm execution gives fluctuating and wrong answers. A user without knowledge of the stability constraints could specify a step size for a fixed step size algorithm that is too large and get an unstable solution. The stability constraint is especially concerning with stiff systems[5] where one term in the solution may place a much higher constraint on the step size than other terms in the solution.

We return to the Euler method and an exponentially decaying ODE to find a stability constraint related to h, the distance between consecutive evaluation points of ODE solvers [65]. With $y' = -a\,y$, where $a > 0$, $y(0) = 1$, the exact solution is $y(t) = e^{-a\,t}$. A property of $y(t)$ that needs to be preserved is that $y(t) \to 0$ as $t \to \infty$.

Referring to section 7.5.1, The forward Euler solution is $y_{k+1} = y_k - h\,a\,y_k$, or $y_{k+1} = (1 - ha)\,y_k$. By induction,

$$y_k = (1 - ha)^k\,y_0. \tag{7.6}$$

A stability constraint on equation (7.6) is needed to ensure that the property of $y_k \to 0$ as $k \to \infty$ is preserved.

$$|1 - h\,a| < 1$$
$$-1 < (1 - h\,a) < 1$$
$$-2 < -h\,a < 0$$
$$h < \frac{2}{a}$$

In the following example $2/a = 2/0.5 = 4$, but $h = 5$[6] and the results are unstable.

```
>> fexp = @(t, y) -0.5*y;
>> [t, y_unstable] = eulerODE(fexp, [0 20], 1, 5)
t =
        0      5     10     15     20
y_unstable =
        1.0000   -1.5000    2.2500   -3.3750    5.0625
```

---

[5]Stiff ODEs are addressed in section 7.5.9.
[6]h = (20 - 0)/(n - 1), $h = 5$, so $n = 5$.

The stability constraint for Euler's method is sufficient for other fixed step size algorithms.

7.5.6 Implicit Backward Euler Method

The implicit backward Euler (IBE) method seeks to remove the stability constraint of fixed step size ODE solvers. To see how the IBE solver works, we will look at equation (7.2) from the perspective of finding y_k from the yet unknown y_{k+1}. So the terms of equation (7.2) are reordered and the limits of the integral are reversed.

$$y_k = y_{k+1} + \int_{t_{k+1}}^{t_k} f(t, y) dt \tag{7.7}$$

Again, we will use the ODE $y' = -a\,y$, where $a > 0$, and $y(0) = 1$. The exact solution is $y(t) = e^{-a\,t}$. As before, we want to find any stability constraints that are required to ensure that $y_k \to 0$ as $k \to \infty$. We can solve the decaying exponential ODE problem numerically by applying Euler's method to equation (7.7).

$$y_k = y_{k+1} + h\,a\,y_{k+1}$$
$$y_k = (1 + h\,a)\,y_{k+1}$$
$$y_{k+1} = \frac{1}{1 + h\,a}\,y_k$$

$$y_{k+1} = \left(\frac{1}{1 + h\,a}\right)^k y_0 \tag{7.8}$$

The constraint to ensure stability of equation (7.8) as $k \to \infty$ requires that

$$\left|\frac{1}{1 + h\,a}\right| < 1,$$

which always holds because both h and a are positive numbers. So the implicit backward Euler method is unconditionally stable.

7.5.7 Adaptive Algorithms

Numerical ODE solvers need to use appropriate step sizes between evaluation steps to achieve high levels of accuracy while also quickly finding solutions across the evaluation interval. Adaptive ODE solvers reduce the step size, h, where needed to meet the accuracy requirements. They also increase the step size to improve performance when the accuracy requirements can be satisfied with a larger step size.

One of two strategies can be used to determine the step size adjustments. As was done with adaptive numerical methods for computing integrals in section 7.4.3, the solution from two half steps could be compared to the solution from a full step. Based on the comparison, the step size can be reduced, left the same, or increased. However, points of an ODE solution are found sequentially, so the strategy of splitting a wide region into successively smaller regions would be computationally slow. Instead, the current step size is used to find the next solution (y_{k+1}) using two algorithms where one algorithm is regarded as more accurate than the other. The guiding principle is that if the solutions computed by the two algorithms are close to the same, then the step size is small enough or could even be increased. But if the difference between the solutions is larger than a threshold, then the step size should be reduced. If the solutions differ significantly, then the current evaluations are recalculated with a smaller step size. The algorithm advances to the next evaluation step only when the differences between the solutions is less than a specified tolerance.

7.5.7.1   The RKF45 Method

The Runge-Kutta-Fehlberg method (denoted as RKF45) uses fourth and fifth order Runge-Kutta ODE solvers to adjust the step size at each evaluation step. The fourth order equation used is more complex than is used by the fourth order classic Runge-Kutta algorithm. But the computed coefficients for the fourth order equation are also used for the fifth order equation, which only requires one additional coefficient. In the `rkf45ODE` function, the fourth order solution values (RK4) are stored in the `z` array, and the fifth order solution values (RKF45) are stored in the `y` array, which is returned from the function. The coefficients and solution values are calculated using the following equations. Note that the s_2 coefficient is used to find later coefficients, but is not part of either the fourth or fifth order solution equations.

There are variances in the literature on the implementation details of the RKF45 method. The `rkf45ODE` function listed in code 7.17 on page 357 is adapted from the description of the algorithm in the numerical methods text by Mathews and Fink [53].

$$z_{k+1} = y_k + \frac{25}{216}s_1 + + \frac{1408}{2565}s_3 + + \frac{2197}{4101}s_4 - \frac{1}{5}s_5$$

$$y_{k+1} = y_k + \frac{16}{135}s_1 + + \frac{6656}{12825}s_3 + + \frac{28561}{56430}s_4 - \frac{9}{50}s_5 + \frac{2}{55}s_6$$

$$s_1 = h\,f\left(t_k, y_k\right)$$

$$s_2 = h\,f\left(t_k + \frac{1}{4}h, y_k + \frac{1}{4}s_1\right)$$

$$s_3 = h f \left(t_k + \frac{3}{8} h, y_k + \frac{3}{32} s_1 + \frac{9}{32} s_2 \right)$$

$$s_4 = h f \left(t_k + \frac{12}{13} h, y_k + \frac{1932}{2197} s_1 - \frac{7200}{2197} s_2 + \frac{7296}{2197} s_3 \right)$$

$$s_5 = h f \left(t_k + h, y_k + \frac{439}{216} s_1 - 8 s_2 + \frac{3680}{513} s_3 - \frac{845}{4104} s_4 \right)$$

$$s_6 = h f \left(t_k + \frac{1}{2} h, y_k - \frac{8}{27} s_1 + 2 s_2 - \frac{3544}{2565} s_3 + \frac{1859}{4104} s_4 - \frac{11}{40} s_5 \right)$$

Some adaptive ODE algorithms make course changes to the step size such as either cutting it in half or doubling it. The rkf45ODE function first finds a scalar value, s, from a tolerance value and the difference between the solutions from the two algorithms. In the implementation of rkf45ODE in code 7.17, the new step size comes from multiplying the current step size by the scalar. The scalar will be one if the difference between the solutions is half of the tolerance. The algorithm advances to the next step if the difference between the two solutions is less than the tolerance value, which occurs when the scalar is greater than 0.84. Otherwise, the current evaluation is recalculated with a smaller step size. The multiplier is restricted to not be more than two. So the step size will never increase to more than twice its current size. The tolerance and the threshold for advancing to the next step are tunable parameters.

$$s = \left(\frac{tolerance}{2 \, \| z_{k+1} - y_{k+1} \|} \right)^{\frac{1}{4}}$$

The initial value (y_0) passed to the rkf45ODE function can be either a scalar for a single equation ODE or a column vector for a system of ODEs. Since the number of steps to be taken is not known in advance, the initial step size is a function argument.

In the example plotted in Figure 7.18 the adaptive Runge-Kutta-Fehlberg (RKF45) method is more accurate than the classic Runge-Kutta method (RK4) with the initial step size of RKF45 being the same as the fixed step size used by RK4. For most ODEs, the ability of RKF45 to reduce the step size where needed likely contributes more to its improved accuracy than can be attributed to using a higher order equation.

```
>> % Code for figure 7.18
>> f = @(t,y) -y;
>> [t, y_rkf45] = rkf45ODE(f, [0 4], 1, 0.5);
>> y_exact = exp(-t);
>> hold on
>> plot(t, y_exact), plot(t, y_rkf45, ':')
>> [t, y_rk4] = rk4ODE(f, [0 4], 1, 8);
>> plot(t, y_rk4, '--')
>> hold off
```

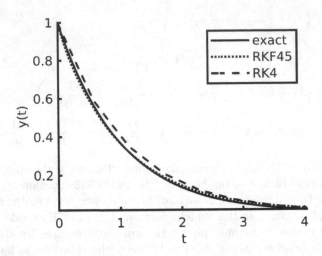

FIGURE 7.18: The adjustable step size of the RKF45 algorithm gives accurate results with good performance.

7.5.7.2 MATLAB's ode45

Although the adaptive `rkf45ODE` function has good results, it is not as accurate as MATLAB's `ode45` function. Using the same comparison test shown in figure 7.18, the `ode45` function had a maximum error of 6×10^{-6}, and a plot of its solution appears to be directly on top of the exact solution plot. MATLAB's `ode45` is the ODE solver that should be tried first for most problems. It has good accuracy and performance for ODEs that are not stiff.[7] Like the RKF45 method, it uses an adaptive step size algorithm based on the fourth and fifth order Runge-Kutta formula.

The `ode45` function is called with the same arguments as the ODE solvers previously described, except neither the initial step size nor the number of evaluation points is given. The MATLAB ODE solvers also offer additional flexibility regarding the evaluation points. If the `tspan` argument contains more than two values, then `tspan` is taken to be a vector listing the evaluation points to use.

As before, we often save two return values from `ode45`, `[t, y] = ode45(f, tspan, y0)`. The return values are stored in column vectors rather than row vectors as was the case with the simple functions previously used. If no return values are specified, `ode45` will make a plot of the y values against the t axis.

---

[7]Stiff ODEs are addressed in section 7.5.9.

The MATLAB ODE solvers have several additional options and tunable parameters. Rather than passing those individually to the solver, a data structure holding the options is passed as an optional argument. The data structure is created, and modified using the **odeset** command with name–value pairs. Refer to the MATLAB documentation for **odeset** for a full list of the options and their descriptions.

```
>> options = odeset('RelTol', 1e-5);        % create
>> options = odeset(options, 'AbsTol',1e-7); % modify
>> [t, y] = ode45(f, tspan, y0, options);   % use
```

7.5.8   The Logistic Equation

This example problem is a model of growth slowed by competition. The competition is within the population for limited resources. The growth gives a y term in the ODE—the growth is proportional to the population size. The competition (y against y) adds a $-y^2$ term.

$$\frac{dy}{dt} = a\,y - b\,y^2.$$

The solution is modeled with positive and negative values for t. The analytic solution is: ([70] page 55)

$$y(t) = \frac{a}{d\,e^{-at} + b}.$$

where, $d = \frac{a}{y(0)} - b$.

In the distant past, ($t = -\infty$), the population size is modeled as zero. The steady state ($t = \infty$) population size is $y(t) = K = \frac{a}{b}$, which is the sustainable population size.

The half way point, can be modeled as $y(0) = a/2b$.

In code 7.15, we will compute the positive half of the solution and then reflect the values to get the negative time values. The solution plot is shown in figure 7.19.

7.5.9   Stiff ODEs

A problem is stiff if the solution contains multiple terms with different step size requirements. A slower moving component may dominate the solution; however, a less significant component of the solution may change much faster. Adaptive step size algorithms must take small steps to maintain accurate results. Because stiff ODEs problems have solutions with multiple terms, they come from either second or higher order ODEs or from systems of ODEs.

```
% File: logistic_diffEq.m
% Logistic Equation: a = 1, b = 1
f = @(t, y) y - y.^2;
tspan = [0 5];
y0 = 1/2;
[t, y] = ode45(f, tspan, y0);
figure, plot(t,y)
t1 = flip(-t);
y1 = flip(1-y);
t2 = [t1;t];
y2 = [y1;y];
analytic = 1./(1 + exp(-t2));
figure, hold on
plot(t2, y2, 'o')
plot(t2, analytic)
hold off, title('Logistic Equation')
xlabel('t'), ylabel('y')
legend('ode45 Numeric Solution','Analytic Solution', ...
    'Location', 'northwest');
```

CODE 7.15: The logistic ODE solved with ode45.

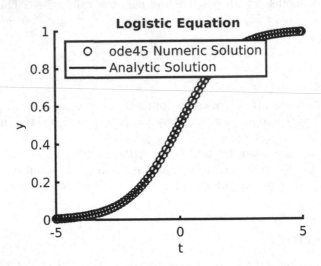

FIGURE 7.19: Numerical and analytical solution to the logistic ODE.

We will now look at an example of a moderately stiff system of ODE equations. Then we will see how implicit solvers solve stiff systems of ODEs and are unconditionally stable. Then we will look at MATLAB's stiff ODE solver, ode15s.

7.5.9.1 Stiff ODE Systems

Here we show a system of ODEs that have a moderately stiff solution. The reader is referred to section 6.7 for a review of systems of first order, linear ODEs.

$$\begin{cases} x' = -25\,x + 24\,y, & x(0) = 2 \\ y' = 24\,x - 25\,y, & y(0) = 1 \end{cases}$$

It is more convenient to express systems of ODEs using vectors and matrices.

$$\boldsymbol{u}(t) = \begin{bmatrix} x(t) \\ y(t) \end{bmatrix} \qquad \mathbf{A} = \begin{bmatrix} -25 & 24 \\ 24 & -25 \end{bmatrix} \qquad \boldsymbol{u}' = \mathbf{A}\,\boldsymbol{u} \qquad \boldsymbol{u}_0 = \begin{bmatrix} 1 \\ 2 \end{bmatrix}$$

The eigenvectors, eigenvalues, and c coefficients are found with some help from MATLAB.

$$\boldsymbol{x}_1 = \frac{1}{\sqrt{2}}\begin{bmatrix} 1 \\ -1 \end{bmatrix} \qquad \boldsymbol{x}_2 = \frac{1}{\sqrt{2}}\begin{bmatrix} 1 \\ 1 \end{bmatrix} \qquad \begin{aligned} \lambda_1 &= -49 \\ \lambda_2 &= -1 \end{aligned} \qquad \begin{aligned} c_1 &= -\sqrt{2}\cdot 0.5 \\ c_2 &= \sqrt{2}\cdot 1.5 \end{aligned}$$

The solution to the system of ODEs then comes from the general solution for systems of first order, linear ODEs as given by equation (6.7) in section 6.7.

$$\boldsymbol{u}(t) = c_1 e^{\lambda_1 t}\boldsymbol{x}_1 + c_2 e^{\lambda_2 t}\boldsymbol{x}_2 \tag{7.9}$$

$$\boldsymbol{u}(t) = \begin{cases} x(t) = -0.5\,e^{-49t} + 1.5\,e^{-t} \\ y(t) = 0.5\,e^{-49t} + 1.5\,e^{-t} \end{cases}$$

This system is stiff because the exact solution has terms with largely different exponent values coming from the eigenvalues of $\mathbf{A}$. The stability constraints on the e^{-49t} terms require $h < 2/49 \approx 0.0408$. From the beginning of the evaluation, those terms will move toward zero much faster than the e^{-t} terms. We can see from figures 7.20 and 7.21 that an adaptive algorithm such as RKF45 or MATLAB's ode45 will significantly reduce the step size from the beginning of the evaluations until the e^{-49t} term has sufficiently gone to zero. Using a longer evaluation time, the errors between the RKF45 solution and the exact solution go to near zero and the step sizes get much larger.

```
>> % Code for figure 7.20 and figure 7.21
>> A = [-25 24; 24 -25]; u0 = [1;2];
>> fA = @(t, u) A*u;
>> [t, u] = rkf45ODE(fA, [0 1], u0, 0.04);
>> x = @(t) -0.5*exp(-49*t) + 1.5*exp(-t);
>> y = @(t) 0.5*exp(-49*t) + 1.5*exp(-t);

>> figure, hold on
>> plot(t, u(1,:), 'o'), plot(t, u(2,:), '+')
>> plot(t, x(t), ':'), plot(t, y(t), '--')
>> hold off
```

```
>> U = [x(t); y(t)];
>> figure, hold on
>> plot(t, vecnorm(U - u))
>> plot(t(1:end-1), diff(t), 'o:')
>> hold off
```

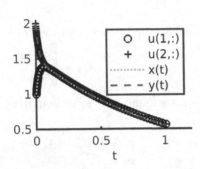

FIGURE 7.20: Stiff ODE system—the $\pm 0.5\,e^{-49\,t}$ terms quickly goes to zero and the $x(t)$ and $y(t)$ solutions approximately merge to $1.5\,e^{-t}$.

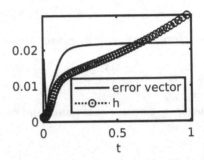

FIGURE 7.21: Error and step size for stiff ODE system. First curve—the norm of the error vector between the exact solution and that found from the RKF45 solver. Second curve—the step size, h, for the stiff ODE system.

7.5.9.2   Implicit Solvers for Systems of ODEs

The implicit backward Euler (IBE) method is used here to solve stiff systems of ODEs. The initial discussion parallels that of section 7.5.6 but the equations are extended to systems of ODEs. As with the implicit solver for a single ODE equation, the IBE method for systems of ODEs is also unconditionally stable.

We will again use the decaying exponential ODE from the previous section. The $\mathbf{A}$ matrix is required to be a symmetric matrix with real, negative eigenvalues.

$$\mathbf{u}'(t) = \mathbf{A}\,\mathbf{u}(t) \qquad \mathbf{u}(0) = \mathbf{u}_0$$

The solution is $\mathbf{u}(t) = \mathbf{u}_0 e^{\mathbf{A}\,t}$. Refer to section 6.7 and appendix B.2 for more information about how we get from a solution with a matrix in the exponent to equation (7.9).

Notice that we do not have a negative sign in front of the $\mathbf{A}$ matrix as we previously saw with a in the scalar ODE from section 7.5.5. But it is a decaying exponential that goes to zero when t goes to infinity because the eigenvalues of $\mathbf{A}$ are negative.

We get the starting equation for the numerical solution by applying equation (7.7).

$$u_k = u_{k+1} - h\,\mathbf{A}\,u_{k+1} = (\mathbf{I} - h\,\mathbf{A})\,u_{k+1}$$

$$u_{k+1} = (\mathbf{I} - h\,\mathbf{A})^{-1}\,u_k \qquad (7.10)$$

Equation (7.10) is an implicit equation for solving systems of ODEs numerically, and is implemented in the IBEsysODE function listed in code 7.16. Notice that the implicit approach requires solving an equation at every step. In this case, the equation is a linear system of equations with a fixed matrix, so the IBEsysODE function uses LU decomposition once and calls on MATLAB's triangular solver via the left-divide operator to quickly find a solution in each loop iteration. If this were a nonlinear problem, then numerical methods such as Newton's method or the secant method described in section 7.1.1 would be needed, which could add significant computation. For this reason, implicit methods are not recommended for problems that are either not stiff or nonlinear [65].

```
function [t, u] = IBEsysODE(A, tspan, u0, n)
% IBESYSODE - ODE solver using the Implicit Euler Backward
%             method for systems of ODEs.
%
% A - matrix defining the system coefficients
% tspan - row vector [t0 tf] for initial and final time
% y0 - initial value of y
% n - number of y values to find
  h = (tspan(2) - tspan(1))/n;
  t = linspace(tspan(1), tspan(2), n);
  m = size(A, 1);
  u = zeros(m, n);
  u(:,1) = u0;
  [L, U, P] = lu(eye(m) - h*A);
  for k = 1:n-1
      u(:,k+1) = U\(L\(P*u(:,k)));
  end
```

CODE 7.16: A function implementing the implicit backward Euler method for systems of ODEs.

Figure 7.22 shows the numerical solution that the IBEsysODE function found to our stiff system of ODEs. Since this is a fixed step size solution, the step size was decreased to 0.02 to track the quickly changing e^{-49t} terms in the solution. Of course, the small step is not needed once the e^{-49t} term is sufficiently close to zero. Remember that there is a difference between accuracy and stability. A lack of stability causes the solution to be wrong as $k \to \infty$. A lack of accuracy just means that the exact solution is not well modeled by the numerical solution, which is often solved by reducing the step size.

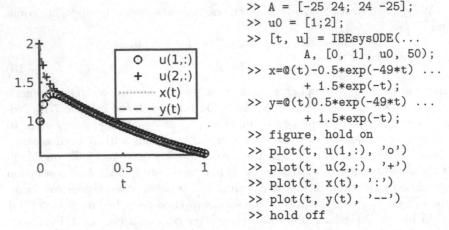

```
>> A = [-25 24; 24 -25];
>> u0 = [1;2];
>> [t, u] = IBEsysODE(...
         A, [0, 1], u0, 50);
>> x=@(t)-0.5*exp(-49*t) ...
         + 1.5*exp(-t);
>> y=@(t)0.5*exp(-49*t) ...
         + 1.5*exp(-t);
>> figure, hold on
>> plot(t, u(1,:), 'o')
>> plot(t, u(2,:), '+')
>> plot(t, x(t), ':')
>> plot(t, y(t), '--')
>> hold off
```

FIGURE 7.22: With enough data points, the `IBEsysODE` function is able to to reasonably well track the rapidly changing solution of the e^{-49t} terms.

To evaluate the stability of the IBE method, we start by using induction to express the steady state solution in terms of the initial value.

$$\boldsymbol{u}_k = \left[(\mathbf{I} - h\,\mathbf{A})^{-1}\right]^k \boldsymbol{u}_0$$

The stability constraint is that $\boldsymbol{u}_k \to 0$ as $k \to \infty$. So we need that

$$\left\|(\mathbf{I} - h\,\mathbf{A})^{-1}\right\|_2 < 1, \quad \text{for all } h.$$

Since our $\mathbf{A}$ matrix is symmetric, we can use the property of $\|\cdot\|_2$ matrix norms that $\|\mathbf{A}\|_2 = \max_i |\lambda_i(\mathbf{A})|$, where $\lambda_i(\mathbf{A})$ means the ith eigenvalue of $\mathbf{A}$ (appendix A.1.3).

$$\max_i \left|\lambda_i\left((\mathbf{I} - h\,\mathbf{A})^{-1}\right)\right| < 1$$

Now we can make use of the eigenvalue property for symmetric matrices that $\lambda_i(\mathbf{A}^{-1}) = \frac{1}{\lambda_i(\mathbf{A})}$ (equation (6.1) in section 6.4).

$$\max_i \left| \frac{1}{\lambda_i(I - h\,\mathbf{A})} \right| < 1$$

$$\min_i |\lambda_i(I - h\,\mathbf{A})| = \min_i |1 - h\,\lambda_i(\mathbf{A})| > 1 \tag{7.11}$$

Equation (7.11) holds for any symmetric $\mathbf{A}$ matrix with negative eigenvalues. So the implicit backward Euler method is unconditionally stable for systems of ODEs.

Unconditionally stable ODE solving algorithms remove the concern about having an unstable numerical solution because the step size is too large, but as we saw in figure 7.22, accuracy requirements may still require a reduced step size. The more complex and robust algorithms of MATLAB's ODE solvers for stiff systems are able to adjust the step size enough to find accurate solutions without being excessively slowed by a stiff system.

7.5.9.3 MATLAB's ode15s

MATLAB has several stiff ODE solvers. The ode15s function should be tried first for stiff ODEs. The primary authors of MATLAB's suite of ODE solvers, Shampine and Reichelt, describe ode15s as an implicit quasi-constant step size algorithm. They further give the following suggestion regarding picking which of their ODE solver to use [64].

> ... Except in special circumstances, ode45 should be the code tried first. If there is reason to believe the problem to be stiff, or if the problem turns out to be unexpectedly difficult for ode45, the ode15s code should be tried.

Figure 7.23 shows how ode15s handles the stiff system of ODEs used in the previous examples.

```
>> % Code for figure 7.23
>> A = [-25 24; 24 -25]; u0 = [1;2];
>> fA = @(t, u) A*u;
>> [t, y_ode15s] = ode15s(fA, [0 1], u0);
>> t = t'; u = y_ode15s';
>> x = @(t) -0.5*exp(-49*t) + 1.5*exp(-t);
>> y = @(t) 0.5*exp(-49*t) + 1.5*exp(-t);
>> figure, hold on
>> plot(t, u(1,:)', 'o'), plot(t, u(2,:)', '+')
>> plot(t, x(t), ':'), plot(t, y(t), '--')
>> hold off
```

7.5.10 MATLAB's Suite of ODE Solvers

MATLAB's ode45 and ode15s are the nonstiff and stiff ODE solvers to use for most problems. If those do not give satisfactory results then ode23 and ode113 might be tried for nonstiff problems. For stiff problems, ode23s should be the next function tried after ode15s. Additional documentation and research may be needed to select the best solver to use for some problems. The MathWorks' web pages have quite a bit of documentation on the suite of ODE solvers. The *MATLAB Guide* by Higham and Higham [34] has extended coverage MATLAB's suite of ODE solvers. Table 7.1 lists MATLAB's suite of ODE solvers.

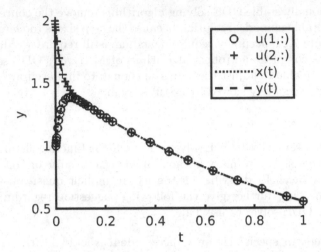

FIGURE 7.23: The ode15s ODE solver is designed for stiff systems, so the less significant, but rapidly changing terms do not significantly reduce the step size.

TABLE 7.1: The MATLAB ODE solvers

Solver	Problem type	Description
ode45	Nonstiff	Explicit Runge-Kutta, orders 4 and 5
ode23	Nonstiff	Explicit Runge-Kutta, orders 2 and 3
ode113	Nonstiff	Explicit linear multistep, orders 1–13
ode15s	Stiff	Implicit linear multistep, orders 1–5
ode15i	Fully implicit	Implicit linear multistep, orders 1–5
ode23s	Stiff	Modified Rosenbrock pair, orders 2 and 3
ode23t	Mildly stiff	Implicit trapezoid rule, orders 2 and 3
ode23tb	Stiff	Implicit Runge-Kutta, orders 2 and 3

Online Resources

- The MathWorks website[a] has some good tutorial videos about numerical solutions to differential equations. Although, the ODE solvers in MATLAB are more complex than the examples in the first three videos, they explain the basic concepts.

- MATLAB's ODE solvers are reviewed on the MathWorks website[b].

[a]https://www.mathworks.com/videos/series/solving-odes-in-matlab-117658.html

[b]https://www.mathworks.com/help/matlab/math/choose-an-ode-solver.html

```
function [t, y] = rkf45ODE(f, tspan, y0, h0)
% rk5ODE - Adaptive interval ODE solver using the
%           Runge-Kutta-Fehlberg method.
%
%   f - vectorized derivative function, y' = f(t, y)
%   tspan - row vector [t0 tf] for initial and final time
%   y0 - initial value of y, scalar or column vector
%   h0 - initial step size
    tol = 2e-5;
    h = h0;
    nEst = ceil((tspan(2) - tspan(1))/h);
    m = size(y0, 1);
    t = zeros(1, 2*nEst);  % estimated preallocation
    y = zeros(m, 2*nEst);
    z = zeros(m, 2*nEst);
    t(1) = tspan(1);
    y(:,1) = y0;        % RKF45 results
    z(:,1) = y0;        % RK4 results
    k = 1;
    while t(k) < tspan(2)
        s1 = h*f(t(k), y(:,k));
        s2 = h*f(t(k) + h/4, y(:,k) + s1/4);
        s3 = h*f(t(k) + h*3/8, y(:,k) + s1*3/32 + s2*9/32);
        s4 = h*f(t(k) + h*12/13, y(:,k) + s1*1932/2197 ...
            - s2*7200/2197 + s3*7296/2197);
        s5 = h*f(t(k) + h, y(:,k) + s1*439/216 - s2*8 ...
            + s3*3680/513 - s4*845/4104);
        s6 = h*f(t(k) + h/2, y(:,k) - s1*8/27 + s2*2 ...
            - s3*3544/2565 + s4*1859/4104 - s5*11/40);
        z(:,k+1) = y(:,k) + s1*25/216 + s3*1408/2565 ...
            + s4*2197/4101 - s5/5;  % RK4
        y(:,k+1) = y(:,k) + s1*16/135 + s3*6656/12825 ...
            + s4*28561/56430 - s5*9/50 + s6*2/55; % RKF5
        % no divide by 0, h at most doubles
        algDiff = max(tol/32, norm(z(:,k+1) - y(:,k+1)));
        s = (tol/(2*algDiff))^0.25;
        h = h*s;
        if s > 0.84
            t(k+1) = t(k) + h;
            k = k + 1;
        end
    end
    % delete unused preallocated memory
    t(k+1:end) = []; y(:,k+1:end) = [];
```

CODE 7.17: A function implementing the Runge-Kutta-Fehlberg method ODE solver.

7.6   Exercises

Exercise 7.1   Roots of a Function

1. Use the `fzero` function to find where $f(x) = 5 + 10x - x^2$ is equal to zero in the range $-5 \le x \le 15$.

2. Write a MATLAB function that uses the Newton-Raphson method to find the cube root of K, $x_r = \sqrt[3]{K}$, where $K > 0$. The function should take K as an argument and return x_r. You may use the `newtSqrt` function in code 7.1 as a guide.

3. Use the `bisectRoot` function (code 7.3) to find the point x_a where $f(x_a) = \cos^2 x_a - \sin^2 x_a = 0.25$. Note that you need to alter the function so that x_a is a root of the function. Use `fplot` to find a search range for `bisectRoot`. Find all x_a where $f(x_a) = 0.25$, $0 \le x_a < 2\pi$.

4. Use the `tic` and `toc` functions to measure the execution times of `bisectRoot` and `secantRoot` (code 7.4) for several functions. Which function is faster?

Exercise 7.2   Numerical Minimum

1. Use the `fminbnd` function to find the maximum value of $f(x) = 5x^2e^{-3x}$, in the range $0 < x < 2$. Use `fplot` to narrow the search range before using `fminbnd`. Recall that since we are looking for a maximum value that we need to search the negative of $f(x)$.

2. Use the `fminsearch` function to find the location of the minimum of the surface plots shown in section 2.3.1. Let $w = (x, y)$ be a vector of the x and y values of the surface. Then the height of the surface is defined by

$$z(w) = (w(2) - w(1)) \, e^{(-0.12(w(1)^2 + w(2)^2))}.$$

Exercise 7.3   Data Interpolation

Use the following code to make 8 (x, y) points. Then use the `interp1` function to expand the data to 32 data points. Generate a 2×2 subplot with the smooth function, 8 point plot, 32 point plots with *linear* and *pchip* interpolation.

```
>> f = @(x) cos(x) + 0.7*sin(x);
>> x = linspace(-pi,pi,8);
>> y = f(x);
>> plot(x,y)
```

Exercise 7.4   Numerical Derivative

Let $y(t) = \cos(t) + \frac{1}{2}\cos 2t$, for $0 \leq t \leq 2\pi$, then $\frac{dy(t)}{dt} = -\sin(t) - \sin(2t)$. Make a plot over the range $0 \leq t \leq 2\pi$ showing three data plots: the analytic derivative, the Euler derivative with $h = \pi/64$ and the spectral derivative.

Exercise 7.5   Numerical Integration

Using the trapezoid rule, Simpson rule, the adaptive integral algorithm presented in section 7.4.3, and MATLAB's `integral` function compute the definite integral $\int_0^{10} x^2 e^{-x}dx$

Exercise 7.6   Numerical Differential Equations

1. Using the anonymous function `f = @(t,y) t.^2 .* exp(-t);`, let $y'(t) = f(t)$, and $y(0) = 0$ define a differential equation. Use `ode45` to plot $y(t)$, for $0 \leq t \leq 20$.

2. Modify the `rkf45ODE` function listed in code 7.17 on page 357 to take on of three actions after each evelution step.

   - Cut the step size in half and repeat the calculation for that step.

   - Leave the step size as is and advance to the next evaluation step.

   - Double the step size and advance to the next evaluation step.

   You will want to establish thresholds for the variable s to decide which action to take. Test and compare your program with plots to the function listed in code 7.17 and to `ode45`.

Appendix A

Linear Algebra Appendix

A.1   Norms

MATLAB includes a function called norm for the purpose of find the length of vectors or matrices. The most frequent usage is to find the Euclidean length of a vector, which is called a l_2-norm. It comes direct from the Pythagorean theorem—the square root of the sum of the squares. The length of a vector is a familiar concept, but the length of a matrix feels somewhat mysterious. As always, the norm function is well documented in the MATLAB documentation. However, there are different measures of length and properties of norms that should be reviewed. Moreover, the names, symbols, and application of the various norms could use some clarification.

In technical literature, the most common mathematics symbol for a norm is a pair of double bars around the variable name with a subscript for the type of norm, $\|v\|_2$. If the subscript is left off, then it is assumed to be 2. The generic name given to the type of norm for vectors is the italics letter l with a subscript of the type, l_2. You may sometimes see the type given as a super-script instead of a subscript.

Cardinality

SYMBOLS $\|v\|_0$, $\|\mathbf{A}\|_0$, l_0

DESCRIPTION The l_0 norm is the number of nonzero elements of either a vector or a matrix. The l_0 norm does not fit the properties that are normally expected of a norm, so it is not always classified as being a norm. It has utility for sparse (a lot of zeros) vectors and matrices.

MATLAB EXAMPLE

```
>> v = [1; 2; 0; 4];
>> v_card = nnz(v)
v_card =
     3
```

A.1.1   Vector Norms

General Vector Norm, p-Norm

The `norm` function takes a second argument, p, that specifies the order of the calculation as follows.

$$\|\boldsymbol{v}\|_p = \text{norm}(v,\ p) = \left[\sum_{k=1}^{N} |v_k|^p\right]^{1/p}$$

Figure A.1 shows plots of the x and y values that satisfy $\|[x\ y]\|_p = 1$ for various values of p. Keep these plots in mind as you read the description of the l_1, l_2, and l_∞ norms. We see straight lines in the $p = 1$ plot because the value of the norm is the sum of the absolute values of the elements. The plot of a circle for $p = 2$ relates to the l_2 norm being the length of a vector by Pythagorean theorem. As the p values get larger, the plots approach the shape of a square, which models the l_∞ norm where the norm takes the value of the largest absolute value of the elements.

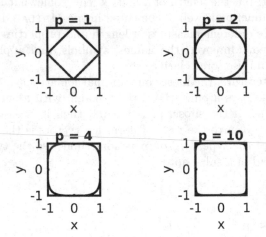

FIGURE A.1: Plots of $\|[x\ y]\|_p = 1$.

Taxicab Norm, Manhattan Distance, or City Block Distance

SYMBOLS $\|\boldsymbol{v}\|_1$, l_1

DESCRIPTION The l_1 norm is the sum of the absolute values of the elements. It is the distance that a car would drive between two points on a rectangular grid of streets.

$$\|\boldsymbol{v}\|_1 = \sum_{k=1}^{N} |v_k|$$

The l_1 norm has application to compressed sensing, which seeks to combine data collection and compression into a single algorithm ([7], pages 88-91). The l_1 norm has also shown to improve linear regression, compared to the l_2 norm, when the data contains outlier values.

MATLAB EXAMPLE

```
>> v = [1; 2; 0; 4];
>> v_l1 = norm(v, 1)
v_l1 =
      7
```

Euclidean Norm

SYMBOLS $\|v\|$, $\|v\|_2$, l_2

DESCRIPTION The l_2 norm of a vector is by far the most frequently used norm calculation. It finds the length of a vector by the same means that the Pythagorean theorem finds the length of the hypotenuse of a right triangle.

$$\|v\|_2 = \sqrt{\sum_{k=1}^{N} v_k^2} = \sqrt{v_1^2 + v_2^2 + \cdots + v_N^2}$$

MATLAB EXAMPLE

```
>> v = [1; 2; 0; 4];        >> norm(v, 2)
>> v_len = norm(v)          ans =
v_len =                        4.5826
      4.5826                >> sqrt(sum(v.^2))
                            ans =
                               4.5826
```

Infinity Norm

SYMBOLS $\|v\|_\infty$, $\|v\|_{-\infty}$, l_∞ $l_{-\infty}$

DESCRIPTION The $l_{\pm\infty}$ is the maximum or minimum absolute value of the vector elements.

$$\|v\|_\infty = \max\left(|v|\right)$$

$$\|v\|_{-\infty} = \min\left(|v|\right)$$

MATLAB EXAMPLE

```
>> v = [1; 2; 0; 4];          >> norm(v, -Inf)
>> v_infty = norm(v, Inf)     ans =
v_infty =                          0
     4                        >> min(abs(v))
>> max(abs(v))                ans =
ans =                              0
     4
```

A.1.2   vecnorm function

MATLAB has a handy function called vecnorm that will compute the p–norms of the columns of a matrix. It takes the matrix variable name as an argument. Optional arguments specify p (l_2-norm is the default), and the dimension for computing the norm (columns is the default). The command to compute the l_2-norm of the rows is: vecnorm(A, 2, 2).

Here is an example that computes the length of the columns of a matrix.

```
>> A = randi(20, 3) - randi(10, 3)
A =
      7      9      4
     17      8      6
     -7     -7     10
>> Alen = vecnorm(A)
Alen =
    19.6723   13.9284   12.3288

>> A1 = A./Alen       % unit length
A1 =
     0.3558    0.6462    0.3244
     0.8642    0.5744    0.4867
    -0.3558   -0.5026    0.8111
>> vecnorm(A1)
ans =
     1      1      1
```

A.1.3   Matrix Norms

The concept of the magnitude of a matrix from its norm seems ambiguous compared to the length of a vector from its norm. The $\|\cdot\|_2$ matrix norm is the most frequently used, but different applications have shown specific matrix norms to perform better than other matrix norm calculations.

Maximum Absolute Column Sum

SYMBOLS $\|\mathbf{A}\|_1$, $\|\cdot\|_1$

DESCRIPTION The $\|\cdot\|_1$-norm is the maximum l_1 norm of the column vectors of the matrix.

MATLAB EXAMPLE

```
>> A                          >> max(sum(abs(A)))
A =                           ans =
    -1     7    -5                17
     4    -3    -5            >> max(vecnorm(A, 1))
    -5     0     7            ans =
>> norm(A, 1)                     17
ans =
    17
```

2-Norm of a Matrix

SYMBOLS $\|\mathbf{A}\|_2$, $\|\cdot\|_2$

DESCRIPTION The focus of the $\|\cdot\|_2$ matrix norm is on the ability of a matrix to stretch a vector rather than on the values of the elements in the matrix. The calculation is the maximum ratio of l_2 vector norms.

$$\|\mathbf{A}\|_2 = \max_{v \neq 0} \frac{\|\mathbf{A}\,v\|}{\|v\|}$$

One might think of the eigenvector corresponding to the largest eigenvalue for the v vector that maximizes the ratio of vector lengths. That is correct if $\mathbf{A}$ is a symmetric matrix, but the general purpose best choice for x comes from the SVD. Recall that for each singular value, we have the relationship $\mathbf{A}\,v_i = \sigma_i\,u_i$. So we want to consider the largest singular value, which is the first value since the SVD sorts the singular values.

$$\|\mathbf{A}\|_2 = \frac{\|\mathbf{A}\,v_1\|}{\|v_1\|} = \frac{\sigma_1\,\|u_1\|}{\|v_1\|} = \sigma_1$$

The final simplification stems from the fact that both u and v are unit vectors. Thus, the $\|\cdot\|_2$ matrix norm is the largest (first) singular value from the SVD.

We learned in section 6.8 that the singular values are the square root of the eigenvalues of $\mathbf{A}^T\mathbf{A}$. If we use the notation

$\lambda_i(\mathbf{A})$ to mean the ith eigenvalue of $\mathbf{A}$, then the $\|\cdot\|_2$ matrix norm is also given by

$$\|\mathbf{A}\|_2 = \max_i \sqrt{|\lambda_i(\mathbf{A}^T\mathbf{A})|}.$$

For symmetric matrices, this simplifies to the maximum eigenvalue of $\mathbf{A}$.

$$\|\mathbf{A}\|_2 = \max_i |\lambda_i(\mathbf{A})|.$$

Other Properties of Matrix 2-Norms

- $\|\mathbf{A}\|_2 > 0$
- $\|\mathbf{I}\|_2 = 1$
- $\|c\mathbf{A}\|_2 = |c|\,\|\mathbf{A}\|_2$, from which we also have $\|-\mathbf{A}\|_2 = \|\mathbf{A}\|_2$.
- $\|\mathbf{A} + \mathbf{B}\|_2 \leq \|\mathbf{A}\|_2 + \|\mathbf{B}\|_2$.
- $\|\mathbf{A}\,\mathbf{B}\|_2 \leq \|\mathbf{A}\|_2\,\|\mathbf{B}\|_2$.

MATLAB EXAMPLE

```
>> A                          >> A_12 = norm(A)
A =                           A_12 =
    -1     7    -5                11.3568
     4    -3    -5            >> S = svd(A);
    -5     0     7            >> A_12 = S(1)
>> A_12 = norm(A, 2)          A_12 =
A_12 =                            11.3568
    11.3568
```

Frobenius Norm

SYMBOLS $\|\mathbf{A}\|_F$, $\|\cdot\|_F$

DESCRIPTION The Frobenius norm is similar to the l_2 vector norm in that it is the square root of the sum of the squared elements. It can also be found from the singular values.

$$\|\mathbf{A}\|_F = \sqrt{\sum_{i=1}^{m}\sum_{j=1}^{n} a_{ij}^2} = \sqrt{\operatorname{trace}(\mathbf{A}^T\mathbf{A})}$$

$$\|\mathbf{A}\|_F = \sqrt{\sum_{i=1}^{\min(m,n)} \sigma_i^2}$$

MATLAB EXAMPLE

```
>> A
A =
    -1    7   -5
     4   -3   -5
    -5    0    7
>> A_f = norm(A, 'fro')
A_f =
    14.1067
>> A_f = sqrt(sum(A(:).^2))
A_f =
    14.1067
>> A_f = sqrt(trace(A'*A))
A_f =
    14.1067
>> A_f = sqrt(sum(svd(A).^2))
A_f =
    14.1067
```

Maximum Absolute Row Sum

SYMBOLS $\|\mathbf{A}\|_\infty$, $\|\cdot\|_\infty$

DESCRIPTION The $\|\cdot\|_\infty$ is similar to the $\|\cdot\|_1$ except it uses the l_1 norm of the rows instead of the columns.

MATLAB EXAMPLE

```
>> A
A =
    -1    7   -5
     4   -3   -5
    -5    0    7
>> norm(A, Inf)
ans =
    13
>> max(sum(abs(A')))
ans =
    13
```

Nuclear Norm

SYMBOLS $\|\mathbf{A}\|_N$, $\|\cdot\|_N$

DESCRIPTION The nuclear norm is the sum of the singular values from the SVD, which is related to the rank because most higher order singular values of matrices holding data, such as an image, are close to zero. It has shown to give good results at improving the robustness of PCA (section 6.9) (RPCA) to outlier data points ([7], pages 107-108).

The nuclear norm gained recent recognition when Netflix held a challenge competition to find the best possible movie recommendation algorithm. It was no surprise that the PCA process was key to the algorithms that contestants developed. However, an interesting discovery was that the nuclear norm performed better than the 2-norm.

$$\|\mathbf{A}\|_N = \sum_{k=1}^{r} \sigma_k$$

MATLAB EXAMPLE

```
>> A
A =
    -1     7    -5
     4    -3    -5
    -5     0     7
>> A_nuc_norm = sum(svd(A))
A_nuc_norm =
   20.4799
```

A.2   Vector Spaces

Here we will define some linear algebra terms and explain some foundational concepts related to vector spaces, dimension, linearly independent vectors, and rank.

A.2.1   Vector Space Definitions

Vector Space

A *vector space* consists of a set of vectors and a set of scalars that are closed under vector addition and scalar multiplication. By saying that they are closed just means that we can add any vectors in the set together and multiply vectors by any scalars in the set and the resulting vectors are still in the vector space.

For example, vectors in our physical 3-dimensional world are said to be in a vector space called $\mathbb{R}^3$. If the basis vectors of the space are the standard orthogonal coordinate frame then the set of vectors in $\mathbb{R}^3$ consist of three real numbers defining their magnitude in the x, y, and z directions. The set of scalars in $\mathbb{R}^3$ is the set of real numbers.

Span

The set of all linear combinations of a collection of vectors is called the *span* of the vectors. If a vector can be expressed as a linear combination of a set of vectors, then the vector is in the span of the set.

Basis

The smallest set of vectors needed to span a vector space forms a *basis* for that vector space. The vectors in the vector space are linear combinations of the basis vectors. For example, the basis vectors used for a Cartesian coordinate frame in the vector space $\mathbb{R}^3$ are:

$$\left\{ \begin{bmatrix} 1 \\ 0 \\ 0 \end{bmatrix}, \begin{bmatrix} 0 \\ 1 \\ 0 \end{bmatrix}, \begin{bmatrix} 0 \\ 0 \\ 1 \end{bmatrix} \right\}$$

Many different basis vectors could be used as long as they are a linearly independent set of vectors, but we have a preference for unit length orthogonal vectors. In addition to the Cartesian coordinates, other sets of basis vectors are sometimes used. For example, some robotics applications may use basis vectors that span a plane corresponding to a work piece. Other applications, such as difference equations, use the set of eigenvectors of a matrix as the basis vectors.

Dimension

The number of vectors in the basis gives the *dimension* of the vector space. For the Cartesian basis vectors, $\mathbb{R}^3$ and $\mathbb{R}^2$, this is also the number of elements in the vectors. However, other vector spaces may have a smaller dimension. For example, the subspace where the z axis is the same as the x axis (plotted in figure A.2) forms a plane with dimension 2. The orthonormal basis vectors for this subspace might be

$$\{v_1, v_2\} = \left\{ \frac{1}{\sqrt{2}} \begin{bmatrix} 1 \\ 0 \\ 1 \end{bmatrix}, \begin{bmatrix} 0 \\ 1 \\ 0 \end{bmatrix} \right\}.$$

We may reference points or vectors in the subspace with 2-dimensional coordinates, (a, b). Linear combinations of the basis vectors are used to find the coordinates of the point in the $\mathbb{R}^3$ world.

$$\begin{bmatrix} a \\ b \end{bmatrix} \longmapsto a\,v_1 + b\,v_2$$

A.2.2 Linearly Independent Vectors

The official definition of linearly independent vectors may at first seem a bit obscure, but it is worth stating along with some explanation and examples.

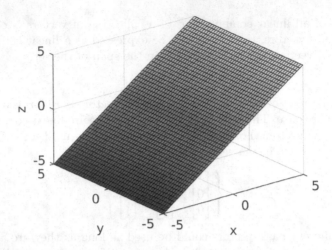

FIGURE A.2: Vector space where $Z = X$, dimension $= 2$.

The set of vectors, $u_1, u_2, \ldots u_n$, are linearly independent if for any scalars $c_1, c_2, \ldots c_n$, the equation $c_1 u_1 + c_2 u_2 + \cdots + c_n u_n = 0$ has only the solution $c_1 = c_2 = \cdots = c_n = 0$.

This means that no vector in the set is a linear combination of other vectors in the set. If one or more vectors in the set can be found by a linear combination of other vectors in the set, then the vectors are not linearly independent and there are nonzero coefficients, c_i, that will satisfy the equation $c_1 u_1 + c_2 u_2 + \ldots + c_n u_n = 0$.

For example, the basis vectors used for a Cartesian coordinate frame in the vector space $\mathbb{R}^3$ are linearly independent.

$$\left\{ \begin{bmatrix} 1 \\ 0 \\ 0 \end{bmatrix}, \begin{bmatrix} 0 \\ 1 \\ 0 \end{bmatrix}, \begin{bmatrix} 0 \\ 0 \\ 1 \end{bmatrix} \right\}$$

Whereas, the following set of vectors are not linearly independent because the last vector is the sum of the first two vectors.

$$\left\{ \begin{bmatrix} 1 \\ 0 \\ 1 \end{bmatrix}, \begin{bmatrix} 0 \\ 1 \\ 0 \end{bmatrix}, \begin{bmatrix} 1 \\ 1 \\ 1 \end{bmatrix} \right\}$$

Some phrases that are used to describe a vector that is linearly dependent on a set of vectors is: "*In the span of*" or "*In the space*". For example, when we evaluate the columns of a matrix, $\mathbf{A}$, with regard to another vector, b, we might say that vector b is in the column space of $\mathbf{A}$, which means that b is a linear combination of the vectors that define the columns of matrix $\mathbf{A}$.

A.2.3 Rank

The rank function can be used to test the linear independence of vectors that make up the columns of a matrix.

The rank of a matrix can be determined by two different methods. The rank of a matrix is the number of nonzero pivots in its row-echelon form, which is achieved by Gaussian elimination. Here is an example illustrating how elimination on a singular matrix results in a pivot value being equal to zero. Because of dependent relationships between the rows and columns, the row operations to change elements below the diagonal to zeros also move the later pivots to zero. In this example, the third column is the sum of the first column and two times the second column.

$$\begin{bmatrix} 3 & 2 & 7 \\ 0 & 3 & 6 \\ 1 & 0 & 1 \end{bmatrix}$$

Add $-1/3$ of *row 1* to *row 3*. The pivot then moves to *row 2*.

$$\begin{bmatrix} 3 & 2 & 7 \\ 0 & \underline{3} & 6 \\ 0 & -2/3 & -4/3 \end{bmatrix}$$

Add $2/9$ of *row 2* to *row 3*.

$$\begin{bmatrix} 3 & 2 & 7 \\ 0 & 3 & 6 \\ 0 & 0 & 0 \end{bmatrix}$$

Thus, with 2 nonzero pivots, the rank of the matrix is 2.

The second method for computing rank is to use the singular value decomposition (SVD) as demonstrated in section 6.8.5.3. The number of singular values above a small tolerance is the rank of the matrix. The MATLAB rank function uses the SVD method.

A.3 Fundamental Matrix Subspaces

There are four possible fundamental vector subspaces of an $m \times n$ matrix **A**. They are called the null space, the column space, the row space, and the left null space. It is shown in section 6.8.5.4 that the subspaces are found from rank determined regions of the SVD factors. The regions are identified in figure A.3. Here we discuss the meaning and application of each subspace in more detail.

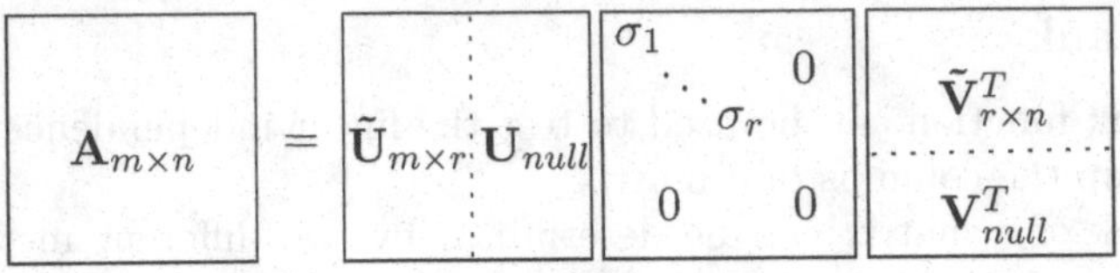

$$
\mathbf{A}_{m\times n} = \begin{array}{|c:c|} \hline & \\ \tilde{\mathbf{U}}_{m\times r} & \mathbf{U}_{null} \\ & \\ \hline \end{array}
\begin{array}{|cc|} \hline \sigma_1 & \\ \ddots & 0 \\ & \sigma_r \\ \hline 0 & 0 \\ \hline \end{array}
\begin{array}{|c|} \hline \tilde{\mathbf{V}}^T_{r\times n} \\ \hdashline \mathbf{V}^T_{null} \\ \hline \end{array}
$$

FIGURE A.3: The columns of $\mathbf{U}$ and rows of $\mathbf{V}^T$ that are multiplied by either nonzero or zero valued singular values define the column space ($\tilde{\mathbf{U}}$), null space ($\mathbf{V}^T_{null}$), row space ($\tilde{\mathbf{V}}^T$), and the left null space ($\mathbf{U}_{null}$) of the matrix.

A.3.1   Column Space

The *column space* of a matrix $\mathbf{A}$, which we denote as $\mathrm{Col}(\mathbf{A})$, is the vector space spanned by all independent columns of $\mathbf{A}$. If $\mathbf{A}$ is a square full rank matrix, then it is all of the columns of $\mathbf{A}$. For singular or under-determined matrices the column space is the set of columns with nonzero pivots during elimination. The vectors in the column space can also be referred to as basis vectors. Of course, an orthogonal set of basis vectors is preferred and can be found using the algorithms mentioned appendix A.4.

Rank and Column Space Dimension

The *rank* of the matrix is the same as the dimension of the column space, which is the number of columns in the column space.

When there is a solution to the equation $\mathbf{A}x = b$, we say that b is "*in the column space of* $\mathbf{A}$" because it is a linear combination of the independent columns of $\mathbf{A}$. The vector x tells us the coefficients for the linear combination of the columns that are needed to get to b.

Here is a singular matrix example, which shows a case where the column space is a subset of the columns of $\mathbf{A}$.

```
>> A1 = randi(10, 3, 2) - randi(5, 3, 2);
>> v = randi(5, 2, 1) - randi(3, 2, 1);
>> A = [A1 A1*v]
A =
      -1      4     -2
       3      0     -6
       6      2    -14
>> rank(A)
ans =
     2
>> rref(A)
ans =
```

```
     1      0     -2
     0      1     -1
     0      0      0

>> [U, S, ~] = svd(A)
U =
     0.1175   -0.9717   -0.2051
     0.3937    0.2352   -0.8887
     0.9117    0.0237    0.4102
S =
    16.8496        0         0
         0    4.2533        0
         0         0    0.0000
```

Notice from the output of rref that the first two columns are pivot columns. Thus we can see, as was designed from constructing the singular matrix, that the first two columns span the column space.

$$\mathrm{Col}(\mathbf{A}) = \mathrm{Span}\left\{ \begin{bmatrix} -1 \\ 3 \\ 6 \end{bmatrix}, \begin{bmatrix} 4 \\ 0 \\ 2 \end{bmatrix} \right\}$$

Then from the SVD, we get a column space from the first two columns of $\mathbf{U}$, which are orthogonal vectors.

A.3.2 Null Space

The *null space* (right null space) of a matrix $\mathbf{A}$, which we denote as Null($\mathbf{A}$), is the vector space spanned by all column vectors x that satisfy the matrix equation $\mathbf{A}x = \mathbf{0}$. Square, singular matrices and under-determined matrices have a null space. The number of vectors in the null space is the number of dependent columns (size(A, 2) - rank(A)). If $\mathbf{A}$ is a square full rank matrix, then the null space is an empty set.

The null function returns the normalized basis vectors of a matrix's null space. MATLAB uses the SVD to find the null space from the last $(n - r)$ columns of $\mathbf{V}$ corresponding to the singular values equal to zero.

The null space vectors can also be found by elimination from the reduced row echelon form of the matrix as was done in section 5.5.3.

Here is an example of a singular matrix.

```
% Find a random, but singular, matrix
>> A1 = randi(10, 3, 2) - randi(5, 3, 2);
>> v = randi(5, 2, 1);
>> A = [A1 A1*v]
A =
     1      7     18
     1      5     14
```

```
          6    -2    20

>> rank(A)
ans =
     2

>> rref(A)
ans =
     1    0    4
     0    1    2
     0    0    0
```

The third column from the RREF shows the dependent relationship of the columns of **A**. The third column is four of the first column plus two of the second column.

$$4\,a_1 + 2\,a_2 - a_3 = 0$$

$$\begin{bmatrix} | & | & | \\ a_1 & a_2 & a_3 \\ | & | & | \end{bmatrix} \begin{bmatrix} 4 \\ 2 \\ -1 \end{bmatrix} = \begin{bmatrix} 0 \\ 0 \\ 0 \end{bmatrix}$$

```
>> x = [4; 2; -1];

>> A*x
ans =
     0
     0
     0

>> x/norm(x)
ans =
     0.8729
     0.4364
    -0.2182

>> null(A)
ans =
     0.8729
     0.4364
    -0.2182

% The last singular value from the SVD is zero, so the
% last column of V is the null space vector.

>> [~,S,V] = svd(A)
S =
    31.1763         0         0
          0    8.0025         0
          0         0    0.0000
```

```
V =
    -0.1583    0.4616    0.8729
    -0.1697   -0.8836    0.4364
    -0.9727    0.0791   -0.2182
```

Here is an example of an under-determined matrix. For this example, the matrix is converted to a symbolic variable to get fractions rather than floating point numbers.

```
>> A = randi(10, 3, 5) - randi(5, 3, 5)
A =
    2   -1   -4    5   -1
    4    5    3   -1    6
    6    3   -3    5   -2

>> rref(sym(A))
ans =
[ 1, 0, 0,  10/3,  50/3]
[ 0, 1, 0, -11/3, -61/3]
[ 0, 0, 1,   4/3,  41/3]
```

Each row from the RREF can be used as an equation of the null vector. The equations can either leave the terms to the left of the equal sign with zeros on the right, or for under-determined equations they can be expressed in parametric form as follows. The difference in the null solution is a multiplication by -1, which is also a correct solution.

$$
\begin{aligned}
x_1 &= -10/3\,x_4 - 50/3\,x_5 \\
x_2 &= 11/3\,x_4 + 61/3\,x_5 \\
x_3 &= -4/3\,x_4 - 41/3\,x_5 \\
x_4 &= x_4 \\
x_5 &= x_5
\end{aligned}
$$

$$
\text{Null}(\mathbf{A}) = \text{span}\left\{ \begin{bmatrix} -10/3 \\ 11/3 \\ -4/3 \\ 1 \\ 0 \end{bmatrix}, \begin{bmatrix} -50/3 \\ 61/3 \\ -41/3 \\ 0 \\ 1 \end{bmatrix} \right\}
$$

Notice that both basis vectors of the null space satisfy $\mathbf{A}\boldsymbol{x} = \mathbf{0}$. They span the null vector space.

```
>> x1 = sym([-10/3; 11/3; -4/3; 1; 0]);
>> x2 = sym([-50/3; 61/3; -41/3; 0; 1]);
>> A*x1
ans =
    0
```

```
        0
        0
>> A*x2
ans =
        0
        0
        0
```

The normalized null space vectors found from the SVD (same result as the null command) are the last two columns of **V**. Remember from section 5.8 that under-determined systems have an infinite set of solutions. The RREF expresses the solution in terms of the general and particular solutions, while the null space from the SVD is the least squares solution. Although, the solutions are different, they are both correct.

```
>> [~, S, V] = svd(A)
S =
   10.7836        0        0        0        0
        0   9.6982        0        0        0
        0        0   2.5806        0        0
V =
    0.5505    0.4663    0.2136   -0.4283    0.5004
    0.1603    0.5742    0.3875    0.2622   -0.6524
   -0.4490    0.3278    0.2592    0.5599    0.5571
    0.6485   -0.1139   -0.3634    0.6508    0.1048
   -0.2215    0.5765   -0.7778   -0.1045   -0.0510

>> nullA = V(:, end-1:end)   % last 2 columns of V
nullA =
   -0.4283    0.5004
    0.2622   -0.6524
    0.5599    0.5571
    0.6508    0.1048
   -0.1045   -0.0510

>> A*nullA    % Not exactly zero due to round-off error
ans =
   1.0e-14 *
   -0.0583    0.0992
    0.0444   -0.0666
   -0.0722   -0.1568
```

A.3.3   Row Space

The *row space* of a matrix **A**, which we denote as Row(**A**), is the column space of A^T. It is the first r columns of **V** from the SVD.

A.3.4   Left Null Space

The *left null space* of a matrix **A** are the set of row vectors, **Y**, that satisfy the relationship **YA** = **0**. It is also the right null space of $\mathbf{A}^T$. The left null space of **A** may be found from the SVD as the transpose of the last $(m-r)$ columns of **U**. It is more common in MATLAB to compute it as the transpose of the right null space of $\mathbf{A}^T$, Y = null(A')'. The left null space exists for singular and over-determined matrices.

A.3.5   Orthogonality of Spaces

The four fundamental vector subspaces form interesting orthogonality relationships. Vectors in the null space are orthogonal to vectors in the row space. Similarly, vectors in the column space are orthogonal to vectors in the left null space.

```
>> A
A =
      -4      2     -3      4      0
       1      8      5      3      2
       1      6      4      5     -4

>> nullSpace = null(A)
nullSpace =
    0.6618   -0.3729
   -0.2290   -0.5357
   -0.2466    0.6319
    0.5914    0.3689
    0.3146    0.1962

>> rank(A)
ans =
     3

% The row Space is all columns of A^T.
>> rowSpace = A';

% Check that the null space vectors are orthogonal to the row space
% vectors. Zero dot product shows orthogonal vectors, which we can
% see from matrix multiplication inner products.

>> nullSpace'*rowSpace
ans =
   1.0e-14 *
         0         0         0
         0   -0.0444    0.1110

% The left null space is empty because the column space vectors
```

```
% are full rank.
>> left_null = null(A')
left_null =
    3x0 empty double matrix
```

Here is a 3×3 matrix of rank 1 to demonstrate orthogonality of the column space and the left null space.

```
>> a = randi(5, 3, 1);        >> colSpace = A(:,1);
>> a2 = a.*2;                 >> left_null = null(A')
>> a3 = a.*3;                 left_null =
>> A = [a a2 a3]                 -0.0276    0.8732
A =                               0.3855   -0.4399
     3     6     9               -0.9223   -0.2100
     5    10    15
     2     4     6             >> colSpace'*left_null
>> rank(A)                    ans =
ans =                            1.0e-15 *
     1                           -0.1943    0.4441
```

A.4   Finding Orthogonal Basis Vectors

We sometimes need to find a set of orthogonal basis vectors for the columns of a matrix. Three such needs are for vector projections (section 5.9.3.2), finding least squares solutions to rectangular systems of equations (section 5.11), and finding the eigenvalues of a matrix (section 6.3 and appendix A.7).

One of three algorithms are typically used to find orthogonal basis vectors. A well known algorithm is the *classic Gram–Schmidt process* (CGS). It was proposed by Laplace (1749 - 1827) and later refined by Gram (1850 - 1916) and Schmidt (1876 - 1959). Gram–Schmidt is often taught in linear algebra courses, usually along with or shortly after the study of vector projections. However, the classic Gram–Schmidt process is known to be numerically unstable when the matrix is poorly conditioned with nearly dependent columns resulting in a loss of orthogonality [18]. Interestingly, the algorithm can be made more reliable by rearranged it into what is know as the *modified Gram–Schmidt process* (MGS) [28]. Although the MGS process is still less accurate for poorly conditioned matrices than the QR algorithm [18]. MATLAB does not include functions for either the CGS or MGS algorithms. Of course CGS is not included because of its lack of numerical stability and the MGS is not essential because MATLAB's qr function is faster, more accurate and returns functionally the same results.

Some implementations of CGS and MGS return both an orthogonal matrix, **Q**, whose columns form basis vectors for the columns of **A**, and an upper triangular matrix, $\mathbf{R}$, such that $\mathbf{A} = \mathbf{Q}\mathbf{R}$. The $\mathbf{Q}$ and $\mathbf{R}$ factors of $\mathbf{A}$ are the same as, or sometimes the negative of, the matrices returned by MATLAB's qr function. The *QR factorization* described in appendix A.5 is the second algorithm for finding orthogonal basis vectors. Although QR returns the same functional results as MGS, it is implemented using a different algorithm and is faster.

The third algorithm for finding orthogonal basis vectors is from the SVD. MATLAB provides a function called orth, which returns the $\mathbf{U}$ matrix from the economy singular value decomposition, [U, S] = svd(A,'econ') (section 6.8.5.2). The SVD based algorithm yields a different set of vectors than either Gram–Schmidt or QR.

A.4.1 The Gram–Schmidt Algorithm

Although MATLAB does not provide a function that implements the Gram–Schmidt process, we will briefly describe it here as a reference for those that encounter it in other linear algebra studies.

The input to the Gram–Schmidt process is a matrix, $\mathbf{A}$, and the output is a matrix, $\mathbf{Q}$, whose columns are an orthogonal basis of $\mathbf{A}$.

The columns of $\mathbf{Q}$ are first formed from vector projections (see section 5.9.1), and then made unit length. Recall that when a vector is projected onto another vector, the vector representing the error between the projection and the original vector are orthogonal to each other. Here we want to find the vector representing the error from projection.

Let matrix $\mathbf{A}$ be formed from column vectors.

$$\mathbf{A} = \begin{bmatrix} | & | & & | \\ a_1 & a_2 & \cdots & a_n \\ | & | & & | \end{bmatrix}$$

We will first find a matrix, $\mathbf{B}$, who's columns are orthogonal and are formed from projections of the columns of $\mathbf{A}$. For each column after the first, we subtract the projection of the column onto the previous columns. Thus we are subtracting away the projection leaving a vector that is orthogonal to all of the previous column vectors.

$$b_1 = a_1$$

$$b_2 = a_2 - \frac{b_1^T a_2}{b_1^T b_1} b_1$$

$$b_3 = a_3 - \frac{b_1^T a_3}{b_1^T b_1} b_1 - \frac{b_2^T a_3}{b_2^T b_2} b_2$$

$$\vdots$$

$$b_n = a_n - \frac{b_1^T a_n}{b_1^T b_1} b_1 - \frac{b_2^T a_n}{b_2^T b_2} b_2 - \cdots - \frac{b_{n-1}^T a_n}{b_{n-1}^T b_{n-1}} b_{n-1}$$

The columns of $\mathbf{Q}$ are then the columns of $\mathbf{B}$ scaled to be unit vectors.

A.4.2   Implementation of Classic Gram–Schmidt

The `gram_schmidt` function listed in code A.1 is an implementation of the classic Gram–Schmidt (CGS) algorithm. Note that CGS is considered to be numerically unstable for some matrices, so it is for educational purposed only. Although, the CGS, MGS, and QR algorithms will return the same $\mathbf{Q}$ matrix for well conditioned input matrices.

A.4.3   Implementation of Modified Gram–Schmidt

The modified Gram–Schmidt (MGS) algorithm improves, but does not completely correct, the stability problems of the CGS process. The algorithm in the `mod_gram_schmidt` function listed in code A.2 is adapted from a code segment in Golub and Van Loan's *MATRIX Computations* text [28]. In this implementation, an upper triangular $\mathbf{R}$ matrix is also returned. As with the $\mathbf{Q}$ and $\mathbf{R}$ matrices returned from MATLAB's `qr` function, $\mathbf{Q}$ and $\mathbf{R}$ are factors of $\mathbf{A}$, $\mathbf{A} = \mathbf{Q}\mathbf{R}$.

A.5   QR Factorization

QR factorization finds sub-matrices $\mathbf{Q}$ and $\mathbf{R}$, where $\mathbf{Q}$ contains orthogonal column vectors, and $\mathbf{R}$ is an upper triangular matrix, such that

$$\mathbf{A} = \mathbf{Q}\mathbf{R}.$$

```
function Q = gram_schmidt(A)
% GRAM_SCHMIDT - Classic Gram-Schmidt Process (CGS)
%    Input - A matrix. The algorithm operates on the columns.
%    Output - unitary matrix - columns are basis for A.

[m, n] = size(A);
Q = zeros(m, n);
Q(:,1) = A(:,1)/norm(A(:,1));       % the first column

% find next orthogonal column and normalize the column
for ii = 2:n
    b = A(:,ii);
    c = b;
    for k = 1:ii-1
        a = Q(:,k);
        c = c - (a*a'*b);
    end
    % now normalize
    Q(:,ii) = c/norm(c);
end
```

CODE A.1: Classic Gram–Schmidt function to find orthogonal column vectors.

For all but poorly conditioned matrices, the **Q** and **R** matrices from QR are nearly the same as found from the modified Gram–Schmidt process of appendix A.4.3. However, the QR algorithm implemented in MATLAB's `qr` function and the `qrFactor` function that follows are faster, especially for larger matrices, and give more numerically accurate results [79, 71].

The QR factorization uses an algorithm based on Householder transformation matrices.[1] The algorithm feels similar to LU decomposition, where products of elementary matrices are used to change matrix elements to zero resulting in an upper triangular matrix. Except the Householder matrices are orthogonal operations on the matrix columns rather than row operations. Each Householder matrix multiplication sets all matrix elements below the main diagonal to zero. The algorithm finds **R** first and then **Q** is the product of the Householder matrices [18, 28].

In the following example, the × symbols represent potentially nonzero matrix elements.

$$\mathbf{H_1 A} = \begin{bmatrix} \times & \times & \times & \times \\ 0 & \times & \times & \times \\ 0 & \times & \times & \times \\ 0 & \times & \times & \times \end{bmatrix}, \qquad \mathbf{H_2 H_1 A} = \begin{bmatrix} \times & \times & \times & \times \\ 0 & \times & \times & \times \\ 0 & 0 & \times & \times \\ 0 & 0 & \times & \times \end{bmatrix},$$

---

[1] The QR algorithm may be accomplished with Givens rotation matrices instead of Householder matrices.

```
function [Q, R] = mod_gram_schmidt(A)
% MOD_GRAM_SCMIDT - Modified Gram-Schmidt Process
%   Variation of the Gram-Schmidt process with
%   improved numerical stability.
%  This function is slower, but functionally
%  equivalent to MATLAB's qr function.
%
%  A - Input matrix
%  Q - Matrix - Unitary basis vectors of columns of A
%  R - Upper triangular matrix, A = Q*R

    [m, n] = size(A);
    Q = zeros(m, n);
    R = zeros(m, n);
    for k = 1:n
        R(k,k) = norm(A(1:m,k));    % start with normalized
        Q(1:m,k) = A(1:m,k)/R(k,k); % column vector
        for j = k + 1:n
            R(k,j) = Q(1:m,k)'*A(1:m, j);
            A(1:m,j) = A(1:m,j) - Q(1:m,k)*R(k,j);
        end
    end
end
```

CODE A.2: Modified Gram–Schmidt function to find QR factorization and orthogonal column vectors.

$$\mathbf{H}_3\mathbf{H}_2\mathbf{H}_1\mathbf{A} = \begin{bmatrix} \times & \times & \times & \times \\ 0 & \times & \times & \times \\ 0 & 0 & \times & \times \\ 0 & 0 & 0 & \times \end{bmatrix}$$

The $\mathbf{R}$ and $\mathbf{Q}$ matrices are found after $n - 1$ Householder matrix multiplications.

$$\mathbf{R} = (\mathbf{H}_{n-1} \dots \mathbf{H}_2\,\mathbf{H}_1)\,\mathbf{A}$$
$$\mathbf{Q} = (\mathbf{H}_{n-1} \dots \mathbf{H}_2\,\mathbf{H}_1)^{-1}$$

Since the Householder matrices are orthogonal, $\mathbf{Q}$ is a product of matrix transposes rather than inverse matrices. Moreover, the Householder matrices are symmetric, so we only reverse the multiplication order to find $\mathbf{Q}$ from a product of Householder matrices.

$$\mathbf{Q} = \mathbf{H}_1\,\mathbf{H}_2 \dots \mathbf{H}_{n-1}$$

Each Householder transformation matrix is a combination of an identity matrix, zeros, and a Householder reflection matrix.

$$\mathbf{H}_1 = \mathbf{H}'_1, \qquad \mathbf{H}_2 = \left[\begin{array}{c|c} 1 & 0 \\ \hline 0 & \mathbf{H}'_2 \end{array}\right], \qquad \mathbf{H}_3 = \left[\begin{array}{cc|c} 1 & 0 & 0 \\ 0 & 1 & \\ \hline & 0 & \mathbf{H}'_3 \end{array}\right]$$

The Householder reflection matrices are designed such that when multiplied by a column vector of elements from the diagonal and below, all vector elements below the first (the one on the diagonal) are set to zero.

$$\mathbf{H}'_2\,\boldsymbol{x} = \mathbf{H}'_2 \begin{bmatrix} \times \\ \times \\ \times \end{bmatrix} = \begin{bmatrix} \times \\ 0 \\ 0 \end{bmatrix}$$

A.5.1   Householder Reflection Matrices

A Householder reflector matrix derives from vector projections as described in section 5.9.1. We find a reflection of vector $\boldsymbol{x}$ about the hyperplane $\mathrm{span}(\boldsymbol{u})^{\perp}$ by multiplying $\boldsymbol{x}$ by a reflector matrix $\mathbf{H}$ as illustrated in figure A.4.

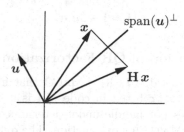

FIGURE A.4: Householder reflector matrix $\mathbf{H}$ times $\boldsymbol{x}$ is a reflection of $\boldsymbol{x}$ about the hyperplane $\mathrm{span}(u)^{\perp}$.

The Householder reflector matrix is found from the following equation [35].

$$\mathbf{H} = \mathbf{I} - \frac{2}{\boldsymbol{u}^T\boldsymbol{u}}\boldsymbol{u}\boldsymbol{u}^T, \qquad \boldsymbol{u} \neq 0, \boldsymbol{u} \in \mathbb{R}^n \qquad (A.1)$$

To use reflector matrices in the QR algorithm, we first find the $\boldsymbol{u}$ vector such that all elements of $\mathbf{H}\boldsymbol{x}$, except for the first, are equal to zero.

$$\mathbf{H}\boldsymbol{x} = \begin{bmatrix} c & 0 & \cdots & 0 \end{bmatrix}^T = c\,\boldsymbol{e}_1, \text{ where } \boldsymbol{e}_1 = \begin{bmatrix} 1 & 0 & \cdots & 0 \end{bmatrix}^T$$

From the geometry and since $\|\boldsymbol{x}\|_2 = \|\mathbf{H}\boldsymbol{x}\|_2 = |c|$, $\boldsymbol{u}$ must be parallel to $\tilde{\boldsymbol{u}} = \boldsymbol{x} \pm \|\boldsymbol{x}\|_2\,\boldsymbol{e}_1$, and $\boldsymbol{u} = \frac{\tilde{\boldsymbol{u}}}{\|\tilde{\boldsymbol{u}}\|_2}$. So we can find $\boldsymbol{u}$ simply from $\boldsymbol{x}$ with $\|\boldsymbol{x}\|_2$ added to x_1 with the same sign as x_1 and then making it a unit vector.

$$\tilde{\boldsymbol{u}} = \begin{bmatrix} x_1 + \mathrm{sign}(x_1)\,\|x\|_2 \\ x_2 \\ \vdots \\ x_n \end{bmatrix}, \qquad \boldsymbol{u} = \frac{\tilde{\boldsymbol{u}}}{\|\tilde{\boldsymbol{u}}\|_2}$$

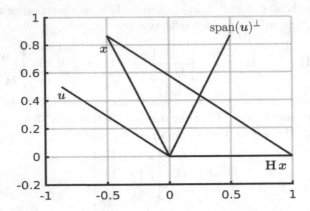

FIGURE A.5: The u vector is set so that only the first element of the Householder reflection $\mathbf{H}\,x$ has a nonzero value.

Since $\|u\|_2 = 1$, the calculation of $\mathbf{H}$ from equation (A.1) is simplified.

$$\mathbf{H} = \mathbf{I} - 2u\,u^T$$

A.5.2   Implementation of QR Factorization

MATLAB has a command, [Q, R] = qr(A) that finds the factors $\mathbf{Q}$ and $\mathbf{R}$ of $\mathbf{A}$. Function qrFactor listed in code A.4 is a simple implementation of the algorithm. It does not handle under-determined matrices, nor is it as flexible or as robust as MATLAB's qr function. The qrFactor function makes use of the House function in code A.3 to find the u vector that is needed to compute the Householder reflector matrix. The code is based on pseudo-code in Demmel's *Applied Numerical Linear Algebra* text [18] and Golub and Van Loan's *Matrix Computations* text [28].

```
function u = House(x)
% HOUSE - Find u for Householder reflector matrix
%   x - Real vector
%   u - Unit length real vector for finding the Householder
%       reflector matrix for QR factorization.
    x(1) = x(1) + sign(x(1))*norm(x);
    u = x/norm(x);
end
```

CODE A.3: Function to find u for Householder reflection.

```
function [Q, R] = qrFactor(A)
% QRFACTOR - Simple QR factorization using Householder reflections.
%
% A - input real matrix, square or over-determined (m-by-n)
% Q - Unitary basis vectors of A (m-by-m)
% R - Upper triangular matrix (m-by-n), A = Q*R
    [m, n] = size(A);
    if m < n
        error('Matrix must be square or over-determined')
    end
    last = min((m-1), n);
    H = zeros(m, m, last);
    for k = 1:last
        u = House(A(k:m, k));
        Hi = eye(m-k+1) - 2*(u*u');
        H(1:k-1,1:k-1,k) = eye(k-1);
        H(k:m,k:m,k) = Hi;
        A(k:m,k:n) = Hi*A(k:m,k:n);
    end
    R = A;
    Q = eye(m);
    for k = 1:last
        Q = Q*H(:,:,k);
    end
end
```

CODE A.4: Function implementing QR factorization using Householder reflections.

A.6 Similar Matrices

Two matrices are said to be *similar* if they have the same eigenvalues.

Aside from comparing the eigenvalues, there is a simple test to verify if two matrices are similar. Matrices **A** and **B** are similar if there exists a matrix **M** for which the following relationship holds.

$$B = M^{-1} A M$$

To show this, we will start with this relationship and the eigenvalue equation for matrix **A** and work toward the eigenvalue equation for the similar matrix **B**.

$$A x = \lambda x$$

We can insert an identity matrix ($M M^{-1} = I$) into the eigenvalue equation.

$$A (M M^{-1}) x = \lambda x$$

Now multiply both sides by $\mathbf{M}^{-1}$.

$$(\mathbf{M}^{-1}\mathbf{A}\,\mathbf{M})\,\mathbf{M}^{-1}\,\boldsymbol{x} = \lambda\,\mathbf{M}^{-1}\,\boldsymbol{x}$$

Substitute for $\mathbf{B}$.

$$\mathbf{B}\,(\mathbf{M}^{-1}\,\boldsymbol{x}) = \lambda\,(\mathbf{M}^{-1}\,\boldsymbol{x})$$

Thus, an eigenvalue of $\mathbf{B}$ is λ, the same as $\mathbf{A}$, and the corresponding eigenvector of $\mathbf{B}$ is $\mathbf{M}^{-1}\,\boldsymbol{x}$.

Some examples of similar matrices are:

- From diagonalization (section 6.5.1), $\mathbf{A}$ and the diagonal matrix $\mathbf{\Lambda}$ are similar: $\mathbf{A} = \mathbf{X}\,\mathbf{\Lambda}\,\mathbf{X}^{-1}$. The eigenvalues of an upper triangular matrix are on the diagonal, and a diagonal matrix is certainly upper triangular.

- From QR, $\mathbf{A}$ and $\mathbf{R}\,\mathbf{Q}$ are similar (appendix A.7).

A.7   QR Eigenvalue Computation Algorithm

The `eig` function in MATLAB is quite complex. It looks at the data and performs pre-conditioning transformations to make the computation faster. It uses different algorithms to compute the eigenvalues depending on the size and relationships of the values in the matrix.

It does not find the eigenvalues by finding the roots of the characteristic polynomial, i.e., find the λs that satisfy the equation $det(\mathbf{A} - \lambda\mathbf{I}) = \mathbf{0}$. We will focus our discussion on the iterative QR algorithm, which is a well known algorithm for finding the real eigenvalues without computing determinants. It is part of a primary algorithm used in MATLAB to compute eigenvalues. What is described here is the basics of the algorithm for purposes of learning.

Prerequisite to the iterative algorithm is the QR matrix factorization, which is described in appendix A.5.

A.7.1   Iterative QR Algorithm

First, use QR factorization, $\mathbf{A} = \mathbf{Q}\,\mathbf{R}$, to find a new matrix $\mathbf{A}_1 = \mathbf{R}\,\mathbf{Q}$. The original $\mathbf{A}$ matrix and the modified $\mathbf{A}_1$ matrix are similar matrices (appendices A.6 and A.7.2). Then QR factorization is repeated on $\mathbf{A}_1$ to find $\mathbf{A}_2 = \mathbf{R}_1\,\mathbf{Q}_1$ continuing until $\mathbf{A}_n$ is upper triangular and additional iterations will not change the result.

As the $\mathbf{A}_k$ matrices converge toward upper triangular at each iteration, the $\mathbf{Q}$ matrices converge toward the identity matrix. When $\mathbf{A}_k$ is upper triangular, the eigenvalues will be on its diagonal.

A more complex question about this algorithm is why does it converge to a stable upper triangular matrix. The complete explanation of convergence is quite advanced and beyond the scope of this text. Texts that are specifically about numerical linear algebra such as that by Demmel [18], and Golub and Van Loan [28] have extensive coverage on the iterative QR algorithm and its convergence. A concise, but interesting proof of convergence is given by Townsend on a web-page posting [74].

One can see from the following example that with repetition, the A1 matrix is converging to an upper triangular matrix with the eigenvalues on the diagonal. More repetitions are needed to complete the algorithm.

```
>> A = [4 5; 3 6]              >> [Q,R] = qr(A1);
A =                            >> A2 = R*Q
    4    5                     A2 =
    3    6                         8.9651    2.1306
>> eig(A)                          0.1306    1.0349
ans =
    1                          >> [Q,R] = qr(A2);
    9                          >> A3 = R*Q
                               A3 =
>> [Q,R] = qr(A);                  8.9963   -2.0146
>> A1 = R*Q                       -0.0146    1.0037
A1 =
    8.5600   -3.0800
   -1.0800    1.4400
```

Three other algorithm refinements have shown to lend to faster convergence. First, the matrix is converted to Hessenberg form, which is a similar matrix with all zeros below the first sub-diagonal (the diagonal below the main diagonal). For a larger matrix, this changes a lot of values in the lower triangular to zero, but the Hessenberg algorithm can not yield a full upper triangular matrix. Secondly, values on the diagonal are shifted so that the last diagonal value is zero before each QR factoring. Later the diagonal values are shifted back. Thirdly, when the last diagonal value has converged to the desired eigenvalue, then for future iterations the size of the matrix is reduced to exclude the last row and the last column. Convergence to the eigenvalue by the last diagonal value is detected when, after QR factoring, the value is less than a tolerance value, which indicates that the shift of the diagonal values holds the eigenvalue.

The function QReig listed in code A.5 is a simple implementation of the iterative QR algorithm that finds real eigenvalues. It works best with symmetric matrices. Note that the code places a limit on the number of iterations used to find each eigenvalue, which is a guard against the algorithm not converging.

```
function eigs = QReig(A)
% QREIG - iterative QR algorithm to find real eigenvalues.
%
% NOTE: The purpose of this code is to illustrate the algorithm.
%   This function does not find complex eigenvalues.
%   It converges reasonably quick for many matrices, but not
%   all matrices. So use MATLAB's eig function when you want
%   to find the eigenvalues of a matrix.
%
% Algorithm that uses the Hessenberg form of matrix A and uses
% a shift of the diagonal for faster convergence.
%
    [m, n] = size(A);
    if m ~= n
        error("Matrix must be square to find eigenvalues")
    end             % Start with Hessenberg - zeros below first
    B = hess(A);    % subdiagonal, but same eigenvalues
    tol = 1e-5;
    count = 0;      % Track the performance
    for k = n:-1:2
        loopcount = 0;
        I = eye(k);
        C = B(1:k,1:k); % sub-matrix
        converg = C(k,k);
        while abs(converg) > tol && loopcount < 1000
            loopcount = loopcount + 1;
            count = count+1;
            shift = C(k,k)*I;
            [Q, R] = qr(C - shift);
            C = R*Q;
            converg = C(k,k);
            C = C + shift;
        end
        B(1:k,1:k) = C; % kth eigenvalue at B(k,k)
    end
    eigs = diag(B);
    fprintf('iterations = %g\n', count);
end
```

CODE A.5: Function to find eigenvalues using the iterative QR algorithm.

A.7.2 Similar Matrices in QR

A key component of the QR algorithm is that after QR factoring ($[Q, R] = qr(A)$), the matrices $\mathbf{A}$ and $\mathbf{R}\,\mathbf{Q}$ are similar (appendix A.6). Since $\mathbf{A} = \mathbf{Q}\,\mathbf{R}$, then $\mathbf{R} = \mathbf{Q}^T\mathbf{A}$ and $\mathbf{R}\,\mathbf{Q} = \mathbf{Q}^T\mathbf{A}\,\mathbf{Q}$. Let $\mathbf{M} = \mathbf{Q}^T$, then $\mathbf{R}\,\mathbf{Q} = \mathbf{M}\,\mathbf{A}\,\mathbf{M}^{-1}$, which establishes similarity with the same eigenvalues.

A.8   Linearly Independent Eigenvectors

When the eigenvalues of a matrix are distinct (unique), the eigenvectors form a set of linearly independent vectors (appendix A.2.2). Here we show a proof that when a matrix has distinct eigenvalues then the eigenvectors are linearly independent. We will first show that pairs of eigenvectors are linearly independent, and then the proof is extended to show general linear independence of all of the eigenvectors.

A.8.1   Pairwise Independence

Let λ_1 and λ_2 be distinct eigenvalues of $\mathbf{A}$, with corresponding eigenvectors $\boldsymbol{x}_1$ and $\boldsymbol{x}_2$. To prove that $\boldsymbol{x}_1$ and $\boldsymbol{x}_2$ are linearly independent, we need to show that if

$$c_1\,\boldsymbol{x}_1 + c_2\,\boldsymbol{x}_2 = \mathbf{0},$$

then it must be that $c_1 = c_2 = 0$. Make a copy of the above equation and multiply one equation on the left by $\mathbf{A}$ and multiply the other equation on the left by λ_2.

$$\mathbf{A}\,(c_1\,\boldsymbol{x}_1 + c_2\,\boldsymbol{x}_2) = c_1\,\lambda_1\,\boldsymbol{x}_1 + c_2\,\lambda_2\,\boldsymbol{x}_2 = \mathbf{0}$$

$$\lambda_2\,(c_1\,\boldsymbol{x}_1 + c_2\,\boldsymbol{x}_2) = c_1\,\lambda_2\,\boldsymbol{x}_1 + c_2\,\lambda_2\,\boldsymbol{x}_2 = \mathbf{0}$$

Now subtract.

$$c_1\,\lambda_1\,\boldsymbol{x}_1 + c_2\,\lambda_2\,\boldsymbol{x}_2 = \mathbf{0}$$
$$-\quad c_1\,\lambda_2\,\boldsymbol{x}_1 + c_2\,\lambda_2\,\boldsymbol{x}_2 = \mathbf{0}$$
$$\overline{\qquad\qquad c_1\,(\lambda_1 - \lambda_2)\,\boldsymbol{x}_1 = \mathbf{0}\qquad\qquad}$$

Since the λ's are different and $\boldsymbol{x}_1 \neq \mathbf{0}$, we conclude that $c_1 = 0$, and similarly $c_2 = 0$. Thus $c_1\,\boldsymbol{x}_1 + c_2\,\boldsymbol{x}_2 = \mathbf{0}$ only when $c_1 = c_2 = 0$. So the eigenvectors $\boldsymbol{x}_1$ and $\boldsymbol{x}_2$ are linearly independent of each other.

A.8.2   General Independence

Given that the eigenpairs $(\lambda_1, \boldsymbol{x}_1)$ and $(\lambda_2, \boldsymbol{x}_2)$ are independent, there can not exist a third eigenpair $(\lambda_3, \boldsymbol{x}_3)$ such that $\boldsymbol{x}_3 = k\,\boldsymbol{x}_1$ or $\boldsymbol{x}_3 = k\,\boldsymbol{x}_2$, for any scalar k. To prove general independence, we have only to show that any third eigenvector can not be a linear combination of other eigenvectors.

This is a proof by contradiction, so we begin by considering an assertion that we will prove to be false.

If eigenvector $\boldsymbol{x}_3$ is a linear combination of $\boldsymbol{x}_1$ and $\boldsymbol{x}_2$, then there must exist constants k_1 and k_2 such that

$$\boldsymbol{x}_3 = k_1\,\boldsymbol{x}_1 + k_2\,\boldsymbol{x}_2. \tag{A.2}$$

Since $\mathbf{A}\,\boldsymbol{x}_3 = \lambda_3\,\boldsymbol{x}_3$,

$$\mathbf{A}\,\boldsymbol{x}_3 = \lambda_3(k_1\,\boldsymbol{x}_1 + k_2\,\boldsymbol{x}_2) = \lambda_3\,k_1\,\boldsymbol{x}_1 + \lambda_3\,k_2\,\boldsymbol{x}_2.$$

Vectors $\boldsymbol{x}_1$ and $\boldsymbol{x}_2$ can be substituted for $\boldsymbol{x}_1 = \frac{1}{\lambda_1}\mathbf{A}\,\boldsymbol{x}_1$ and $\boldsymbol{x}_2 = \frac{1}{\lambda_2}\mathbf{A}\,\boldsymbol{x}_2$.

$$\mathbf{A}\,\boldsymbol{x}_3 = \frac{\lambda_3\,k_1}{\lambda_1}\mathbf{A}\,\boldsymbol{x}_1 + \frac{\lambda_3\,k_2}{\lambda_2}\mathbf{A}\,\boldsymbol{x}_2 \qquad\qquad (A.3)$$

$$\boldsymbol{x}_3 = \frac{\lambda_3\,k_1}{\lambda_1}\boldsymbol{x}_1 + \frac{\lambda_3\,k_2}{\lambda_2}\boldsymbol{x}_2 \qquad\qquad (A.4)$$

Pre-multiplying both sides of equation (A.3) by $\mathbf{A}^{-1}$ removes the $\mathbf{A}$ matrices leaving equations (A.2) and (A.4) as equivalent equations. If equations (A.2) and (A.4) are true, then it must be that $\lambda_3 = \lambda_1$ and $\lambda_3 = \lambda_2$, so $\lambda_1 = \lambda_2 = \lambda_3$, which is a contradiction of the initial statement that each eigenvalue is distinct. Therefore, equation (A.2) is false. If each eigenvalue is distinct, then all eigenvectors are linearly independent.

A.9   Symmetric Matrix Eigenvalues and Eigenvectors

Symmetric matrices ($\mathbf{S}^T = \mathbf{S}$) have nice proprieties. All square, symmetric matrices have real eigenvalues and eigenvectors. The eigenvector matrix is also orthogonal—a square matrix whose columns and rows are orthogonal unit vectors (section 5.2.3.2).

It follows that since symmetric matrices have such nice properties, $\mathbf{A}^T\mathbf{A}$ is often used in eigenvalue problems.

A.9.1   Proof of Real Eigenvalues and Eigenvectors

We will show that the eigenvalues of symmetric matrices are real. The eigenvectors are real when the eigenvalues are real. Our proof allows that an eigenvalue λ and its eigenvector $\boldsymbol{x}$ of a symmetric matrix $\mathbf{S}$ might be complex with complex conjugates $\bar{\lambda}$ and $\bar{\boldsymbol{x}}$ and then shows that the eigenvector equation is only satisfied with real eigenvalues.

We start with the eigenvalue equation and its complex conjugate. We then pre-multiply by the transpose of the same eigenvector and the transpose conjugate. Finally we subtract to see that the eigenvalues must be real.

$$
\begin{array}{rcl}
\bar{\boldsymbol{x}}^T\,\mathbf{S}\,\boldsymbol{x} & = & \lambda\,\bar{\boldsymbol{x}}^T\,\boldsymbol{x} \\
- \quad \boldsymbol{x}^T\,\mathbf{S}\,\bar{\boldsymbol{x}} & = & \bar{\lambda}\,\boldsymbol{x}^T\,\bar{\boldsymbol{x}} \\
\hline
0 & = & (\lambda - \bar{\lambda})\,\boldsymbol{x}^T\,\bar{\boldsymbol{x}}
\end{array}
$$

Because of the symmetry of $\mathbf{S}$, the scalar values on the left hand sides are the same (subtracting to zero). On the right hand side, the dot product is the sum of the squares of the eigenvector $\|\boldsymbol{x}\|^2$ and can not be zero for an nonzero vector. Thus, it must be that $\lambda - \bar{\lambda} = 0$, which is true only when λ is real.

A.9.2 Proof of Orthogonal Eigenvectors

Recall that the vectors of a dot product may be reversed because of the commutative property of the dot product (section 5.1.4). Then because of the symmetry of matrix $\mathbf{S}$, we have the following equality relationship between two eigenvectors and the symmetric matrix.

$$\boldsymbol{x}_i^T \mathbf{S} \boldsymbol{x}_j = \boldsymbol{x}_j^T \mathbf{S} \boldsymbol{x}_i$$

Starting with two eigenvector equations, we can pre-multiply the first equation by the transpose of the eigenvector from the second equation, then pre-multiply the second equation by the transpose of the eigenvector from the first equation and subtract the two equations.

$$
\begin{aligned}
\boldsymbol{x}_j^T \mathbf{S} \boldsymbol{x}_i &= \lambda_i \, \boldsymbol{x}_j^T \boldsymbol{x}_i \\
- \quad \boldsymbol{x}_i^T \mathbf{S} \boldsymbol{x}_j &= \lambda_j \, \boldsymbol{x}_i^T \boldsymbol{x}_j \\
\hline
0 &= (\lambda_i - \lambda_j) \, \boldsymbol{x}_i^T \boldsymbol{x}_j
\end{aligned}
$$

Because the dot products between any two eigenvectors $(i \neq j)$ of a symmetric matrix is zero, the set of eigenvectors must be orthogonal. When, as usual, the eigenvectors are scaled to unit length, the eigenvector matrix, $\mathbf{Q}$, is orthonormal, and orthogonal because it is square.

Recall also from section 5.2.3.2 that from the spectral theorem, that $\mathbf{Q}^T = \mathbf{Q}^{-1}$. The property is illustrated with the following simple example.

```
>> S = [1 2; 2 2]
S =
     1     2
     2     2
>> [Q, Lambda] = eig(S)
Q =
    -0.7882     0.6154
     0.6154     0.7882
Lambda =
    -0.5616          0
          0     3.5616
>> Q'*Q
ans =
     1.0000    -0.0000
    -0.0000     1.0000
```

Appendix B

The Number e

B.1 All About the Number e

This section is devoted to topics related to the irrational number $e \approx$ 2.718281828459046, which is also called the Euler number. We know from section 6.7 that the special number e is important to solutions to ordinary differential equations (ODEs). When e is raised to a complex number the equation represents oscillations, which factor into both control system stability analysis and the conversion of time and spatial domain data into the frequency domain. The number e also factors into the calculation of compound interest rates for savings and loans.

B.1.1 Definition and Derivative

The importance of the number e to ODEs stems from the derivative of exponential equations of e. If a function is defined as the following, where a is a real constant and x is a real variable,

$$y(x) = e^{a\,x},$$

then

$$\frac{dy(x)}{dx} = a\,e^{a\,x} = a\,y(x).$$

Let us now regard exponential equations of e as special cases of a more general class of exponential equations. In doing so, we will see some interesting properties of e and also get a start toward finding a definition of the value of e. In the following equation, k is any positive, real number.

$$y(x) = k^{a\,x}$$

A strategy for finding the derivative of $y(x)$ is to first take the natural logarithm of $y(x)$ (base e logarithm, denoted as $\ln y$). Although we have not yet found the value of e, we know that it is a number and can abstractly use it as the base for a logarithm.

$$\ln y(x) = a\,x\,\ln k$$

$$\frac{d\,[\ln y(x)]}{dx} = \frac{d\,[a\,x\,\ln k]}{dx}$$

The derivative of e^{at}

To find the derivative of e^{at} we can either take the derivative of its Maclaurin[a] series, or use its numeric definition in terms of a limit. The later is used here.

$$e^{at} = \lim_{N \to \infty} \left(1 + \frac{at}{N}\right)^N$$

The chain rule is used to find the derivative. If $f(t) = \left(1 + \frac{at}{N}\right)^N$, then $f'(t) = a \left(1 + \frac{at}{N}\right)^{N-1}$. We see the desired equality then in the limit.

$$e^{at} \quad = \lim_{N \to \infty} f(t)$$

$$\frac{d}{dt}\left(e^{at}\right) \quad = \lim_{N \to \infty} f'(t) = a \lim_{N \to \infty} f(t) = a\, e^{at}$$

---

[a]The Maclaurin series for function f is the Taylor series for function $f(a)$ about the point $(a = 0)$.

The left side of the above equation is the more difficult to find. Implicit differentiation and the chain rule shows that

$$\frac{d\,[\ln y]}{dx} = \frac{1}{y}\frac{dy}{dx}.$$

$$\frac{1}{y}\frac{dy}{dx} = a\,(\ln k)$$

Thus after multiplying by y we have,

$$\frac{dy}{dx} = a\,(\ln k)\, y = a\,(\ln k)\, k^{ax} \tag{B.1}$$

If $k = 2$, $\frac{dy}{dx} = a\,(0.693147)\,2^{ax}$.
If $k = e$, $\frac{dy}{dx} = a\,e^{ax}$.
If $k = 3$, $\frac{dy}{dx} = a\,(1.0986)\,3^{ax}$.

Equation (B.1) is useful, but it assumes that the value of e is already known. We need to use the definition of a derivative to find an equation for the value of e,

$$\frac{dy}{dx} = \lim_{h \to 0} \frac{y(x+h) - y(x)}{h}.$$

We will let $a = 1$ in the remaining equations.

$$\frac{dy}{dx} = \lim_{h\to 0} \frac{k^{(x+h)} - k^x}{h}$$

$$= \lim_{h\to 0} \frac{k^x\, k^h - k^x}{h}$$

$$= \lim_{h\to 0} k^x \frac{(k^h - 1)}{h}$$

Relating the last equation to equation (B.1), we find a limit equation for $\ln k$.

$$\ln(k) = \lim_{h\to 0} \frac{k^h - 1}{h}$$

Let us test this with the natural log of 2 and 3. We need a very small value for h to get a reasonably accurate result.

```
>> h = 0.00000001;
>> ln2 = (2^h - 1)/h
ln2 =
    0.693147184094300
>> log(2)
ans =
    0.693147180559945
>> ln3 = (3^h - 1)/h
ln3 =
    1.098612290029166
>> log(3)
ans =
    1.098612288668110
```

When $k = e$, then $\ln(k = e) = 1$, which we can use to find an equation with relation to the value of e.

$$\ln(e) = 1 = \lim_{h\to 0} \frac{e^h - 1}{h}$$

To solve for e, we need to make a change to the limit.

$$\lim_{h\to 0} h \mapsto \lim_{N\to\infty} \frac{1}{N}$$

$$1 = \lim_{N\to\infty} N\left(e^{1/N} - 1\right)$$

$$\lim_{N\to\infty}\left[\left(e^{1/N} - 1\right) = \frac{1}{N}\right]$$

$$\lim_{N \to \infty} \left[e^{1/N} = \left(1 + \frac{1}{N} \right) \right]$$

Raise both sides to the N power.

$$\boxed{e = \lim_{N \to \infty} \left(1 + \frac{1}{N} \right)^N} \qquad \text{(B.2)}$$

Let us test it with MATLAB. Here, the limited digital resolution of the computer can limit the accuracy if N is too large.

```
>> N = 1E10;
>> (1 + 1/N)^N
ans =
    2.718282053234788
>> exp(1)
ans =
    2.718281828459046
```

The limit can also be used to find powers of e.

$$\boxed{e^x = \lim_{N \to \infty} \left(1 + \frac{x}{N} \right)^N} \qquad \text{(B.3)}$$

```
>> (1 + 3/N)^N
ans =
    20.085541899804120
>> exp(3)
ans =
    20.085536923187668
```

B.1.2   Euler's Complex Exponential Equation

When the exponent of the number e is a complex number, we see the presence of oscillation. This result is especially important to control system analysis and signal processing (i.e., Fourier transform).

Here, we follow the engineering practice of using the variable j for the imaginary number $\sqrt{-1}$ rather than the math practice of using i.

We can see how e^{jx} relates to the $\sin(x)$ and $\cos(x)$ functions by looking at the Maclaurin series for these functions.

The needed Maclaurin series are:

$$\sin x = \sum_{n=0}^{\infty} \frac{(-1)^n x^{2n+1}}{(2n+1)!} = x - \frac{x^3}{3!} + \frac{x^5}{5!} - \frac{x^7}{7!} + \cdots$$

$$\cos x = \sum_{n=0}^{\infty} \frac{(-1)^n x^{2n}}{(2n)!} = 1 - \frac{x^2}{2!} + \frac{x^4}{4!} - \frac{x^6}{6!} + \cdots$$

$$e^x = \sum_{n=0}^{\infty} \frac{x^n}{n!} = 1 + x + \frac{x^2}{2!} + \frac{x^3}{3!} + \cdots$$

Now replace x in the equations for e^x with jx. Remember that $j^2 = -1$.

$$e^{jx} = 1 + jx + \frac{(jx)^2}{2!} + \frac{(jx)^3}{3!} + \frac{(jx)^4}{4!} + \frac{(jx)^5}{5!} + \frac{(jx)^6}{6!} + \frac{(jx)^7}{7!} + \cdots$$

$$= 1 + jx - \frac{x^2}{2!} - \frac{jx^3}{3!} + \frac{x^4}{4!} + \frac{jx^5}{5!} - \frac{x^6}{6!} - \frac{jx^7}{7!} + \cdots$$

$$= \left(1 - \frac{x^2}{2!} + \frac{x^4}{4!} - \frac{x^6}{6!} + \cdots\right) + j\left(x - \frac{x^3}{3!} + \frac{x^5}{5!} - \frac{x^7}{7!} + \cdots\right)$$

$$= \cos x + j \sin x$$

This relationship between complex exponentials of e to the cosine and sine functions in the complex plane ($\mathbb{C}^2$), is often called Euler's formula.

B.1.3 Numerical Verification of Euler's Formula

If the derivation of Euler's formula from the Maclaurin series didn't convince you, we can try some numerical analysis to show that complex exponentials actually produce complex, sinusoidal functions.

We will use the definition of e^x from equation (B.3). We expect to see a circle in $\mathbb{C}^2$, just as we would by plotting $\cos \theta + j \sin \theta$, $0 \le \theta \le 2\pi$. The plot is shown in figure B.1. In the MATLAB script listed in code B.1, we just assign N to be a fairly large number. Since the definition uses a limit as N goes to infinity, the results become more accurate when a larger value for N is used.

```
% File: cmplxEuler.m
% let n = some big number
N = 100000;
z = linspace(0,2*pi); % test 100 numbers between 0 and 2*pi

% Now show that e^jz = cos(z) + j*sin(z)
% Begin with definition of value of e^z.
%    e^z = lim(N = infinity) (1 + z/N)^N
eulerValues = complex(1, z/N).^N;

% This should plot a unit circle, just like
% figure, plot( cos(z), sin(z));
figure, plot(real(eulerValues), imag(eulerValues));
axis equal tight
```

CODE B.1: Script to verify Euler's complex exponential formula

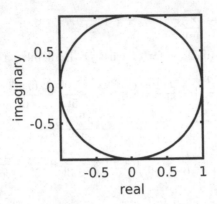

FIGURE B.1: Points along a complex unit circle—numerical verification of Euler's formula.

B.1.4   Compound Interest

Another application of the number e relates to one way to calculate interest on loans or saving accounts.

Loans normally use *simple interest* to track the current principle. The interest paid at each payment is the product of the previous principle, the daily interest rate, and the number of days between payments. The interest is taken out of the payment and the remainder of the payment is applied to the principle.

$$\text{Simple interest} = P \times I \times N$$

Saving accounts normally use *compound interest* where the interest is calculated on the principal amount and also on the accumulated interest of previous periods, and can thus be regarded as "interest on interest."

If you invest P dollars without continued contribution, then the balance of your savings after each compound (t) is described by

$$\text{Balance} = P\left(1 + \frac{I\,t}{N}\right)^N,$$

where I is the annual interest rate and N is the number of times that interest is compounded each year. If interest were compounded continuously, then we have

$$\text{Balance} = P\,e^{I\,t}.$$

If a loan uses compound interest, then the principle can grow multiple times between each payment depending on how often it is compounded. This results in the borrower paying much more over the life of the loan. Fortunately, automobile loans and home mortgages use simple interest. However, credit card debt and student loans are more likely to use compound interest.

The difference between simple and compound interest made for a humorous exchange between George Washington, first president of the United States, and his step son. Washington's wife, Martha, was previously a widow to a wealthy man. One of Martha's sons, John Parke Custis, was not very knowledgeable of business matters when he borrowed money at compound interest to purchase a plantation. When Washington learned of the terms of the loan, he wrote the following to his step son, "No Virginia estate (except a few under the best management) can stand simple interest. How then can they bear compound Interest?" [25].

B.2 A Matrix Exponent and Systems of ODEs

We have already learned from section 6.7 that ordinary differential equations (ODEs) of the form

$$\frac{dy(t)}{dt} = a\, y(t),$$

have the solution

$$y(t) = c\, e^{a\, t}.$$

The same principle applies to systems of ODEs, except that we use vectors and matrices to describe the equations.

$$
\begin{cases}
y_1' = a_{11}\, y_1 + a_{12}\, y_2 + \cdots + a_{1n}\, y_n \\
y_2' = a_{21}\, y_1 + a_{22}\, y_2 + \cdots + a_{2n}\, y_n \\
\quad \vdots \\
y_n' = a_{n1}\, y_1 + a_{n2}\, y_2 + \cdots + a_{nn}\, y_n
\end{cases}
$$

In matrix notation, this is $y' = A\, y$, which has a solution of the form $y(t) = e^{A\, t}\, k$.

We will show here that the matrix form of the solution leads to the same general solution given in section 6.7. First, let us put the matrix exponent e^A into a more manageable form and then apply it to the ODE.

B.2.1 A Matrix in the Exponent

The equation e^A seems unusual. What does it mean to have a matrix in the exponent of the natural number e?

We need to use a Maclaurin series expansion of e^A to get the matrix out of the exponent and then we will use diagonalization from section 6.5.1 to simplify the powers of A that the series expansion gives us.

From the Maclaurin series expansion,

$$e^{\mathbf{A}} = \sum_{n=0}^{\infty} \frac{\mathbf{A}^n}{n!} = \mathbf{I} + \mathbf{A} + \frac{\mathbf{A}^2}{2!} + \frac{\mathbf{A}^3}{3!} + \cdots.$$

From section 6.5, we know that $\mathbf{A}^k = \mathbf{X}\mathbf{\Lambda}^k\mathbf{X}^{-1}$ Where $\mathbf{X}$ and $\mathbf{\Lambda}$ are the eigenvector and diagonal eigenvalue matrices of $\mathbf{A}$.

Thus,

$$
\begin{aligned}
e^{\mathbf{A}} &= \mathbf{I} + \mathbf{X}\mathbf{\Lambda}\mathbf{X}^{-1} + \frac{\mathbf{X}\mathbf{\Lambda}^2\mathbf{X}^{-1}}{2!} + \frac{\mathbf{X}\mathbf{\Lambda}^3\mathbf{X}^{-1}}{3!} + \cdots \\
&= \mathbf{X}\left(\mathbf{I} + \mathbf{\Lambda} + \frac{\mathbf{\Lambda}^2}{2!} + \frac{\mathbf{\Lambda}^3}{3!} + \cdots\right)\mathbf{X}^{-1} \\
&= \mathbf{X}\,e^{\mathbf{\Lambda}}\,\mathbf{X}^{-1}
\end{aligned}
$$

We will call the middle term $\mathbf{\Gamma}$ (Gamma).

$$\mathbf{\Gamma} = e^{\mathbf{\Lambda}} = \left(\mathbf{I} + \mathbf{\Lambda} + \frac{\mathbf{\Lambda}^2}{2!} + \frac{\mathbf{\Lambda}^3}{3!} + \cdots\right)$$

Adding the elements in the Maclaurin series of $\mathbf{\Gamma}$ reveals a Maclaurin series for e^{λ_i} in each diagonal element.

$$
\mathbf{\Gamma} = \begin{bmatrix}
\left(1 + \lambda_1 + \frac{\lambda_1^2}{2!} + \frac{\lambda_1^3}{3!} + \cdots\right) & \cdots & 0 \\
0 & \ddots & 0 \\
\vdots & & \vdots \\
0 & \cdots & \left(1 + \lambda_n + \frac{\lambda_n^2}{2!} + \frac{\lambda_n^3}{3!} + \cdots\right)
\end{bmatrix}
$$

$$
= \begin{bmatrix}
e^{\lambda_1} & 0 & \cdots & 0 \\
0 & e^{\lambda_2} & \cdots & 0 \\
\vdots & & \ddots & \vdots \\
0 & 0 & \cdots & e^{\lambda_n}
\end{bmatrix}
$$

In MATLAB, if variable L (for Lambda) is used for $\mathbf{\Lambda}$, then `Gamma = exp(L) .* eye(n)`, where n is the size of the matrix. Or if we have the variable `e = exp(1)`, then `Gamma = e.^L .* eye(n)`. The expression `Gamma = expm(L)` gives the same result using a built-in MATLAB function.

$$\boxed{e^{\mathbf{A}} = \mathbf{X}\,\mathbf{\Gamma}\,\mathbf{X}^{-1}}$$

B.2.2 Example Matrix Exponent

```
>> A = [10 0; -10 7];          >> % find e^A
>> [X, L] = eig(A)             >> X*Gamma*inv(X)
X =                            ans =
         0    0.2873             1.0e+04 *
    1.0000   -0.9578             2.2026          0
L =                            -6.9766     0.1097
    7    0
    0   10                      >> % verify with MATLAB's
                                >> % expm() function
>> e = exp(1)                   >> expm(A)
e =                            ans =
    2.7183                      1.0e+04 *
>> Gamma = e.^L .* eye(2)        2.2026          0
Gamma =                        -6.9766     0.1097
    1.0e+04 *
    0.1097          0
         0    2.2026
```

B.2.3 Matrix Solution to a System of ODEs

Returning to the problem of systems of ODEs, we know that in matrix notation, the ODE $y' = \mathbf{A}\,y$, has a solution of the form

$$y(t) = e^{\mathbf{A}\,t}k = \mathbf{X}\,\mathbf{\Gamma}^t\,\mathbf{X}^{-1}k, \tag{B.4}$$

where

$$\mathbf{\Gamma}^t = \begin{bmatrix} e^{\lambda_1 t} & 0 & \cdots & 0 \\ 0 & e^{\lambda_2 t} & \cdots & 0 \\ \vdots & \vdots & \ddots & \vdots \\ 0 & 0 & \cdots & e^{\lambda_n t} \end{bmatrix}.$$

Note:

$$
\begin{aligned}
e^{\mathbf{A}} &= \mathbf{X}\,\mathbf{\Gamma}\,\mathbf{X}^{-1} \\
e^{\mathbf{A}\,2} &= (\mathbf{X}\,\mathbf{\Gamma}\,\mathbf{X}^{-1})(\mathbf{X}\,\mathbf{\Gamma}\,\mathbf{X}^{-1}) \\
&= \mathbf{X}\,\mathbf{\Gamma}^2\,\mathbf{X}^{-1} \\
e^{\mathbf{A}\,3} &= (\mathbf{X}\,\mathbf{\Gamma}^2\,\mathbf{X}^{-1})(\mathbf{X}\,\mathbf{\Gamma}\,\mathbf{X}^{-1}) \\
&= \mathbf{X}\,\mathbf{\Gamma}^3\,\mathbf{X}^{-1} \\
&\vdots \\
e^{\mathbf{A}\,t} &= \mathbf{X}\,\mathbf{\Gamma}^t\,\mathbf{X}^{-1}
\end{aligned}
$$

Using equation (B.4), we find k from $y(t)$ when $t = 0$.

$$y(0) = \mathbf{X}\,\mathbf{\Gamma}^0\,\mathbf{X}^{-1}k = \mathbf{X}\,\mathbf{I}\,\mathbf{X}^{-1}k = k$$

Letting $c = \mathbf{X}^{-1} k = \mathbf{X}^{-1} y(0)$ gives the solution used in section 6.7 (equation (6.7) on page 269).

$$y(t) = \mathbf{X}\,\boldsymbol{\Gamma}^t\,\mathbf{X}^{-1}\,y(0) = \mathbf{X}\,\boldsymbol{\Gamma}^t\,c \tag{B.5}$$

$$= c_1 e^{\lambda_1 t} x_1 + c_2 e^{\lambda_2 t} x_2 + \cdots + c_n e^{\lambda_n t} x_n \tag{B.6}$$

B.2.4   Example ODE Matrix Solution

In section 6.7, we solved the following ODE example with initial conditions,

$$\begin{cases} y_1(t)' = -2\,y_1(t) + \ y_2(t), & y_1(0)=6 \\ y_2(t)' = \ \ y_1(t) - 2\,y_2(t), & y_2(0)=2 \end{cases}$$

Here we keep all of the variables as matrices and vectors in MATLAB and find the same solution.

```
>> A = [-2 1; 1 -2]
A =
    -2    1
     1   -2
>> [X, L] = eig(A)
X =
    0.7071    0.7071
   -0.7071    0.7071
L =
    -3    0
     0   -1
>> X = X*2/sqrt(2)     % scaling for appearance - not needed
X =
    1.0000    1.0000
   -1.0000    1.0000
>> y0 = [6; 2]
y0 =
     6
     2
>> c = X\y0
c =
    2.0000
    4.0000
```

$$y(t) = \mathbf{X}\,\boldsymbol{\Gamma}^t\,c = \begin{bmatrix} 1 & 1 \\ -1 & 1 \end{bmatrix}\begin{bmatrix} e^{-3t} & 0 \\ 0 & e^{-t} \end{bmatrix}\begin{bmatrix} 2 \\ 4 \end{bmatrix} = \begin{bmatrix} e^{-3t} & e^{-t} \\ -e^{-3t} & e^{-t} \end{bmatrix}\begin{bmatrix} 2 \\ 4 \end{bmatrix}$$

$$\begin{cases} y_1(t) = \ \ 2\,e^{-3t} + 4\,e^{-t} \\ y_2(t) = -2\,e^{-3t} + 4\,e^{-t} \end{cases}$$

Bibliography

[1] Robert K. Adair. *The Physics of Baseball*. Harper Perennial, New York, NY, 1990.

[2] Falah Alsaqre. Two-dimensional PCA for face recognition. https://www.mathworks.com/matlabcentral/fileexchange/69377-two-dimensional-pca-for-face-recognition), 2019. MATLAB Central File Exchange, Retrieved July 10, 2019.

[3] ANSI/IEEE. IEEE standard for binary floating-point arithmetic. *ANSI/IEEE Std 754-1985*, pages 1–20, 1985.

[4] Stormy Attaway. *MATLAB, A Practical Introduction to Programming and Problem Solving*. Butterworth-Heinemann/Elsevier, Amsterdam, fourth edition, 2017.

[5] A. Azzalini and A. W. Bowman. A look at some data on the Old Faithful geyser. *Journal of the Royal Statistical Society. Series C (Applied Statistics)*, 39(3):357–365, 1990.

[6] Major League Baseball. Mustc: Gordon's clutch home run. https://www.mlb.com/video/must-c-gordon-s-clutch-home-run-c526415283.

[7] Steven Brunton and Nathan Kutz. *Data-Driven Science and Engineering: Machine Learning, Dynamical Systems, and Control*. Cambridge University Press, 2019.

[8] Garrett Buffington. Polar decomposition of a matrix. http://buzzard.ups.edu/courses/2014spring/420projects/math420-UPS-spring-2014-buffington-polar-decomposition.pdf, 2014.

[9] Rizwan Butt. *Introduction to Numerical Analysis using MATLAB*. Jones and Bartlett, 2010.

[10] Andrew Chamberlain. The linear algebra view of the Fibonacci sequence. https://medium.com/@andrew.chamberlain/the-linear-algebra-view-of-the-fibonacci-sequence-4e81f78935a3, 2016.

[11] Stephen J. Chapman. *MATLAB Programming for Engineers*. Cengage Learning, Boston, MA, fifth edition, 2016.

[12] Steven C. Chapra and Raymond L. Canale. *Numerical Methods for Engineers*. McGraw-Hill, seventh edition, 2015.

[13] Mei-Qin Chen. A brief history of linear algebra and matrix theory. `http://www.macs.citadel.edu/chenm/240.dir/12fal.dir/history2.pdf`, 2012.

[14] R.E. Cline and R.J. Plemmons. l_2–solutions to undetermined linear systems. *SIAM Review*, 18(1):92–106, Jan., 1976.

[15] Peter Corke. *Robotics, Vision and Control–Fundamental Algorithms in MATLAB*. Springer, New York, NY, second edition, 2017.

[16] Peter Corke. Spatial math toolbox. `https://petercorke.com/toolboxes/spatial-math-toolbox/`, 2020.

[17] Marc Deisenroth, Aldo Faisal, and Cheng Ong. *Mathematics for Machine Learning*. Cambridge University Press, 02 2020.

[18] James W. Demmel. *Applied Numerical Linear Algebra*. SIAM, Philadelphia, PA, 1997.

[19] Froilan Dopico. Alan Turing and the origins of modern Gaussian elimination. *Arbor*, 189:a084, 12 2014.

[20] Dheeru Dua and Casey Graff. UCI Machine Learning Repository. `http://archive.ics.uci.edu/ml`, 2017.

[21] C. Eckart and G. Young. The approximation of one matrix by another of lower rank. *Psychometrika*, 1(3):211–218, 1936.

[22] Elsie Eigerman. How to import data from spreadsheets and text files without coding. `https://www.mathworks.com/videos/importing-data-from-text-files-interactively-71076.html`.

[23] R. A. Fisher. The use of multiple measurements in taxonomic problems. *Annals of Eugenics*, 7(2):179–188, 1936.

[24] G. E. Forsythe, M. A. Malcolm, and C. B. Moler. *Computer Methods for Mathematical Computations*. Prentice Hall, Englewood Cliffs, NJ, 1976.

[25] National Archives Founders Online. From George Washington to John Parke Curtis, 3 August 1778. `https://founders.archives.gov/documents/Washington/03-16-02-0249`, 2006.

[26] Matan Gavish and David L. Donoho. The optimal hard threshold for singular values is $4/\sqrt{3}$. *IEEE Transactions on Information Theory*, 60(8):5040–5053, 2014.

[27] G. Golub and C. Reinsch. Singular value decomposition and least squares solution. *Numerische Mathematik*, 14:403–420, 1970.

[28] Gene H. Golub and Charles F. Van Loan. *Matrix Computations.* Johns Hopkins University Press, Baltimore, MD, fourth edition, 2013.

[29] Michael C. Grant and Stephen P. Boyd. CVX: MATLAB software for disciplined convex programming. `http://cvxr.com/cvx/`, 2020. CVX Research, Inc.

[30] John V. Guttag. *Introduction to Computation and Programming Using Python: With Application to Understanding Data.* The MIT Press, second edition, 2016.

[31] Gabriel Ha. Creating a basic plot interactively. `https://www.mathwork s.com/videos/creating-a-basic-plot-interactively-68978.html`.

[32] Brian D. Hahn and Daniel T. Valentine. *Essential MATLAB for Engineers and Scientists.* Academic Press/Elsevier, Amsterdam, sixth edition, 2017.

[33] E. Cuyler Hammond and Daniel Horn. The relationship between human smoking habits and death rates: A follow-up study of 187,766 men. *Journal of the American Medical Association,* 155(15):1316–1328, Aug. 1954.

[34] Desmond J. Higham and Nicholas J. Higham. *MATLAB Guide.* SIAM, Philadelphia, PA, third edition, 2017.

[35] Nicholas J. Higham. *Accuracy and Stability of Numerical Algorithms.* SIAM, Philadelphia, PA, second edition, 2002.

[36] Nicholas J. Higham. Gaussian elimination. *Wiley Interdisciplinary Reviews: Computational Statistics,* 3:23–238, 2011.

[37] Robert V. Hogg and Allen T. Craig. *Introduction to Mathematical Statistics.* Macmillan, London, fourth edition, 1978.

[38] The MathWorks Inc. Create 2-d line plots. `https://www.mathworks.co m/help/matlab/creating_plots/using-high-level-plotting-fun ctions.html`.

[39] The MathWorks Inc. Greek letters and special characters in chart text. `https://www.mathworks.com/help/matlab/creating_plots/g reek-letters-and-special-characters-in-graph-text.html`.

[40] The MathWorks Inc. Line properties. `https://www.mathworks.com/he lp/matlab/ref/matlab.graphics.chart.primitive.line-propertie s.html`.

[41] The MathWorks Inc. MATLAB for new users. `https://www.mathwork s.com/videos/matlab-for-new-users-1487714181074.html`.

[42] The MathWorks Inc. Supported file formats for import and export. `https://www.mathworks.com/help/matlab/import_export/supporte` `d-file-formats-for-import-and-export.html`.

[43] The MathWorks Inc. Two-D and Three-D plots. `https://www.mathwo` `rks.com/help/matlab/learn_matlab/plots.html`.

[44] The MathWorks Inc. MATLAB fundamentals. `https://matlabacadem` `y.mathworks.com/`, 2017.

[45] The MathWorks Inc. Characters and strings. `https://www.mathworks.` `com/help/matlab/characters-and-strings.html`, 2021.

[46] The MathWorks Inc. Systems of linear equations. `https://www.math` `works.com/help/matlab/math/systems-of-linear-equations.html`, 2021.

[47] J. Kautsky, N.K. Nichols, and P. Van Dooren. Robust pole assignment in linear state feedback. *International Journal of Control*, 41(5):1129–1155, 1985.

[48] Philip N. Klein. *Coding the Matrix: Linear Algebra through Applications to Computer Science*. Newtonian Press, 2013.

[49] Jose Nathan Kutz. *Data-Driven Modeling & Scientific Computation: Methods for Complex Systems & Big Data*. Oxford University Press, 2013.

[50] J. C. Lagarias, J. A. Reeds, M. H. Wright, and P. E. Wright. Convergence properties of the Nelder-Mead simplex method in low dimensions. *SIAM Journal of Optimization*, 9(1):112–147, 1998.

[51] James Lambers. CME 335 lecture 6 notes. `https://web.stanford.edu` `/class/cme335/lecture6.pdf`, 2010.

[52] Cris Luengo. Boxplot. `https://www.mathworks.com/matlabcentral/` `fileexchange/51134-boxplot`, 2015. MATLAB Central File Exchange. Retrieved September 3, 2020.

[53] John H. Mathews and Kurtis D. Fink. *Numerical Methods Using Matlab*. Pearson Prentice Hall, fourth edition, 2005.

[54] J.L. Meriam. *Engineering Mechanics Statics and Dynamics*. John Wiley & Sons, Inc. New York, NY, 1978.

[55] Cleve Moler. *Numerical Computing with MATLAB*. SIAM, Philadelphia, PA, 2004.

[56] Cleve Moler. Professor svd. `https://www.mathworks.com/company/ne` `wsletters/articles/professor-svd.html`, 2006. A blog post in the MathWorks' Technical Articles and Newsletters.

[57] Cleve Moler. Gil Strang and the cr matrix factorization. `https://blogs.mathworks.com/cleve/2020/10/23/gil-strang-and-the-cr-matrix-factorization/`, 2020. Blog: Cleve's Corner: Cleve Moler on Mathematics and Computing.

[58] David S. Moore, William Notz, and Michael Fligner. *The Basic Practice of Statistics*. W.H. Freeman and Co., New York, 2018.

[59] Dan Mullen. How to use basic plotting functions. `https://www.mathworks.com/videos/using-basic-plotting-functions-69018.html`.

[60] Alexander D. Poularikas and Samuel Seely. *Signals and Systems*. Krieger Pub. Co., second edition, 1994.

[61] Sergio Obando Quintero. Beyond Excel: Enhancing your data analysis with MATLAB. `https://www.mathworks.com/videos/beyond-excel-enhancing-your-data-analysis-with-matlab-1503081232623.html`.

[62] Karl J. Åström and Richard M. Murray. *Feedback Systems: An Introduction for Scientists and Engineers*. Princeton University Press, 2012.

[63] National Park Service. Predicting Old Faithful geyser. `https://www.nps.gov/features/yell/ofvec/exhibits/eruption/prediction/predict8.htm`, 2021.

[64] Lawrence F. Shampine and Mark W. Reichelt. The MATLAB ODE suite. *SIAM Sci. Comput.*, 18(1):1–22, 1997.

[65] Wen Shen. *An Introduction to Numerical Computation*. World Scientific Publishing Company, 2015.

[66] Timmy Siauw and Alexandre M. Bayen. *An Introduction to MATLAB Programming and Numerical Methods: For Engineers*. Academic Press, 2015.

[67] G. W. Stewart. On the early history of the singular value decomposition. *SIAM Review*, 35(4):551–566, 1993.

[68] Gilbert Strang. *Computational Science and Engineering*. Wellesly-Cambridge Press, Wellesley, MA, 2007.

[69] Gilbert Strang. Linear algebra, MIT course 18.06. `https://ocw.mit.edu/courses/mathematics/18-06-linear-algebra-spring-2010/video-lectures/`, 2010.

[70] Gilbert Strang. *Differential Equations and Linear Algebra*. Wellesly-Cambridge Press, Wellesley, MA, 2014.

[71] Gilbert Strang. *Introduction to Linear Algebra*. Wellesly-Cambridge Press, Wellesley, MA, fifth edition, 2016.

[72] Gilbert Strang. Res.18-010 a 2020 vision of linear algebra. `https://oc w.mit.edu`, 2020.

[73] Alaa Tharwat. PCA (principal component analysis). `https://www. mathworks.com/matlabcentral/fileexchange/30792-pca-princip al-component-analysis`. MATLAB Central File Exchange. Retrieved July 12, 2019.

[74] Alex Townsend. The qr algorithm. `http://pi.math.cornell.edu/~web 6140/TopTenAlgorithms/QRalgorithm.html`, 2019.

[75] Alan. Tucker. The growing importance of linear algebra in undergraduate mathematics. *The College Mathematics Journal*, 24(1):3–9, 1993.

[76] A. M. Turing. Rounding-off errors in matrix processes. *The Quarterly Journal of Mechanics and Applied Mathematics*, 1(1):287–308, 1948.

[77] Gareth Williams. *Linear Algebra with Applications*. Jones & Bartlett Learning, Burlington, MA, eighth edition, 2014.

[78] Won Y. Yang, Wenwu Cao, Tae-Sang Chung, and John Morris. *Applied Numerical Methods Using MATLAB*. John Wiley & Sons, Inc. Hoboken, NJ, 2005.

[79] Kalidas Yeturu. Chapter 3 - machine learning algorithms, applications, and practices in data science. In Arni S.R. Srinivasa Rao and C.R. Rao, editors, *Principles and Methods for Data Science*, volume 43 of *Handbook of Statistics*, pages 81–206. Elsevier, 2020.

Index

Printed in the United States
by Baker & Taylor Publisher Services

Printed in the United States
by Baker & Taylor Publisher Services